Introduction to Sustainability

Introduction to Sustainability

Robert Brinkmann

Second Edition

WILEY Blackwell

This edition first published 2021

Edition History
Wiley Blackwell (1e, 2016)

Registered Offices
John Wiley & Sons, Inc., 111 River Street, Hoboken, NJ 07030, USA
John Wiley & Sons Ltd, the Atrium, Southern Gate, Chichester, West Sussex, PO19 8SQ, UK

Editorial Office
9600 Garsington Road, Oxford, OX4 2DQ, UK

For details of our global editorial offices, customer services, and more information about Wiley products visit us at www.wiley.com.

Wiley also publishes its books in a variety of electronic formats and by print-on-demand. Some content that appears in standard print versions of this book may not be available in other formats.

Library of Congress Cataloging-in-Publication Data

Names: Brinkmann, Robert, 1961- author.
Title: Introduction to sustainability / Robert Brinkmann.
Description: Second edition. | Hoboken, NJ : Wiley Blackwell, 2021. |
Includes bibliographical references and index.
Identifiers: LCCN 2020029249 (print) | LCCN 2020029250 (ebook) | ISBN 9781119675464 (paperback) | ISBN 9781119675488 (adobe pdf) | ISBN 9781119675495 (epub)
Subjects: LCSH: Sustainable development. | Economic development–Environmental aspects.
Classification: LCC HC79.E5 B743 2021 (print) | LCC HC79.E5 (ebook) | DDC 338.9/27–dc23
LC record available at https://lccn.loc.gov/2020029249
LC ebook record available at https://lccn.loc.gov/2020029250

Cover Design: Wiley
Cover Image: © Siegfried Haasch/Getty Images
Printed and bound by CPI Group (UK) Ltd, Croydon, CR0 4YY

C9781119675464_161023

Contents

Acknowledgments *xv*
About the Author *xvii*
About the Companion Website *xix*

1 **Roots of the Modern Sustainability Movement** *1*
Meaning of Sustainability *1*
Nineteenth Century Environmentalism *3*
Pinchot, Roosevelt, and Muir *5*
Aldo Leopold and the Land Ethic *6*
Better Living Through Chemistry, The Great Smog of 1952, and Rachel Carson *8*
Environmental Activism of the 1960s and 1970s and the Development of Environmental Policy *10*
The Growth of Environmental Laws in the 1960s and 1970s *13*
The First Earth Day *14*
International Concerns *14*
Ozone and the World Comes Together *15*
Globalization and the Brundtland Report *16*
Deep Ecology *18*
Environmental Justice *19*
Measuring Sustainability *21*
The Climate Change Challenge *23*
The Road Ahead *24*
Organization *26*

2 **Understanding Natural Systems** *29*
The Earth, its Layers, and the Rock Cycle *29*
The Rock Cycle *32*
Biogeochemical Cycles *33*
Water and the Water Cycle *34*
The Carbon Cycle and Global Climate Change *37*
Global Climate Change and the Carbon Cycle *38*
The Sulfur Cycle *40*

The Nitrogen and Phosphorus Cycles *42*
Nitrate Pollution of groundwater *45*
Organisms and Ecosystems *46*
Urban Ecosystems *49*
Understanding the Anthropocene *53*

3 **Measuring Sustainability** *57*
The United Nations Millennium Goals *58*
The United Nations Sustainable Development Goals *59*
National sustainability planning *60*
Canada *64*
Bhutan *67*
Regional sustainability planning *69*
Local sustainability measurement *73*
Green local governments in Florida *75*
Specific community plans *79*
PlaNYC *79*
London and sustainability *81*
Small towns and sustainability *84*
Business sustainability *85*

4 **Energy** *89*
World Energy Production and Consumption *89*
Traditional or "Dirty" Energy Resources *91*
Oil *91*
Oil shale and tar sands *93*
Natural gas *95*
Coal *97*
Coal mining *98*
Pollution from coal *99*
Green energy *100*
Biomass *100*
Biomass: wood, manure, peat, and other organic sources *100*
Burning of garbage: waste-to-energy *101*
Conversion of biomass to liquid or gas fuel *102*
Wind energy *103*
Solar energy *105*
Passive solar energy *105*
Active solar energy *106*
Concentrated solar power *107*
Critiques of solar power *107*
Nuclear energy *107*
Other innovations *110*
Energy efficiency *110*
Living off the grid *112*

5 **Global Climate Change and Greenhouse Gas Management** *113*
The end of nature? *113*
The science of global climate change: The greenhouse effect *114*
Water vapor *116*
Carbon dioxide *116*
Methane *118*
Sinks of carbon *120*
Forests *120*
Reefs *120*
The IPCC and evidence for climate change, and the future of our planet *121*
Ocean acidification *122*
Phenological changes *123*
Conducting greenhouse gas inventories *124*
Step 1 Setting boundaries *125*
Step 2 Defining scope *125*
Step 3 Choosing a quantitative approach *126*
Step 4 Setting a baseline year *126*
Step 5 Engaging stakeholders *126*
Step 6 Procuring certification *127*
Greenhouse gas equivalents used in greenhouse gas accounting *127*
Greenhouse gas emission scopes *128*
De minimis emissions *129*
Computing greenhouse gas credits *129*
Climate action plans *129*
Religion and climate change *135*
Evangelical Environmental Network *136*
Young Evangelicals for Climate Action *136*
Catholic Climate Covenant *136*
Jewish Climate Change Campaign *137*
The International Muslim Conference on Climate Change *138*
Buddhist Declaration on Climate Change *138*
Hindu Declaration on Climate Change *138*
Art, culture, and climate change *139*
Swoon *139*
Raul Cardenas Osuna and Toro Labs *139*
Isaac Cordal *140*

6 **Water** *143*
Sources of water *143*
Consumption trends *148*
Sources of water pollution *150*
Agricultural pollution *150*
Industrial pollution *150*
Storm water pollution *151*
Sewage *152*

Leaking underground tanks *153*
Landfills *153*
Water management and conservation *155*
National and regional water conservation and management *155*
Water as a tool for regional development *156*
Water supply management *157*
Hard path water management *157*
Soft path water management *158*
Water management and innovation *159*
Water quality *161*
Understanding drainage basins *168*
Drainage basins out of synch *169*
Drainage basin pollution *169*
Stream profile and base level *169*
Lakes *169*
Seas *171*
Oceans *171*

7 **Food and Agriculture** *173*
Development of modern agriculture *173*
Meat production *177*
Piggeries *178*
Feed lots *179*
Chicken houses *179*
World agricultural statistics *181*
Food deserts and obesity *182*
Sustainable alternatives to the industrial food movement *185*
Vegetarianism and veganism *185*
Organic farming *186*
Small farm movement *186*
Locavores *188*
Farm to table *189*
Community sponsored agriculture *191*
Community gardens *193*
Farmers' markets *193*
Beekeeping *195*
The urban chicken movement *196*
Guerilla gardening, freegans, and other radical approaches to food *196*

8 **Green Building** *201*
LEED rating systems *201*
Site selection *204*
Brownfield development *204*
Other aspects of sustainable building siting *207*
Water use *207*

Energy and atmospheric health 208
Materials and resources 210
Material re-use 211
Recycled content of construction material 211
Locally derived materials 211
Renewable materials and certified sustainable wood 212
Waste management 212
Summary 213
Indoor environmental quality 213
Ventilation and air delivery monitoring 213
Construction indoor air quality management 214
Use of low-emitting materials 214
Indoor chemical and pollution source control 215
Controllability and design of lighting and temperature systems 215
Access to daylight 215
Summary 215
Innovation 215
Regional priorities 216
Expansion of green building technology 216
Other green building rating systems 216
BREEAM 217
PassivHaus 219
Green building policy 220
Critiques of green building 221
The greenest building and historic preservation 222
Small house movement 226
Further reading 229

9 **Transportation** 231
Transportation options 232
Vehicles 232
Cars 234
Trucks 234
Vehicles and fuels 235
Electric cars 237
Automated Vehicles 238
Rail 238
Ship transport 239
Bulk carriers 239
Container ships 239
Tankers 240
Refrigerated ships 240
Roll-on/roll-off ships 240
Environmental issues associated with ship transport 240
Air transport 241

Space travel 243
Roads 245
Environmental issues with roads 246
Storm water pollution management 246
Street sweeping 250
Ground stability 250
Mass transit 252
Forms of mass transit 252
Railways 252
Light rail 253
Buses 253
Bus rapid transit 254
Ferries 254
Transit hubs and transit-oriented development 254
The future 255

10 Pollution and Waste 259
Pollution 259
Chemical pollution 259
Metals 259
Organic compounds 260
Nutrients 261
Radioactive Pollutants 262
Pharmaceutical pollutants 263
Heat pollution 263
Light pollution 264
Noise pollution 265
Visual pollution 265
Littering 266
Understanding pollution distribution 266
The US approach to pollution 268
Clean Air Act 268
Clean Water Act 270
National Environmental Policy Act 271
Superfund 272
Sewage treatment 274
Sewage and sustainability 277
Garbage and recycling 277
Garbage composition 278
Managing garbage 278
Landfills 279
Reducing waste 280
Composting 281
Recycling 281

11 **Environmental Justice** *287*
Social justice *287*
Civil rights and the modern environmental movement in the United States *290*
Lead pollution and the growth of the urban environmental justice movement *291*
Environmental racism in the United States *293*
Brownfields, community re-development, and environmental justice *295*
US EPA and environmental justice *297*
Indigenous people and environmental justice *299*
Exporting environmental problems *300*
Environmental justice around the world *301*
Environmental justice in Europe *302*
Environmental justice in Asia and the Pacific *302*
The Three Gorges Dam *302*
Bhopal and environmental justice in India *303*
Tuvalu and global climate change *304*
Environmental justice in Africa *305*
Environmental justice in Latin America and the Caribbean: oil pollution in Ecuador *306*
Environmental justice in a Globalized World *308*

12 **Sustainability Planning and Governance** *313*
Local governments and their structure *313*
The role of citizens and stakeholders in local government *314*
Community stakeholders *315*
Boundaries and types of local governments *316*
Leadership *319*
Efforts to aid local governments on sustainability issues *319*
Scale and local governments *321*
Green regional development *322*
Sustainable development *326*
Globalization *327*
Development of globalization *328*
Drivers of globalization *329*
Internet and communications *329*
Transportation *330*
Economic development *331*
Transnational organizations *332*
War and sustainability *339*
Further reading *342*

13 **Sustainability, Economics, and the Global Commons** *343*
The global commons *343*
Economic processes that put the Earth out of balance *345*
Social and economic theories *346*

Neoclassical economics *346*
Environmental critiques of neoliberalism *347*
Environmental economics *349*
Cost-benefit analysis and its application in environmental economics *349*
Environmental impact assessment *351*
Environmental ethics *352*
Green economics *352*
Non-capitalistic economies *353*
Deep ecology *353*
Ecofeminism *356*
Destruction regardless of theory *356*
Environmental economics: externalities *357*
Measuring the economy *358*
Green jobs *362*

14 Corporate and Organizational Sustainability Management *371*
Cognitive dissonance *371*
Why are businesses concerned with sustainability? *372*
Profit *372*
Public relations *372*
Altruism *372*
Concern over the long-term sustainability of the industry *373*
Professional standards and norms *373*
Total quality management and sustainability *373*
People, planet, and profits *374*
Ray Anderson, the father of the green corporation and the growth of green corporate environmentalism *379*
Anderson's legacy *380*
Greenwashing in the corporate world *380*
Green consumers *380*
Global Reporting Initiative *382*
Sustainability reporting in the S & P 500 *382*
Dow Jones Sustainability Index *385*
Sustainability reporting *388*
International Organization for Standardization (ISO): ISO 14000 and ISO 26000 *388*
ISO 14000 *388*
ISO 26000 *388*
Case studies of sustainability at the corporate level *389*
Walmart *391*
Unilever *393*
Lessons from Walmart and Unilever *395*
Can businesses with unsustainable products be sustainable? *396*

15 Sustainability at Universities, Colleges, and Schools *401*
Curriculum at colleges and universities *401*
Sustainability curriculum at K-12 schools *403*
External benchmarking *405*
American Association for Sustainability in Higher Education *405*
Presidents' Climate Leadership Commitments *406*
Other external benchmarking organizations *408*
Internal initiatives *409*
Sustainability officers *410*
Sustainability committees *411*
Food service *411*
Student and faculty activism *414*
Building your own case study *417*
Sustainability at Oxford: a campus commitment *418*
Making school lunches healthier in the United States *419*
The cow powered carbon neutral campus *421*
Whitman College builds wind turbines on campus farm *421*
Stanford University: dumping the car for bikes *422*
Green fleets: The University of South Florida's biodiesel Bullrunner *422*
Community engagement at Portland State University *423*
Green buildings on college campuses: University of Florida goes for gold *424*
Native and sustainable landscaping at one of the largest schools in the nation: Valencia College *425*
Campus archaeology at Michigan State University *425*

Index *427*

Acknowledgments

I am grateful to everyone at John Wiley & Sons, Ltd. for their guidance throughout the production of this book. They were professional, patient, and extremely helpful. I am also deeply appreciative of the reviewers that provided insightful initial comments on the text.

There are many who helped me in one way or another as I worked on this book. I cannot name everyone, but please know that all of my friends, family members, colleagues, and former students have my thanks. Yet there are some who deserve some special thanks for inspiration or assistance. They are: Mario Jose Gomez, Jim Brinkmann, Charlie Brinkmann, Fenda Akiwumi, Kamal Alsharif, J. Bret Bennington, Elizabeth Bird, David Brinkmann, James Brinkmann, John Brinkmann, Rose Brinkmann, Jennifer Collins, Michelle and Craig DeBruyn, Andrea Del Toro, Lauren D'Orsa, Joni Downs, Emma Farmer, Bernard Firestone, Sandra Garren, Jody and Erik Gartzke, Lynne Goldstein, Antonio Gomez, Carolina Gomez, Elis Grecia Pulido Gomez, Elis Vera de Gomez, Charleen Gonzalez, Karla Gonzalez, Mark Hafen, Grant Harley, Nancy Heller, Adriane Hoff, Sharon and Bob Hoff, Randy Honig, Heidi Hutner, Robin Jones, Rafael Jaramillo, Bhavani Jaroff, Sophia Kasselakis, Ina Katz, Beth Larson, Lawrence Levy, Burrell Montz, Keshanti Nandlall, Christophser Niedt, Gitfah Niles, Joanne Norris, Leslie North, Juan Penso, Lisa Marie Pierre, Jason Polk, Phil Reeder, Norma Camero Reno, Geary Schindel, Patricia and Nelson Sohns, Elizabeth Strom, Graham Tobin, Maya Trotz, Naimish Upadhyay, Phil Van Beynen, George Veni, and Laurie Walker.

I would also like to thank all of the photographers who submitted photographs for this book. Finally, I would like to thank all of my current and former students for the inspiration you give me to have hope for the future of our planet.

About the Author

Robert (Bob) Brinkmann, Ph.D. is the Dean of the College of Liberal Arts and Sciences at Northern Illinois University where he is also a Professor in the Department of Geology and Environment. He was born in 1961 in rural Wisconsin and was greatly influenced by his experiences growing up in a quaint, small-town environment. As a child he spent many hours in nature hiking, fishing, and canoeing, especially in the wilderness of northern Wisconsin. In 1979, he entered the geology program at the University of Wisconsin at Oshkosh. There, he earned a Bachelor of Science with a focus on lithology, mineralogy, and field geology. During this period, he travelled throughout North America and participated in a geology field school in Alberta, British Columbia, and the Yukon. His first publication, on the formation of the Berlin Rhyolite, was published in 1982.

After graduation, Brinkmann attended the University of Wisconsin-Milwaukee where he earned an MS in Geology in 1986 and a Ph.D. in Geography in 1989. During this period, he worked in diamond exploration, ice crystallography, and soil chemistry. It was while conducting fieldwork in diamond exploration that Brinkmann began to be influenced by sustainability issues. He started to take courses with the late Forest Stearns, one of the first ecologists to call for research on urban ecosystems, and the late Robert Eidt, a soil scientist noted for his definition and interpretation of anthrosols, or humanly modified soils. Brinkmann began to study a number of topics including heavy metal geochemistry of garden soils in cities, ancient agricultural soils in the Arabian Peninsula, and soil and sediment erosion in mountainous regions. He also took courses with cave and karst expert, Michael J. Day, and noted archaeologist, Lynne Goldstein.

In 1990, Brinkmann became an Assistant Professor at the University of South Florida (USF) where he continued his research on urban sustainability, particularly as associated with soil and sediment pollution in urban and suburban areas and cave and karst research. He published numerous articles and books including the only book on the science, policy, and management of urban street sweeping (with Graham Tobin) and the only book on sinkholes in Florida. He became a Full Professor in 2000 and the first Chair of USF's Department of Environmental Science and Policy. He also served as Chair of the Department of Geography and as Associate Dean for Faculty Development in the 2000s. He arrived at Hofstra University in 2011 to start a new sustainability studies program. The undergraduate program offers a BS, BA, and MA in sustainability. From 2016 until 2019 Brinkmann served as Vice Provost for Research and Dean of Graduate Studies at Hofstra University.

Over the years, he has designed a number of courses, including ones on sustainability management, wetlands, and community-based sustainability. Brinkmann has served as an elected officer with a number of national, regional, and local organizations and has appeared on a number of national news outlets as an expert on geologic and environmental issues, including CBS News and CNN. His blog, *On the Brink,* which focuses on environmental and sustainability issues, gets thousands of hits a day. He also has a regular column on *Huffington-post*, and his opinion pieces have appeared on *Newsday* and CNN.com.

About the Companion Website

Don't forget to visit the companion website for this book:

www.wiley.com/go/Brinkmann/IntroductiontoSustainability

There you will find valuable material designed to enhance your learning, including:

- Learning Outcomes for all chapters
- Color version of figures
- Exercises for all chapters

Scan this QR code to visit the companion website

1

Roots of the Modern Sustainability Movement

In the summer of 2019, Iceland held a funeral for Okjukull, a glacier that melted as a result of global climate change. The entire country is losing ice at a rate of 11 billion tonnes (1 tonne is equivalent to 2204 pounds) a year. A plaque at the site of the ancient glacier states, "OK is the first Icelandic glacier to lose its status as a glacier. In the next 200 years all our glaciers are expected to follow the same path. This monument is to acknowledge that we know what is happening and what needs to be done. Only you know if we did it." (Figure 1.1).

But what action has been taken in the world to try to solve the climate change problem? How did we get to the point that human population is knowingly changing the world's climate? What historical developments have gotten us to this point? While there have always been waves of dramatic climate change over the history of our planet, what specific actions have caused the dramatic changes we have seen over the last 100 years?

Climate change isn't the only problem we face. It is but one of many issues in sustainability that prompt us to take a deeper look at our interaction with the environment. As you will see as you progress through this book, we face many problems. However, we have developed many solutions and there is reason to hope that we can make the appropriate changes to make our world more sustainable in the future.

The purpose of this chapter is to review the development of the modern sustainability movement from its roots in the nineteenth century to the development of international efforts to improve our world's environment. However, prior to getting us to this point, it is worthwhile to define the meaning of sustainability.

Meaning of Sustainability

Sustainability can be succinctly defined as doing what we can now to preserve the environment for future generations. However, in practice the word has a much deeper meaning. There are three components of sustainability: environment, equity, and economy. The environment is an obvious part of sustainability in that we are striving to preserve and protect the environment. Equity focuses on ensuring that fairness in environmental decision-making are front and center as we move forward in the future. And the economy component of sustainability focuses on the reality that we need to ensure that livelihoods are protected and enhanced as we strive to protect the environment for future generations.

Introduction to Sustainability, Second Edition. Robert Brinkmann.

Companion website: www.wiley.com/go/Brinkmann/IntroductiontoSustainability

Figure 1.1 If global climate change is not stopped, Iceland may lose its glaciers (Photo by Peter de Rueter).

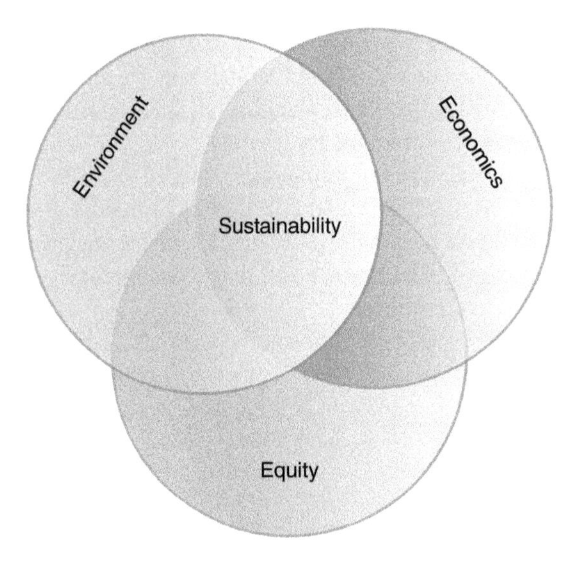

Figure 1.2 The three "E"s of sustainability: Environment, Economics, and Equity. Sustainability is achieved when the three are taken into consideration.

These three components: environment, equity, and economy, are often called the three pillars or three e's of sustainability. All three should be part of any decision making to ensure sustainable development for the future (Figure 1.2).

When businesses and green entrepreneurs think about sustainability, they use similar terms, but call them the triple bottom line: people, planet, and profits. For a business to be sustainable in the modern world, the profits are not the only consideration that must be taken. The impacts of actions on people and the planet are part of the mix. Businesses that embrace the tenets of modern sustainability are often considered green businesses.

Figure 1.3 This is the largest green roof in the United States. It covers the Rouge Factory that manufactures the Ford F-150, one of the least fuel-efficient personal vehicles on the market. Is this greenwashing or a real attempt at corporate sustainability?

Some businesses and other organizations try to embrace the popular environmental practices of our day and promote their efforts as green but in reality follow unsustainable practices. Such discordant behavior is called greenwashing (Figure 1.3).

In our modern world, it is difficult to avoid impacting the planet in some way. The study of sustainability teaches us how we as individuals, organizations, or societies can lessen our impacts so that we can leave our planet in better shape for the future.

As we will see in the next section, while the development of sustainability emerges out of the broad environmental movement of the nineteenth and twentieth centuries, it is deeply informed by the development of global economic and environmental agreements that caused deep concerns in the 1980's.

Nineteenth Century Environmentalism

It could be argued that prior to the western enlightenment and industrial revolution, most people in the world had an intimate relationship with nature. This was due, in part, to necessity. Most of us were farmers or found some way to feed ourselves off of the land and the bounty of nature. At the time, the earth had a larger spiritual role in humanity. The cycles of the moon and stars were more dominant in the non-electrified night sky and the life/death/rebirth annual patterns of nature provided metaphors for daily experiential existence in all of the major religions of the world. Such intimacy exists today in some corners of our continents where the impacts of our modern age are light. Yet, for most of us, the seventeenth century enlightenment put our societies on a path of environmental decline and destruction while at the same time providing amazing technological advances and allowing the creation of the middle class.

The growth in technology during the industrial revolution (lasting roughly from the middle of the eighteenth century to the middle of the nineteenth century) transformed the world in tremendous ways. Urbanization increased and cities grew. At the same time, migration to industrial centers expanded and new markets throughout the world were sought. Europe and North America expanded their spheres of influence.

In the midst of this, many around the world started to question the value of the industrial revolution. Life was grim in many cities and the world started to see mass destruction of natural resources and the decline of air and water quality.

In North America, this critique emerged within the romantic and transcendental movements, particularly in the writing of Henry David Thoreau, author of *Walden* (published in 1854). The romantic and transcendental movements of the nineteenth century idealized nature. Adherents believed that nature helped to transcend the meaning of an ordinary life.

The art of the era is exemplified by the Hudson River school of art that showed humans as observers of grand scenes in nature (this effect was later utilized by Ansel Adams' extraordinary images of the American west in the middle twentieth century). Many of the romantic images of the time were of the northeastern United States or Canada.

This approach to art certainly grew out of other traditions of landscape art found throughout the world, but it uniquely influenced North American thinkers by elevating nature in glorious ways. Nature was depicted as wholly good and as a path to greater enlightenment.

It is this enlightenment that Thoreau sought when he decided to move to a cabin on the property of noted romantic poet, Ralph Waldo Emerson on the edge of Concord, Massachusetts, from his comfortable house in town. He lived simply and contemplated the meaning of life, largely away from distractions of others.

His romantic view of the simple life is one that has been replicated by others for millennia—whether the hermit or the sage of the mountain. There is something innately human about seeking solace in nature. Thoreau, however, placed this experience squarely in the consciousness of the times by writing eloquently about it.

His work certainly influenced many others. John Muir, a Scottish born American naturalist, was perhaps the person who most put Thoreau's writings into practice.

Muir was only eleven when his family moved from Europe to a farm in Wisconsin. He entered the University of Wisconsin when he was in his early 20's and quickly became exposed to the writing of Thoreau. While he never graduated, he took a number of courses in a variety of scientific areas including geology, botany, and chemistry.

His strong religious background and his experience in the beautiful landscape of south-central Wisconsin certainly provided ample opportunity for him to see the hand of God in the works of nature. But as a young man, he set out and saw the world.

Muir completed a number of well-documented travels including a walk to the Gulf Coast of the United States in 1867 and a trip to California in 1868 where he was one of the first western explorers of the Sierra Nevada mountain region—including areas around Yosemite. It was there that he met Ralph Waldo Emerson, the leader of the romantic and transcendental movement. At the time, Emerson was rather elderly and in a slow decline of health. But each had a strong impact on each other.

With time, Muir became known for his own writing and essays documenting the wonders of the west and the beauty of nature found there. He strongly advocated for the preservation of Yosemite in order to preserve its unique natural beauty. His recommendations were followed when Yosemite became a national park in 1890.

The first US national park was Yellowstone and was established by Ulysses S. Grant in 1871, and Canada's first national park (Banff) was established in 1885. Several other nations developed national parks in the same era after the establishment of Yellowstone.

However, it is Yosemite that holds the greatest significance in the history of the sustainability movement, because it is here that we see the development of Muir's ideas about the importance of nature in our life and in providing solace for mankind. He helped to found the influential Sierra Club in 1892, which still works to preserve natural, lands and promote responsible use of the earth's resources.

Pinchot, Roosevelt, and Muir

Muir strongly believed in the total preservation of national parks. He did not believe that the activities of man should interrupt the peace of nature. In 1891, a different type of public land was established—the National Forest. They were established with the intention of providing opportunity for economic development of the resources on public lands (Figure 1.4).

The establishment of this type of public land challenged Muir's preservationist tendencies. The rapid expansion of the west through railroad and shipping lanes brought new settlers and new challenges. Most of the land was public and great economic good could

Figure 1.4 Forested lands like this one in northern Wisconsin provide a sense of peace and tranquillity. However, they can also be seen as holding resources that could be developed.

be obtained from it. There was great demand for timber, for ranching land, and for mining. The establishment of the National Forests allowed the use of the land while maintaining some type of public control.

Gifford Pinchot most articulated this approach to public lands. He became the first head of the US National Forest Service and greatly influenced the future direction of land management on public lands. His family was in the timber business and he knew the impact of poor forest practices on the environment. He decided to learn all he could about how best to protect the land for long-term forest yields. He also developed a strong belief that there should be a national policy around forest management in order to preserve them.

Yet, Pinchot also believed that forests should be utilized to extract the greatest good from them. He developed a conservation ethic that focused on producing the greatest yield possible from the land with as minimal disruption as possible. He is often seen as the father of the *conservation* movement that advocates the wise use of land in order to allow economic gain while preserving it for future generations.

Pinchot and Muir knew each other, but fell out when Pinchot promoted grazing on public lands in 1897. Muir believed that grazing severely damaged land for generations and created unsustainable conditions in forests.

Perhaps the most influential person to embody Muir's and Pinchot's visions was the American President (1901–1909), Theodore "Teddy" Roosevelt (Figure 1.5). An avid outdoorsman, Roosevelt loved being outside in nature and believed that the US should have a distinct conservation policy that protected wild lands. In this respect, he was influenced by John Muir. He had read Muir's writings and even travelled to California to meet with him in 1903. He believed strongly in setting aside public land in perpetuity for the enjoyment of future generations.

At the same time, Roosevelt was influenced by the work of Pinchot. He appointed him to the job as chief of the National Forest service in 1905 and they were friends. In the efforts of Roosevelt we see the realization of the two great visionaries for American public lands: Muir the preservationist and Pinchot the conservationist. Roosevelt found a way to expand both ideals through his conservation efforts.

Aldo Leopold and the Land Ethic

The work of Roosevelt, Muir, and Pinchot were early efforts that provided a framework for managing public space. Yet, there was little understanding about how to manage vast properties effectively. In the early twentieth century, schools of forestry were established and efforts were made to educate a new type of professional forester that was not only interested in seeking timber yields, but also in protecting forests for their intrinsic value.

The most influential of these new foresters was Aldo Leopold. Originally from Wisconsin, he graduated from the Yale School of forestry (one of its first graduates) in 1909 and started his career in land management in the southwestern United States. He developed the first management plan for the Grand Canyon. He eventually became the nation's first Professor of Game Management at the University of Wisconsin in 1933.

Figure 1.5 Teddy Roosevelt was one of the major leaders of the modern conservation movement. This is a statue of him near his home in Oyster Bay, New York.

While in Wisconsin, he purchased a piece of land that was highly impacted by poor agricultural practices in order to try to return it to natural conditions. Part experiment, and part labor of love, this effort provided a fundamental framework for his groundbreaking writing in *A Sand County Almanac,* which was published in 1949 after his death.

This book advocated the development of a land ethic based on ecosystems. Leopold understood that preservation or conservation efforts were somehow flawed. They were not informed by how nature actually worked. Leopold understood through his work in the southwest and Wisconsin that nature was highly impacted by man intentionally or unintentionally. He saw ecosystems, living organisms, and their environment, as the foundation for truly understanding nature and how it should be managed.

In his land ethic, he saw that land should not just be set aside or managed for economic gain as advocated by preservationist or conservationists. Instead, human society needed to understand the components of nature—things such as water, soil, air, and organisms—in order to fully grasp how it worked. If one truly wanted to preserve nature, an ethical system must be developed around the components of the ecosystem. The soil, individual plants, the air, etc. were as important as the land itself.

Better Living Through Chemistry, The Great Smog of 1952, and Rachel Carson

While the large debates over the wise use of public lands moved forward and evolved into Leopold's land ethic, American private land was utilized to advance a new technological revolution centered on chemistry. The world began to see that we were in a new chemical age at the end of World War II, brought about in part by the dropping of atomic bombs over Hiroshima and Nagasaki. The building blocks of nature were discovered and could be transformed in new, exciting, and sometimes deadly ways.

We started to understand the potential of using chemical building blocks for creating new chemicals and products. While there was some concern over the use and management of these emerging products, the world saw the advances as miracles in areas such as fertilizers, pest control, plastics, and fuels. The chemical age saw the transformation of peoples' lives in unimaginable ways (Figure 1.6).

While there was still concern about nuclear war, there was also space exploration, the interstate highway system, and suburban development. The post-war world was a very different place from the pre-war dreariness. We had a sense that we could do anything.

Yet, we started to see that there were impacts from our use of chemicals on the landscape. We saw new forms of pollution, destruction of ecosystems, and new health concerns emerge. One of the first important visible impacts of this new age was the Great Smog of 1952 in London.

This event occurred during a windless period in early December. The stagnant air allowed the buildup of coal smoke that permeated not only the streets of London, but homes and businesses as well. At least 4000 died from the event and tens of thousands became ill. The smog caused great concern among the public about the impacts of air pollution and efforts were made to develop rules to control coal smoke. Eventually, the Parliament of the United Kingdom passed the Clean Air Act of 1956.

While not the first air pollution act in Europe, it was the most important one in that it was the first to develop effective mechanisms for improving public health through the

Figure 1.6 The 1950s and 1960s saw the world change in unimaginable ways. This was my family's dining room in that era. How is your dining room different? How does this change the chemistry around us?

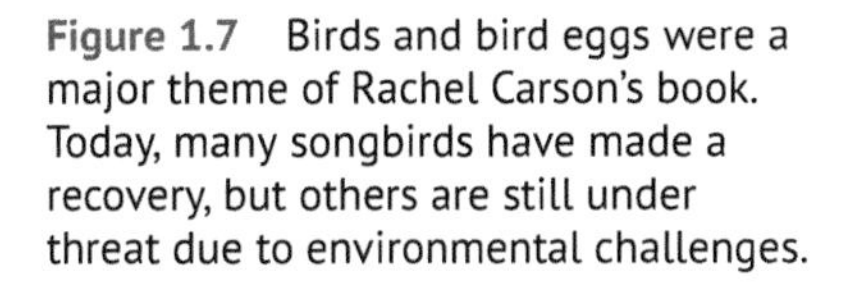

Figure 1.7 Birds and bird eggs were a major theme of Rachel Carson's book. Today, many songbirds have made a recovery, but others are still under threat due to environmental challenges.

development of non-coal fuels and via the regulation of the actions of individual households. The UK's Clean Air Act set the stage for the regulation of a variety of pollutants by national governments around the world.

While the Great Smog was an important event in the development of air pollution controls, the person who is most credited with nudging the world into the understanding of broader pollution issues is Rachel Carson, noted author of the book *Silent Spring*.

Rachel Carson was a nature writer who was connected with some of the leading government and university researchers in the 1950's and early 1960's. Through their work, she learned that there was growing concern in the scientific community about the dangers of unregulated chemical usage to human health and the broader environment. The chemical industry at the time largely acted with impunity and was able to release emissions and create products regardless of their broader impact.

The title of the book, *Silent Spring*, references concerns over the fact that some of the pesticides in use at the time were causing the death of large numbers of birds, thereby causing a silent spring (Figure 1.7). The impact of Carson's writing was significant to the development of the world's understanding that industrialization had distinct costs associated with it. The very existence of nature was impacted by our ability to create new commercial chemicals out of atomic building blocks.

Around the world, people started to question the use of these new organic and inorganic chemicals and investigate their impacts. For example, the investigation of illnesses near a mercury mine in Minamata, Japan led to the discovery of an uncommon illness named for the city, Minamata disease. Water that contained mercury emitted from industrial wastewater systems entered local ecosystems, most importantly the local bay. The mercury accumulated in fish and shellfish that were major sources of food for the local population.

And Minamata was only one of hundreds of cases that were discovered. There were many examples of the environmental impacts of mines and factories on the environment and public health. In addition, there was growing concern over the use of the widespread application of pesticides and fertilizers. Plus, the expansion of the widespread use of coal-burning power plants and individual automobiles in the suburban age began to be felt.

Citizens all over the world started to demand action by their governments to create new laws to protect themselves and their environments from the dangers of environmental pollution. An age of new environmental activism was born.

Environmental Activism of the 1960s and 1970s and the Development of Environmental Policy

New activism emerged in the 1960s as a direct result of widespread environmental contamination and devastation. People began to see the impact of industrialization in their local communities. There was widespread pollution of waterways, soil, and air. In addition, the development of packaging, fast food, and the mobile family caused widespread littering along roadways. The everyday experience of most people was vastly different from that dreamed of by the transcendentalists. Nature was no longer inspirational for most. It was a problem.

Several important works were published that influenced that way we think about the environment and sustainability today. Ralph Nader, long an advocate for the environment and consumer protection published *Unsafe at Any Speed* (1965) about the reluctance of the automobile industry to develop safety measures in cars—things like safety belts that we take for granted today.

The year 1968 was a time of change for much of the world. It also saw the publication of three seminal pieces of writing: *The Population Bomb* by Paul Ehrlich, *Desert Solitaire* by Edward Abbey, and the essay, *The Tragedy of the Commons* by Garret Hardin. Ehrlich's book highlighted concerns of global overpopulation and the impact of large populations on the Earth's environment. While the Earth has been able to accommodate significant population growth since the publication of *The Population Bomb*, concerns remain as to the overall impact of increasing populations on the planet. In similar fashion, Hardin's essay brought to light via metaphor the impacts of economic self-interest on shared resources. It is impossible for us to get everything we want without causing some degree of environmental stress in shared places. This problem is exacerbated when populations increase and resources are stressed. In contrast to the previous two works that focus on population growth and the depletion of resources, Edward Abbey's book brings forward some of the ideas espoused by Thoreau, Muir, and Leopold.

Edward Abbey was a writer and a seasonal park ranger influenced by the American west where he spent most of his adult life. He wrote a number of books including *Desert Solitaire* (1968). This book is a loving tribute to the American West and a memoir reminiscent of the transcendental writers. However, it is also highly critical of the park system and the promotion of tourism within the parks.

His later work, *The Monkey Wrench Gang* (1975) became influential to the philosophy of Earth First! which was founded in 1979 to promote a radical environmental agenda that advocated that there was no compromise in the defense of the environment. Some of their members have been accused of "monkey wrenching", or sabotage, to harm equipment, like bulldozers, that could be used to destroy a natural landscape. They were also accused of

Figure 1.8 How much growth is enough? These developments in Caracas, Venezuela create challenges for infrastructure and also for environmental management.

spiking trees by hammering metal stakes into trunks that could cause breakage of a chainsaw and harm to the operator.

In 1971, Barry Commoner published the book *The Closing Circle* which focused on framing an economy that was less harmful to the environment than traditional open capitalism which he saw as the cause of many environmental problems. He also did not believe that overpopulation was the major problem causing environmental degradation (in contrast to Paul Ehrlich's *Population Bomb* thesis). Instead, he believed that the issues were inherent in the way that people lived within modern capitalistic societies focused on consumption.

Shortly after the release of Commoner's book, the Club of Rome published a book by several authors called *The Limits to Growth* (1972) which returned the conversation to issues of population growth and the impacts of growth on the environment (Figure 1.8). The book, using revolutionary computer models, predicted that the earth's population was running out of resources and that many of the contemporary economic practices were highly unsustainable over the long-term.

The consciousness of the times returned to considering the implications of economic activity within the E.F. Schumacher's important book, *Small is Beautiful: Economics as if People Mattered* (1973). The idea of expansive economic growth and consumption, coupled with the idea that bigger was better was highly critiqued in this work. The question of enoughness is raised. When do we have enough stuff to be happy? Does it take more and more stuff? The implications of wanting more are significant on the environment. Schumacher argues that there are limits to the capacity of the planet to absorb the pollution and provide resources for rampant economic growth. Instead, we should focus on the smallness of what we need and on limiting impacts of economic development.

In 1979, James Lovelock published *Gaia: A New Look at Life on Earth* that claimed that the earth could be seen as a living organism. What happens in one part of the planet will impact other parts in sometimes unpredictable ways. Lovelock helps us view the Earth in more holistic ways. He clearly outlines how the different systems of the earth, such as energy or nutrient cycling interact with each other. Thus, damage done to the planet by humans will likely have unexpected impacts that may end up harming humans in the long term.

The 1960s and 1970s also saw the founding of several key organizations including the World Wildlife Fund, that promotes wildlife conservation, the Environmental Defense Fund, which strives to preserve ecosystems, Greenpeace, an independent group which focuses on direct action to confront environmental problems, and Friends of the Earth, an international environmental advocacy group which protects the environment within social, environmental, political, and human rights contexts; Friends of the Everglades; Worldwatch Institute that examines worldwide data to develop environmentally sustainable solutions; The Land Institute which focuses on sustainable agriculture; and Earth First! that focuses on interventions to protect nature.

The decades also saw the development of key environmental advocates that had important voices within the emerging pop culture of the time that was rife with environmentalism and the overall environmental sentiment brought about by the Age of Aquarius and the widespread hippie movement. While many such individuals could be discussed, three will be highlighted: Lady Bird Johnson, Jacques Cousteau, and Pete Seeger.

Lady Bird Johnson was the wife of US President Lyndon Johnson who served in office after the assassination of President John F. Kennedy. Johnson is considered by many to be the architect of the Federal response to the Civil Rights Movement of the 1960s and was a strong advocate for equality. His wife was very supportive of his efforts. However, she also became very concerned over the state of litter and the environment in the nation. The US interstate highway system was relatively new and many areas were becoming covered with trash. Americans did not have an ethical way to manage garbage along roadways. She took highway beautification on as a major cause in a way similar to how Michelle Obama, the wife of Barack Obama, took on food and health as a platform.

The Highway Beautification Act, sometimes called Lady Bird's Bill, was passed in 1965 and provided rules for billboards, fencing, and the type of development that could be promoted adjacent to roadways. She also sought to plant native flowing plants along roads to improve the overall experience for drivers and passengers. If you see a patch of flowers blooming on a trash-free and billboard-free US interstate, you have Lady Bird Johnson to thank. There is no doubt that her efforts helped to educate a generation of Americans on the importance of trash removal and the aesthetic role of nature in our everyday life. Her celebrity status as the wife of the most powerful man in the US gave her a platform to promote a practical environmentalism that continues to this day.

Jacques Cousteau is another celebrity of the era. He was a French explorer and filmmaker who advocated marine conservation. He became active in the modern environmental movement when the French government planned to dump radioactive waste into the ocean. He was a very vocal critic of such practices. However, he is best known for a series of films that he made that had critical acclaim. He won the Palme D'Or at the Cannes film festival in 1956 for his first film, a documentary, *Silent World*. After that he created a number of films until his death in the 1997.

His films documenting the underwater world were a sensation in the 1960s. They provided a fresh look at a landscape that had been looked at as largely unknown. The seas were convenient dumping grounds because the impact of our waste was largely unknowable. No longer. With Cousteau's work, the seas were seen as amazingly diverse and connected with the broader surface ecology. He became a worldwide celebrity and strongly advocated for marine protection.

There are many American musicians and artists who became active on environmental issues in the 1960s and 1970s. It was a period of activism with actors like Robert Redford lending their celebrity clout to causes. Many songwriters and singers were particularly effective. Some were noted for their regional celebration of places such as John Denver's *Rocky Mountain High.* Others provided important public support for environmentalists advocating for change in local or national policy. The individual who best represents this group of musician activists is the folk singer Pete Seeger.

Seeger was involved with folk music throughout his very long career. He was born in 1919 and died in 2013. He was very involved with labor and civil rights music and protest prior to the Vietnam War. During the war, his music became much more political and edgy, particularly as the Civil Rights movement became much more vital and violent.

Like many artists of the era, he also became involved with environmental causes, most notably the clean-up and preservation of the Hudson River. His music inspired many of the times to get involved in society and seek to change it for the better. The improvements of the Hudson River ecosystem are the result, in part, of Seeger's effort.

The Growth of Environmental Laws in the 1960s and 1970s

In response to public pressure, a slew of new laws emerged during throughout the 1960s and 1970s. In the US alone, hundreds of federal, state, and local laws emerged that strove to protect the environment. Important federal rules include: The Clean Air Act (1963); The National Emissions Standards Act (1965); The Solid Waste Disposal Act (1965); The National Environmental Policy Act (1969); The Formation of the Environmental Protection Agency (EPA-1970); The Formation of the National Oceanic and Atmospheric Administration (NOAA-1970); The Williams-Steiger Occupational Safety and Health Act (creating OSHA-1970); The Lead-Based Point Poisoning Prevention Act (1970); The Marine Protection, Research, and Sanctuaries Act of 1972 (1972); The Endangered Species Act (1973); The Safe Drinking Water Act (1974); The Hazardous materials Transportation Act 1975); The Resource Conservation and Recovery Act (RCRA-1976); The Toxic Substances Control Act (1976); and The Surface Mining Control and reclamation Act (1977). Some of these will be discussed in upcoming chapters.

Environmental rule making continued after the 1970's, but this two-decade period is as important to the environmental policy movement as the 1770s and 1780s were to the growth of democracy in North America and Europe.

Each of these federal laws provided new protections for different areas of the environment. They are among the most significant suite of environmental laws passed by any federal government in the world. While new rules have been created since these laws were passed in the 1960s and 1970s, this time period was the most important era for the development of public policy focused on protection of the environment. Many governments around the world followed the lead of the US federal government by instating rules to protect local, state, and national environments and resources.

The First Earth Day

In the midst of this tumultuous time, there were many events that tried to capture the public's focus on the environment. The most important of these was Earth Day, first held in April of 1970. To many, the event signaled the start of the modern environmental movement.

Earth Day was the brainchild of several individuals, but is often credited to Wisconsin Senator Gaylord Nelson. He was disturbed by pollution problems around the United States and decided to develop a "teach-in" to educate people about the environment.

Since the first Earth Day in 1970, it has been celebrated every year on April 22. The movement has grown from a US event to a broader international event with coordinated activities, themes, and special events each year.

The focus of Earth Day, however, is always on education. Over the years, the themes have changed and new issues have emerged. But, the success of Earth Day is due largely to the flexible nature of the event with its broad focus on educational themes and public ceremonies and entertainment. The spring season, with its elements of rebirth and the signal of hope provide fitting seasonal background to Earth Day events.

International Concerns

The 1960s and 1970s was a time of significant global conflict. Several important events included the Vietnam War, the Arab–Israeli Conflicts, The Arab Oil Embargo, Coups in Africa and Latin America, and the Iran Hostage Crisis. At the same time, the United Nations emerged as an important body that could assist international cooperation and promote peace.

Increasing population led to higher death tolls from natural disasters. For example, a tropical cyclone in Bangladesh killed 300 000 in 1970—a shocking number of fatalities. In addition, there were cases of drought induced starvation or thirst, often a result of poor rural development throughout many parts of the world.

Some of these problems, including conflict, death from natural disasters, and poor development, were seen within the context of the environment—and the environmental issues were often beyond the capability of a single nation to manage. Thus, the United Nations was seen as an organization that could facilitate not only interventions in environmental crises, but also promote international agreements to ensure the protection and preservation of the environment.

The UN and other organizations were able to craft some key agreements. Some of them are discussed below.

Ramsar Convention on Wetlands in 1971—this UN treaty focuses on maintaining the ecology of wetlands and promoting sustainable uses of them.

Establishment of the UN Protectorate of the Environment in 1972—the mission of this group is "To provide leadership and encourage partnership in caring for the environment by inspiring, informing, and enabling nations and peoples to improve their quality of life without compromising that of future generations." The protectorate essentially manages most of the environmental initiatives of the United Nations.

Declaration of the United Nations Conference on the Human Environment (Stockholm Declaration) 1972—the first agreement to recognize the right to an "environment of a quality that permits a life of dignity and well-being..."

Convention for the Protection of World Cultural and Natural Heritage 1972—this agreement established protections for both important cultural and natural sites around the world, thereby elevating historic sites into the previous efforts on the preservation of wilderness.

Establishment of the Convention on International Trade in Endangered Species of Wild Fauna and Flora (CITES)—this 1973 voluntary agreement was organized by the World Conservation Union. There are now 183 signatories to the agreement.

Each of these early agreements helped the world recognize that efforts could be made to establish important links between the developed world and the developing world in order to improve conditions for all. The agreements also helped to establish the general notion that what happens to one group of people has implications for the rest of the world's population. We started to understand that the actions of one person, organization, business, or government could unknowingly make a positive or negative impact on the lives of others around the planet.

Ozone and the World Comes Together

Within this context, alarm bells started to ring within the scientific community about problems associated with ozone depletion in the upper atmosphere. Ozone, or O3, is a chemical that occurs naturally within the atmosphere. It can form from atmospheric pollution in the lower atmosphere to cause significant respiratory stress on high-smog days in urban settings. However, in the upper atmosphere, ozone shields us from ultraviolet radiation coming from the sun.

In the 1970s scientists started to recognize that a group of chemicals called chlorofluorocarbons were destroying the ozone within the stratosphere. There was significant concern about the damage that the unfiltered ultraviolet radiation would do to ecosystems—particularly humans. Excess radiation is known to cause skin cancer. The locations of the thinnest parts of the ozonosphere were found to be at the polls. A number of maps were created to outline the extent of the "ozone holes" year by year. Yet, there was very little policy advocated for the removal of CFCs from use.

Although CFCs are damaging to the upper layers of the atmosphere, they are very useful chemicals. They are effective in refrigerant systems, are excellent propellants in aerosol sprays, and work well as solvents. So, there was significant resistance to ban or reduce these very effective chemicals in everyday life.

However, as the impact of CFCs increased in the 1970s and 1980s, and the ozone holes over the poles grew, something had to be done. Two important events occurred in the mid

1980s to try to address these issues. The first was the Vienna Convention for the Protection of the Ozone Layer (1985) and the second was the Montreal Protocol on Substances that deplete the Ozone Layer (1987). The first is an international agreement to protect the ozonosphere and the second is a legally binding international agreement to reduce ozone-depleting chemicals.

The advent of the Montreal Protocol in 1987 is extremely significant. For the first time, the world realized that it needed to act together to reduce the impact of harmful pollution produced by most nations of the world. One nation could not act unilaterally to solve the problem. Instead, agreements about reductions needed to be reached that all nations could agree on to try to solve the dangerous problems of ozone depletion.

Fortunately, the agreement worked and the chemicals associated with ozone destruction in the upper atmosphere are declining for the most part. The successful development of the Montreal Protocol demonstrated to the world that international agreements could be forged to solve real environmental problems. Unfortunately, ozone was but one major problem confronting the world in the 1980s.

Globalization and the Brundtland Report

The term globalization is used to describe the process by which exchanges between countries lead to a sameness of culture or attitudes, often through an integration of economic or transportation systems. Globalization has been taking place for centuries. For example, when the Romans expanded their empire and "Romanized" their conquered territories, they brought with them to the new lands their architecture, arts, language, and other forms of culture. That is why there are Roman roads in England and Roman theaters in North Africa. But, globalization is a two-way street. Christianity, which had its birth in western Asia, found a willing home in Rome within a short time. Plus, there were extensive biological and cultural exchanges that occurred during the Columbian Exchange in the 1500s between the Americas and the "Old World" of Europe, Africa, and Asia.

Certainly we can find many examples of globalization within the history of our culture. But, the modern form of globalization started to accelerate in the 1980s.

Prior to this time, there certainly was international trade, communications, and cultural exchanges. However, with the advent of modern transportation systems, open trade agreements, and telecommunications and computing, the world changed. Resources could be moved around quickly. Factories could locate in areas of cheap labor, and decisions made in financial capitals like Tokyo, Paris, Cairo, or New York could impact rural areas in Brazil, Yemen, Gabon, or Myanmar.

Plus, it became clear that globalization connects us in unexpected ways. We immediately saw the devastation of environmental disasters like the chemical leak in Bhopal India that killed thousands in 1984, the Chernobyl nuclear plant devastation in 1986, and the 1988 forest fires of in Yellowstone that burned almost 800 000 acres. Each of these environmental events, while not impacting everyone across the world, were felt emotionally by everyone at the time due to the magnitude of the impact—particularly since many of them were caused by worldwide trends of globalization.

In our modern era, globalization is something we take for granted. Many of us buy products in our local stores that were manufactured in far distant places. We have very little connection to the means of production. Many of us live in consumer societies separate from the impacts of factories, shipping, or resource extraction. Goods are routinely shipped around the world in expanding networks of global trade.

Even our food is shipped all over the world. We can buy fruits and vegetables that are out of season in our area from areas of the world where they are in season. We live in amazing times.

In the 1980s the process of globalization was a relatively new trend. While there certainly were international trade and global connections, there was not the current expansive trade we see today.

As globalization started to expand, many started to become concerned about the impacts of globalization on the environment and the world's cultures. The concern framed itself within broad discussion around sustainable development due to the vast evidence for the deterioration of environmental and social conditions in the wake of globalization. In addition, there were clear winners and losers as a result of this economic global change and economic disparities were becoming more evident.

Since its creation, the United Nations has been concerned about these issues. However, in the late 1970s and the early 1980s there was evidence that conditions were deteriorating and many felt it was important to try to address the problem.

Thus, the United Nations in 1983 established the World Commission on Environment and Development. The group was chaired by Gro Harlem Brundtland, the former Prime Minister of Norway. The group was charged with finding ways to develop strategies for global sustainable development. Their groundbreaking work, the *Report of the World Commission on Environment and Development: Our Common Future*, often called the Brundtland Report, became a key document outlining the future of international sustainable development.

The Brundtland Report was the first to provide a concise definition of sustainability within the context of sustainable development, "Development that meets the needs of the present without compromising the ability of future generations to meet their own needs." This definition is commonly used today.

But, the Brundtland Report was not just important for providing one of the first important definitions of sustainability. It also changed the way we look at environmental issues. The report detailed that most environmental problems are not the result of just the action of one person or industry. Instead, most environmental issues are the result of an integrated set of actions and situations that lead to environmental decline. Key to this understanding is the recognition that problems with issues like population, food, species loss, energy, industry, urbanization, and human settlement are connected. We could no longer think of environmental issues as easy to solve by just regulating a pollutant or industry. These issues were cultural and social issues that required deeper understanding of the root of the problem.

The report also detailed the impacts of inequality to the environment. Between nations the beneficiaries of industrialization do not often feel the negative effects of industrialization. Even within cultures, differences in class, gender, education, or age can lead to differences of environmental experiences and harms.

The report also noted that there are distinct limits to growth due of industrialization due to lack of resources to support long-term growth. The report is linked here and should be perused to get a sense of the breadth of content: http://www.un-documents.net/wced-ocf.htm For example, the report discusses at great length the state of energy options across the world and challenges for the future. Similar types of summaries are made for food, population, endangered species, industrialization, and urbanization.

The report is often praised for clearly outlining the problems the world was facing in the 1980s. Yet, many have been disappointed with the lack of progress on many of the issues that were outlined.

The weakness of the report is within the recommendations section. It essentially suggests strengthening national laws, international agreements, and improving global institutions to manage the issues. Yet the report does not provide any specific measurable outcomes by which to measure success. As we will see, this emerges as a key initiative after the 1990s.

Although the Brundtland Report has been critiqued for not having enough clear solutions for the problems it identifies, it stands as one of the most important documents in the sustainability literature because (1) it clearly defines sustainable development, (2) it recognizes that environmental problems are linked with social issues such as governance, poverty, class, and gender; and (3) it highlights that there are limits to industrialization due to diminishing resources.

Globalization continues to be an issue of great concern within the environmental community for a number of reasons and the Brundtland Report only started the conversation. One of the leading voices providing concern about the impacts of globalization on society and the environment is Vandana Shiva. She is a leading critical voice in the anti-globalization movement—a movement that rejects the negative aspects of globalization while embracing the positive developments of international cooperation. She believes strongly in the importance of local, traditional practices, over unsustainable global enterprises.

Shiva has worked in a number of important areas. For example, she has been a critic of the impacts of genetically modified food production. Large agricultural corporations patent their seeds and make farmers dependent upon not only the seeds, but also the required maintenance and fertilizers recommended by the companies. This takes farming away from traditional practices and hurts the ability of farmers to save seeds for future generations. The winners in this practice are the western corporations and the losers are the local farmers who become dependent upon a globalized system of agriculture.

Deep Ecology

About the same time that globalization was emerging as an environmental issue, the philosophical movement known as deep ecology evolved near the end of the 1970s as somewhat of a critique of the bureaucratic response to environmental decline and the overall lack of progress toward environmental improvement. The movement was in some ways founded on the writings of Norwegian Arne Naess and was deeply informed by the works of Leopold, Carson, and Abbey. Many other contemporary authors promote deep ecology principles and it is still an important approach advocated by some in the environmental movement today.

The basic tenet of deep ecology is that the environment and nature as a whole has value regardless of its utility. It should therefore be protected not for its utility, but for its very right to exist. Man is seen not as a protector of nature, but as a problem for it. Thus, deep ecologists often advocate simple living, lowering of human population, and deeper connections with nature. Some have developed alternative communities such as Dancing Rabbit Ecovillage (see www.dancingrabbit.org).

In many ways, deep ecologists have strong connections with the historic Muir/Pinchot dialectic, clearly residing in the Muir camp. They have a discomfort with the idea of utilizing nature for the greatest good of mankind. Instead, they advocate preservation of nature and treading lightly upon it. As we will see, the deep ecology movement greatly influences the modern sustainability movement. However, many are critical of the view of sustainability as a field that promotes some of the traditional notions of conservation and nature as a source of economic development.

Deep ecology also greatly impacted the world of environmental activism. Earth First!, for example, believes in resorting to a variety of approaches to try to stop environmental decline. Deep ecologists garnered significant attention in the 1990s as they tried to stop deforestation of old growth forests. They conducted a number of protests and events to raise attention to the cutting down of old trees. Some resorted to climbing trees to live in them to prevent their felling. The most famous of these "tree sitters" is perhaps Julia Butterfly Hill. On December 10, 1997, she climbed an at-risk tree she later named Luna, and stayed in it for 738 days. She prevented the tree from destruction, and other trees in a 200-foot buffer, as part of her agreement to leave the tree. Hill's effort inspired a variety of non-violent activism around the environment throughout the globe.

While encouraged by the efforts of individuals like Julia Butterfly Hill, deep ecologists become frustrated with the general lack of measurable action on the environment and become frustrated with efforts to preferentially accommodate businesses or individuals over the environment.

Environmental Justice

While the deep ecologists were focused on the preservation of natural landscapes, a new group of activists and researchers began to raise concerns over the preservation of their communities and cultures (Figure 1.9).

The environmental community has traditionally been dominated by white men. As the movement evolved into the 1970s and 1980s there was growing criticism that environmentalists were willing to turn a blind eye toward the environmental issues in communities of color in favor of natural landscapes devoid of people. Several cases highlighted this criticism. Hazel Johnson, often considered the mother of the environmental justice movement, provides a point of departure for discussing one of the most famous of these cases.

Hazel Johnson was an activist on the South Side of Chicago. She became concerned with environmental issues after she started to notice that several people she knew became ill for no apparent reason in the early 1980s. She started to document cases of cancers and other sicknesses and became convinced that there were environmental reasons for the illnesses.

Figure 1.9 This image of street art in Brooklyn gives one a sense of the challenges associated with diversity and fairness in urban issues.

As she conducted her work, she found that her neighborhood had the highest rates of cancer in Chicago. She also learned that the community was surrounded by dozens of landfills and dozens of leaking underground storage tanks. She found that the drinking water for many in the community was contaminated with chemicals leaking from these underground sources.

Johnson's efforts, and those of others such as Robert Bullard (who will we read about Chapter 11) helped define environmental justice as one of the most important new areas to emerge in our era into the modern environmental movement. Eventually, President Clinton in 1994 signed an executive order to ensure environmental justice was taken into consideration in federal efforts. The law also directed the government to take action within minority and low-income communities to ensure that efforts were made to improve environmental conditions. Also, the EPA started the Office of Environmental Equity in 1992 (now the Office of Environmental Justice).

Since then, environmental justice has been applied rather broadly to encompass a variety of different types of regions and situations. For example, environmental justice can be a consideration when examining the impacts of consumer societies on the developing world or when examining the impacts of international conflict on communities. As we will see in Chapter 11, this new approach provides opportunities for community empowerment, renewal, and revitalization.

Measuring Sustainability

One of the ways in which we have looked to address the failures of the past is by developing measurable sustainability goals for the future and assessing success after a period of time via an outcomes assessment process. Thus, it is not enough to set goals, one has to develop procedures to achieve goals, measure the success toward reaching the goals, and evaluate how to improve practices to achieve unmet goals. This approach toward measuring sustainability is an outcomes-based approach that is the cornerstone of the modern environmental movement. It is no longer enough to do small things toward the environment in a feel-good way. One has to measure the outcomes in order to assess whether or not one is truly achieving appropriate goals.

Measurement of goals allows comparisons year to year, from place to place, or from organization to organization. This comparison allows "benchmarking" of outcomes to assess success and best practices. Two states, for example, could compare efforts to reduce greenhouse gas pollution. The state that is most successful could serve as an example to the other state in order to find pathways toward positive outcomes.

This idea of sustainability measurement has been around for a long time. For example, many have tried to lower pollution levels for decades prior to the modern push for sustainability measurement. However, the comprehensive approach toward measuring sustainability indicators expanded greatly in the late 1990s and into the present era. We will discuss sustainability measurement in great detail in Chapter 3, but it is worth reviewing briefly here.

One of the first important international sustainability indicators is the UN Millennium Goals, which evolved out of the 2000 United Nations Millennium Summit. All of the nations of the world signed up to the goals of the summit, which arguably can be traced to the Brundtland Report. The intent of the summit was to achieve the goals by 2015. The major eight goals are:

1. Eradicating extreme poverty and hunger
2. Achieving universal primary education
3. Promoting gender equality and empowering women
4. Reducing child mortality rates
5. Improving maternal health
6. Combating HIV/AIDS, malaria, and other diseases
7. Ensuring environmental sustainability
8. Developing a global partnership for development

Each of these goals has set targets and a timeline for meeting them. For example, Goal 7, Ensure environmental sustainability, has four sub-goals with distinct measurable outcomes:

7A. Integrate the principles of sustainable development into country policies and programs; reverse loss of environmental resources

7B. Reduce biodiversity loss, achieving, by 2010, a significant reduction in the rate of loss
- Proportion of land area covered by forest
- CO^2 emissions, total, per capita and per $1 Gross Domestic Product
- Consumption of ozone-depleting substances
- Proportion of fish stocks within safe biological limits
- Proportion of total water resources used
- Proportion of terrestrial and marine areas protected
- Proportion of species threatened with extinction

7C. Halve, by 2015, the proportion of the population without suitable access to safe drinking water and basic sanitation
- Proportion of population with sustainable access to an improved water source, urban and rural
- Proportion of urban population with access to improved sanitation

7D. By 2020, to have achieved a significant improvement in the lives of at least 100 million slum-dwellers
- Proportion of urban population living in slums

Each of the sub-goals of Goal 7, ensure environmental sustainability, has distinct measurable outcomes associated with it. No longer was it enough to say that species would be protected or that countries would do all they could to provide safe drinking water to their populations. Now, the world was expecting real significant outcomes to improve the lives of humans and a stronger protection of the environment.

Since the development of the Millennium Development Goals, the United Nations approved the Sustainable Development Goals. They will be discussed in more detail in Chapter 3. But it is important to note that there are a number of other measurable goals that have been developed to assist with sustainability management. The LEED green building rating system provides another example of the types of outcomes-based approaches that have emerged in recent years.

LEED stands for leadership in energy and environmental design. The LEED green building rating system was developed by the US Green Building Council in the late 1990s as a way to rate the "greenness" of buildings based on a variety of measurable criteria.

LEED is a point-based system that allows buildings to be rated as silver, gold or platinum based on performance in a number of areas. Plus, LEED certification can be done on new construction, existing buildings, commercial interiors, building cores, retail buildings, schools, homes, and neighborhoods.

In new construction, for example, points can be earned for site selection, development density, brownfield development, transportation infrastructure, habitat protection, storm water design, light pollution reduction, water use reduction, landscaping choices, wastewater management, energy systems, waste management, materials reuse, and indoor air quality. While not a comprehensive list, the previous list demonstrates the complexity of green building certification.

Green buildings are becoming more common and there are thousands of them that have been certified since certification started in the late 1990s. Some notable LEED buildings include the Empire State Building (green renovation) and Target Field in Minneapolis (home of the Minnesota Twins).

As we will see, this type of green certification and outcomes-based sustainability measurement is very common and a source of many jobs for students interested in sustainability careers.

Some have critiqued the sustainability benchmarking movement as too simplistic. For example, the LEED rating system provides excellent ways to measure the green technology used in a building, but does not evaluate how a building is to be used. In other words, you could build a giant mansion with all kinds of green bells and whistles. You could have it run totally on solar energy and utilize composting toilets in all 13 bathrooms. However, if the mansion were for you and your small family, it would not nearly be as green as living within a small conventional apartment.

Thus, it is important to evaluate the nature of the green rating system to examine the values behind it. The US Green Building Association is clearly most concerned about advancing green technologies in order to improve the overall impact of the building on the planet. For the most part, they do not concern themselves with the use of the building after it is built.

Some rating systems may also hide the complexities of an issue. Take for instance the push for the use of recycled paper. Imagine that you are a paper company that utilizes pulp from paper derived from sustainably managed forests in Canada. Your paper mills are fueled by hydroelectric power which is considered a green source of energy. In some rating systems, this would be considered a non-sustainable paper product because the paper does not have recycled content.

The issue can become more complex. Imagine you are a paper purchaser near the paper company noted above. If your firm decides to purchase only recycled paper, this requires you to import paper from a recycling mill that may be hundreds of miles away from the local source. By the measure of many sustainability indicators this would be the appropriate decision. However, when taking a number of variables into consideration, such as carbon footprints, transportation emissions, and community impacts, the decision to follow a sustainability measurement scheme becomes rather complex.

Thus, while sustainability benchmarking systems seem very basic at first glance, they are actually rather complex systems. It is important to understand the purpose of the scheme and understand the implications of choosing to rely on a benchmarking system in any organization. We will discuss this in more detail in Chapter 3.

The Climate Change Challenge

At the start of this chapter, I noted that Iceland held a funeral for the loss of its first glacier. All around us there is evidence for climate change. For many, climate change is the key sustainability problem. There is no doubt that climate change is linked to many of the topical issues covered in the upcoming chapters. It is anticipated that without rapid action on reducing greenhouse gases the world will see many more problems.

Hurricane Sandy was a superstorm that devastated the coast of the Mid-Atlantic region of the United States in October of 2012. It hit the most populated coastal region of the US near the New York Metropolitan area. Striking at night with deadly force, many were surprised by the strength of the storm and its far-reaching impacts.

In Manhattan, one of the five boroughs of New York City, flooding occurred in many areas including the famous neighborhoods of the Lower East Side, Chinatown, and The Battery. Subway tunnels flooded and residents were without power for days. In Staten Island and Queens, other boroughs of the city, a storm surge destroyed entire neighborhoods and caused dozens of deaths.

Was this storm caused by global climate change? In just two years, the New York region was hit by two sizable hurricanes. This has not happened in the history of the city which more typically experiences a sizable storm every half century or so.

After the storm, the Mayor of the City, Michael Bloomberg stated, "Our climate is changing. And while the increase in extreme weather we have experienced in New York City and around the world may or may not be the result of it, the risk that it may be—given the devastation it is wreaking—should be enough to compel all elected leaders to take immediate action."

This book is a textbook, but it is also a call to action. As you learn more about sustainability, think about how you can make a difference to make the world a better place for future generations.

The Road Ahead

The remainder of the book is divided into a total of 13 chapters. Each provides a distinct view of a major theme of sustainability. Along the way, I will introduce you to different experts on sustainability who are making a difference in research, community activism, economic development, or environmental protection. Prior to outlining the book, I think you should know a little bit about me, your author.

I am originally from a small village in southeastern Wisconsin called Waterford (Figure 1.10). This location is right in the middle of North American and in the region of the United States sometimes called the Upper Midwest. By car, my family could get to Chicago in about an hour and a half or to Milwaukee in about an hour. However, we did not travel to these places very much. Instead, we spent much of our free time in the woods of northern Wisconsin near the border of the Upper Peninsula of Michigan where we had a cabin. We would spend weekends and vacations in the woods and on the water. While we had all the comforts of a modern middle-class family, we were certainly more outdoorsy than most.

After graduating from high school I attended the University of Wisconsin at Oshkosh where I majored in Geology. I fell in love with mineralogy and with geologic fieldwork, especially after taking a field school that took me all over northwestern Canada in the Yukon, British Columbia, and Alberta. I wanted to become a mining geologist. I went on for my masters in geology at the University of Wisconsin at Milwaukee where my thesis was on the glacial geology of a portion of western Wisconsin near the Mississippi River. During this time, I worked for a mineral exploration company looking for particular minerals in stream sediment. The idea was to try to find trace minerals associated with gemstones in order to try to find deposits of rare gem minerals.

While conducting this work in rural areas throughout the Midwest, I realized that the surface of the earth was much more altered than most geologists thought it was. The world

Figure 1.10 Your author hiking on our family property in northern Wisconsin in 1968. How did your upbringing impact the way you think about the environment?

was not the simple layers of geology modified by earth forces. The planet was changed by the forces of its human population. Since my epiphany, scientists have caught on to my observations and now call the current geologic age the Anthropocene.

As I reflected on the implications of my discovery, I realized that there was scant information about the role of humans on altering the earth's surface. I was aware of the work of people like Leopold and Carson, but there were few conducting systematic work on the alteration of the surface of the earth. Eventually, I decided to make that the major theme of the rest of my career. I have worked at two major universities, the University of South Florida, and my current home, Hofstra University, to conduct this work.

In order to make that happen, I decided to work with noted soil scientist, Robert Eidt. He developed the term, anthrosol—a soil that is modified by the actions of man. I completed my dissertation on the distribution of lead pollution in soils in the late 1980s.

Over the last three decades, I continued to work on soil and sediment pollution issues. I've studied lead pollution in soils in Florida and I also studied the distribution and magnitude of pollution in street sweeping sediment, storm water, and street debris. I have published a number of pieces on these topics.

In recent years, I have become more and more concerned with problems associated with greenhouse gas pollution and global climate change. I was educated on some of the policy issues when I took a summer course at the Vermont Law School with Patrick Parenteau.

Since then, I've written widely on greenhouse gas policy and a variety of other sustainability issues.

Of course, my work has also stayed close to my roots in the geosciences. I deeply love nature and the field of geology. I continue to conduct research on natural systems and have published works on caves and sinkholes.

I also love to garden and spend time in nature. I have a blog that you can visit called *On the Brink* that includes my informal writing on the environment (http://bobbrinkmann .blogspot.com). I've travelled to all continents but Australia and Antarctica and have been to all of the US states except Oregon. I love to take walks in nature and love parks. Whenever I travel, I try to visit as many parks as I can to get a sense for the landscape and to better understand how different regions of the world organize their public lands.

As a gardener, I am also interested in food. We have greatly transformed our agricultural system in my lifetime and I am interested how this impacts us not only from a health perspective, but also from an ethical standpoint. Most of us have very different relationships with the plants and animals that provide us food than our parents and grandparent had.

During my lifetime the world has changed greatly. In some ways, it has changed for the better. In others, it has changed for the worse. Readers of this book will likely have the same experiences. We have serious energy and water issues, populations continue to expand, and global climate change is likely to vex us for decades.

My purpose in writing this book is to provide an organizational framework for understanding these and other issues within the new field of sustainability. I've had the opportunity to work as a researcher as the field developed and I think my approach to the discipline is a suitable starting off point for someone interested in learning about sustainability. In this book, you'll find that my approach is systematic in that the organization of the text takes us from one theme to another and build upon each other. However, I try to bring in real-world applications of the content and introduce you to people who are at the cutting edge of the sustainability field. I have also provided a series of interesting Weblinks within each chapter that can take you in new directions and that provide real-world information about the work of others in the field.

Organization

This book is divided into four parts, each with three or more chapters. Part 1 focuses on the roots of sustainability. In this section, we will explore the historic background of the field (as we have done in this chapter), provide a basic understanding of earth systems, and review ways that sustainability is measured. Part 2 examines sustainability and natural resources. There are chapters on energy, greenhouse gas management and climate change, water, and food and agriculture. Part 3 brings us toward the understanding of sustainability and communities with information on green building, transportation, waste, and environmental justice. Finally, the book concludes with Part 4 with sections on planning, sustainable development, green economies, business sustainability, and sustainability at colleges and university.

I hope that you enjoy this book. If you have any suggestions for improvement, please do not hesitate to contact me at Robert.brinkmann@hofstra.edu

When I tell people that I teach and write in the sustainability field, their first question, after a bemused look is to ask me, "what's that?!" Some of the sustainability majors in my department have told me that their friends do not fully understand the major or the discipline and that they have a hard time explaining the major to their parents. Here are some of the majors in the environmental sustainability field and roughly what they study.

Environmental Science. An environmental science major takes courses on the major earth systems sciences such as ecosystems (biology), water and other resources (geology, meteorology, hydrology, and geography), environmental chemistry, and environmental engineering. Students in this field often take a number of supporting science and math courses. Students taking an environmental science degree typically earn a bachelor of science (BS) degree.

Environmental Policy. Environmental policy majors focus their studies on a range of issues related to the development and management of environmental policies. They take a number of liberal arts and applied courses in areas that may include government and policy, planning, sociology, economics, and business. Students that earn an environmental policy degree usually obtain a bachelor of the arts.

Environmental Studies. An environmental studies major usually looks at environmental issues within the humanities. They will take courses in literature, philosophy, and art. Students that earn this degree often receive a bachelor of science.

Sustainability Studies. A sustainability major often links science and policy in order to find ways to improve overall environmental, social, or economic sustainability. Due to the diversity of the field, sustainability curricula vary widely. However, sustainability tends to be much more applied than any of the previously mentioned degrees in that the focus of the degree is solving real-world problems via achieving measurable improvements.

There are many other types of environmental degrees that are offered throughout the world including environmental management, environmental planning, environmental economics, environmental geology, and urban ecology. Each offers differing perspectives on the field of the environment.

What environmental courses and degrees are offered at your university or school?

2

Understanding Natural Systems

The world around us works through a variety of organized systematic natural processes (Figure 2.1). It is the goal of this chapter to review them. Starting with understanding the Earth's rock system, we will progress into the hydrosphere, a variety of chemical cycles, including the important carbon cycle, and conclude with a review of the life cycle and the distribution and importance of ecosystems on the planet.

The Earth, its Layers, and the Rock Cycle

The Earth is approximately 4.5 billion years old. Located in a lonely portion of the universe, our planet and our associated solar system consists of a variety of silicon and iron based objects including other planets, moons, and of course, our sun.

The Earth itself consists of a series of layers known as the core, the mantle and the crust. The core consists of two layers, a solid inner layer and a molten outer layer. Both layers are made up of iron and nickel and exist at pressures and temperatures unimaginable to us living at the Earth's surface. Above the core, again at tremendous heat and pressure, is the mantle. This layer is rather thick and (2900 kilometers thick) acts somewhat like a solid and somewhat like a plastic. Because of these properties, the mantle has distinct currents like extremely slow boiling water that moves the mantle around. The currents carry the thinnest layer of the Earth, the crust, around across the planet at very slow rates. It is these movements that are responsible for Earthquakes and most volcanoes.

But it is the Earth's crust that serves as the home to all life on our planet. There really are two major types of crustal systems: oceanic crust and continental crust. These crustal types are divided into several distinct plates that slowly move across the planet as a result of the deep mantle currents in a process called plate tectonics. The oceanic crust is denser and consists largely of volcanic materials extruded from undersea volcanoes in great undersea rift valleys sometimes called spreading centers. The continental crust is lighter and made of a great variety of rock types that formed in a number of different conditions.

The main building blocks of the Earth are minerals, which are naturally occurring substances with distinct chemical, physical, and crystalline properties. There are nearly 5000 types of minerals that have been identified, but most of the surface of the Earth is made up of some commonly occurring minerals. Broadly, minerals can be divided into two classes: silicates and non-silicates. Some of the most commonly occurring minerals are within the

Introduction to Sustainability, Second Edition. Robert Brinkmann.

Companion website: www.wiley.com/go/Brinkmann/IntroductiontoSustainability

Figure 2.1 This chapter focuses on understanding natural systems. While much of the world has been altered by humans, it is important to understand the basics of how the major Earth systems work.

broad silicate mineral group. Approximately 90% of the Earth's crust is made up of them. They are among the most diverse of mineral types and include minerals like quartz, feldspar, clays, hornblende, biotite, zircon, and pyroxene.

The non-silicates, while not as abundant as the silicates, are also important minerals and include halides like halite (salt), carbonate minerals like calcite or dolomite, sulfate minerals like gypsum, phosphate minerals like apatite (a major source of phosphate fertilizer), native element minerals like gold and copper, sulfide minerals that are the main ore or many metals, and oxide minerals like bauxite, hematite, and corundum. Groups of one or more minerals together make up rocks.

There are three broad classes of rocks: igneous, sedimentary, and metamorphic rocks. Igneous rocks form from the solidification of molten rock. There are two ways that this can happen: from the solidification of molten rock deep within the ground or when liquid rock in the form of lava makes its way to the surface. Let's take a deeper look at these two forms of igneous rocks. Then we will move forward to discuss sedimentary and metamorphic rocks.

Igneous rocks that solidify from magma deep within the Earth are known as plutonic rocks or intrusive igneous rocks. Often, the resulting rock is a large solid body that is known as a batholith. Igneous rocks that solidify from molten rock extruding from a volcano are known as volcanic rocks or extrusive igneous rocks. When such materials extrude from a volcano they create lava flows or ash fields that turn into local or regional rock bodies.

While the two rock types can have similar chemistry, they have very different appearances. Intrusive igneous rocks can take tens of thousands of years to cool. Crystals in the

Table 2.1 Intrusive and extrusive rocks of similar chemistry have distinctly different properties and names.

	Intrusive	Extrusive
High Silica	Granite	Rhyolite
Intermediate	Diorite	Andesite
Low Silica	Gabbro	Basalt

rock form slowly and can be rather large. In contrast, the crystals in extrusive igneous rocks, when present, are small due to the rapid rate of cooling at the surface. Due to the violent nature of some of the volcanic events that produce extrusive igneous rocks, some of the rocks can be in the form of ash or a solidified bubbly froth known as pumice. In some cases, the rocks freeze so quickly upon extrusion from the Earth that they have no crystalline structure at all and turn to a glass known as obsidian.

The two types of igneous rocks can be classified based on their chemistry and mineralogy. Table 2.1 lists several igneous rock types based on their silica content and mode of formation (intrusive or extrusive).

Sedimentary rocks, in contrast, form from deposition of mineral matter at the surface of the Earth, either on land or in water. There are three main forms of sedimentary rocks: clastic, biochemical, and chemical sedimentary rocks. Each will be discussed below.

Prior to reviewing the rock types, it is important to discuss chemical and physical rock weathering. When rocks are exposed at the surface, they react to the climatic conditions. A variety of physical processes can impact the rocks: freezing and thawing, salt crystal growth, and biological effects such as tree root growth. At the same time, a number of chemical processes can dissolve or transform minerals into other minerals. When this happens, ionic materials, like calcium, sodium, potassium, or bicarbonate can be released into solution where they can enter water systems and become part of geochemical cycles such as the nutrient cycles for plants.

Clastic sedimentary rocks are rocks that form from other pieces of rock that somehow eroded from another rock type. These sediments can be transported by wind or water in great quantity into some type of depositional environment, often a low or deep area. With time, the sediment can solidify in a process called lithification.

Clastic sedimentary rocks are classified based on the size of the sediment. Three main sizes are recognized (although each is subdivided by sedimentary rock specialists): gravel (> 2mm in diameter), sand (0.0625 mm–2mm); and mud (<0.0625mm). Clastic sedimentary rocks made of rounded gravel are called conglomerates while angular gravel rocks are called breccias. Rocks that are made of sand are unsurprisingly called sandstone. Rocks made of mud are called mudstone or shale. It is worth noting that there is no chemical or mineral designation made in the classification of clastic sedimentary rocks. However, many minerals do not survive the erosion or transportation process. Thus, some clastic sedimentary rock types are mineralogically homogenous. Most sandstone, for example, is made almost entirely of quartz.

Biochemical sedimentary rocks are a series of interesting rocks that form through biogenic processes. Most of us have eaten shellfish like oysters or mussels. The shells that remain after eating the protein are a mineral substance, usually calcite that can turn into a sedimentary rock. Coral and oyster reefs are examples of living ecosystems that constantly create new mineral material. By far, the most common biochemical sedimentary rock is limestone. When thin sections of this rock are seen under a microscope, one can see large to small fragments of shell and calcium rich feces. Coal and chert are other biogenic rocks.

Chemical sedimentary rocks are among the rarest of the common sedimentary rocks. They typically form in depositional environments like lakes or shallow seas where the water becomes saturated with dissolved minerals. The Great Salt Lake in Utah and the Dead Sea in the Middle East are examples of places where these conditions currently exist. In these locations there is more evaporation of water than input of water. The water becomes more and more saturated with dissolved minerals. Eventually, the water becomes supersaturated and precipitation of minerals begins. A suite of rocks, called evaporitic rocks, forms. These include halite and gypsum.

Metamorphic rocks are rocks that form from heat and pressure. They are transformed from other rock types, thus they undergo a metamorphosis to become something entirely new. They typically form where there is great pressure deep under the Earth, usually in places where plate tectonics is active. For example, many metamorphic rocks are found in the Appalachian region of the United States. Here, nearly 500 million years ago, the North American Plate collided with portions of the African plate to create a towering mountain system that would have rivaled the Himalayan chain. For many millennia great pressure was exerted on the rocks deep under the mountains. Today, these metamorphic rocks are exposed in portions of our current Appalachian Mountains. Indeed, metamorphic rocks are often found in the great mountain ranges of the world.

Metamorphism not only transforms the rocks, but it also transforms minerals. Unique mineral assemblages can be created including garnet, asbestos, and andalusite. Metamorphic action can modify the structure of the rock to create unique foliation or banding. Sometimes, the link to the parent rock is obvious. For example, metamorphosed limestone becomes marble, metamorphosed sandstone becomes quartzite, and metamorphosed shale becomes slate. However, the origins of highly metamorphosed rocks, particularly gneiss and schist, are difficult to discern.

The Rock Cycle

Key to the field of geology is the concept of the rock cycle that demonstrates that any rock can turn into any other rock via the processes outlined above (Figure 2.2). An igneous rock can turn into a sedimentary rock via erosion, transport, sedimentation, and subsequent lithification. It can also turn into a metamorphic rock if it is subject to heat and pressure. It can also melt and turn back into an igneous rock. The same is true for metamorphic and igneous rocks. Each can transform into the other over millions and millions of years.

The rock cycle demonstrates that the Earth is constantly changing. Without human action, it is always reacting to surface and subsurface processes that transform our crust. Also, natural processes at the surface of the Earth contribute to the process via chemical

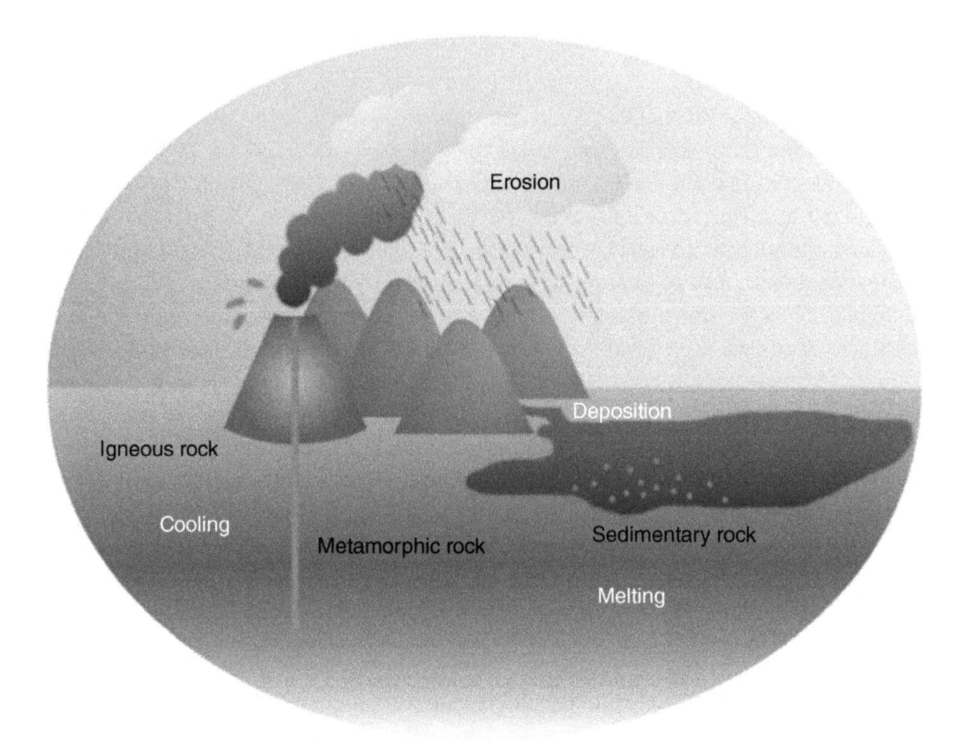

Figure 2.2 The rock cycle showing the major forms of rock and their formation.

and physical weather, erosion, transportation, and deposition. As we will see, many of the chemical and biological cycles associated with the Earth depend upon the rock cycle and the constant recycling of Earth material.

But, the rock cycle is significantly altered in the present era due to the high rate of environmental change. Geologists call this new geologic time the Anthropocene, which roughly extends from the industrial revolution to our current time. Key evidence used by scientists to demonstrate that we are in a new time period is the fact that our climate is changing, that we are in a period of rapid extinctions, and that we have significantly altered biogeochemical cycles, most notably by releasing trace elements into the environment, such as lead or nuclear particles.

Biogeochemical Cycles

There are a number of biogeochemical cycles that are intertwined with each other to create our current environment. Biogeochemical cycles are ways in which important environmental chemicals travel through different areas of the environment (lithosphere, hydrosphere, and atmosphere) and how they are interrelated. As we will see, each biogeochemical cycle described here is important to the regulation of ecosystems and each has in some way been altered in our current era. In this section, we will review water, carbon, sulfur, nitrogen, and phosphorus.

Water and the Water Cycle

Water is a simple compound, H_2O, or hydrogen dioxide. It exists within the crust and the atmosphere in an interconnected system called the hydrosphere.

Water is an interesting chemical because it is one of the few chemicals on the surface of the Earth that exists within three natural phases: gas, liquid, and solid. When it transforms from one phase or another, then energy is lost or gained. This thermodynamic process refers to latent heat or the amount of heat released or absorbed by a body when it goes through a phase change. Thus the transference of energy is a key role of water in our environment. The latent heat of fusion (ice to water or water to ice) is 334 kJ/kg and the latent heat of vaporization of water (water to vapor, vapor to water) is 2260kJ/kg. Thus, as water changes from the phases solid, liquid, or gas, heat is released or gained.

Due to the high latent heat of vaporization, a great deal of energy is required to evaporate water from the surface of the Earth. Since we have a tremendous amount of water in the atmosphere, there is a great deal of heat energy stored there. The evaporation of water helps to cool surfaces thereby transferring the heat into the atmosphere. When the water condenses in the form of rain, the heat is released, thereby warming the planet. These heat exchanges help to transfer heat from equatorial regions poleward via moving air masses and our normal weather systems.

The transformation of water into its different phases also helps to purify water. When it evaporates, it leaves behind impurities. Most rainwater is relatively pure and safe to drink. Thus, the natural system of evaporation and subsequent condensation in the form of rain or snow helps to regulate natural systems.

The movement of water across the planet is known as the water cycle (Figure 2.3). While the water cycle is a planetary scale system, each location on the Earth has a unique relationship with the water cycle due to its individual climate and geologic setting.

The driving force behind the water cycle is the Sun. It provides the energy to cause evaporation from water bodies or the soil. Oceans cover 71% of the Earth's surface and hold 97% of its water. Warm ocean water is a major source of atmospheric moisture. NASA has mapped the worldwide distribution of atmospheric water vapor and they found that most of it is located near the equator. This is not particularly surprising since the world's oceans are warmest between the Tropic of Cancer and the Tropic of Capricorn around the equator.

The amount of energy stored in the tropical air is huge. This is clearly evidenced in the great number of hurricanes and typhoons that occur regularly in these areas. In recent years, devastating tropical storms hit Asia, Africa, and North America. Sometimes, these storms head out of the tropics and bring tropical energy northward where it is released to dramatic effect. One such storm, Hurricane Sandy, devastated portions of the New York City area to cause loss of life and significant damage to many coastal areas that were highly developed.

Water that evaporates in the atmosphere can condense and fall as rain or snow. Worldwide precipitation maps show that most of the precipitation falls within tropical zones where atmospheric moisture is highest. However, many coastal areas north or south of the tropics do receive substantial amounts of moisture. The driest areas of the world tend to be near the poles and the interiors of continents away from sources of moisture.

Much of the rain that falls in the tropics falls due to convectional storms. These storms are near daily occurrences in some tropical areas. During the day, water is evaporated from

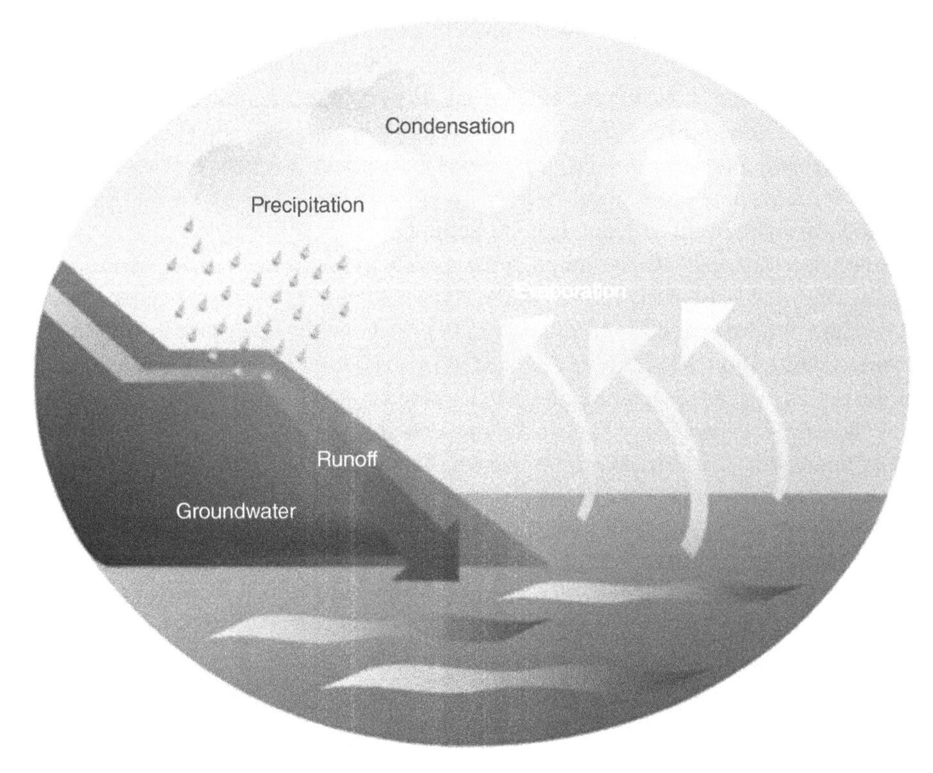

Figure 2.3 The water cycle. How do we alter it in our daily lives?

the land and sea. As the day progresses, large thunderstorms occur in the late afternoon when clouds build to towering heights. At night, the atmosphere cools and clears. This daily cycle of rain is in contrast to continental frontal rainfall systems that occur over many non-tropical portions of the world. In these locations, rainfall typically occurs along frontal boundaries that separate warm air masses originating in the tropics from cold air masses originating in the poles. Sometimes the differences between the air masses are minimal and the resulting storms and rainfall are mild. When the differences are significant, severe weather can occur that results in thunderstorms or tornadoes.

Once moisture reaches the surface of the Earth, many things can happen to it. It can evaporate back into the atmosphere, it can runoff into streams, it can become stored at the surface in ice, lakes or oceans, or it can infiltrate into the ground.

Water stored in ice accounts for approximately 1.75% of the Earth's water and most (70%) of the Earth's freshwater reserves. Most of this ice is fossil water, meaning that it contains precipitation that fell in the past. The amount of additional new ice each year is relatively small compared to the total ice stored on the planet. Plus, as our planet warms, we are losing ice in many areas of the world.

The runoff of water into streams is an important way that water moves across our planet. Moving water also helps to erode the Earth and transport sediment. Streams are also important habitats. Only a very small percentage of the Earth's water is in streams (0.0002%).

When rains fall on the Earth's surface, it moves through a process called overland flow, or sheet flow, prior to entering a channelized system. These streams can also be fed by groundwater. Streams that flow all year long are called perennial streams. Those that flow part of the year are called intermittent streams. Those that flow only after a rainfall are called ephemeral streams. Streams go by a variety of names such as rivers, creeks, coulees, brooks, and bayous. There are a huge variety of regional names for streams all over the world. Terms like gulch, wadi, kill, and run are examples.

As noted earlier, water can also be stored at the surface in lakes and oceans (Figure 2.3). While the oceans hold most of the Earth's water, lakes hold only 0.007% of the Earth's total water. There are hundreds of millions of lakes in the world, with most of them located in the northern latitudes in areas that were impacted by the action of glaciers. Giant bodies of ice in the last Ice Age left behind many small lakes throughout North America, Europe, and Asia. There are also some very large lakes, most notably the Great Lakes of the United States and Canada that were carved out by glaciers.

Water can also be stored in groundwater systems called aquifers. Approximately 1.7% of the Earth's water is stored in groundwater systems. When water is at the surface of the Earth, it can infiltrate into the soil. When it enters, it becomes part of the vadose zone above an aquifer. The vadose zone is an unsaturated area that stores moisture, but also contains gases. The plain separating the vadose zone from an aquifer is called the water table.

Water infiltrates into the subsurface through the forces of gravity and due to capillary action (the ability of water to move against the forces of gravity in narrow pore spaces due to surface forces and adhesion). While rainfall helps to supply moisture to the subsurface, there must be a suitable host sedimentary body or rock to store water. In order to store and transmit water, the subsurface materials must have pore space and permeability (the ability to transmit water). Some of the most productive aquifers in the world are in karst limestone systems in tropical or subtropical regions.

Karst is a term broadly applied to landscapes of limestone terrain (Figure 2.4). Approximately 20% of the world is covered by karst landscapes and 40% of the world's groundwater is derived from karst systems. The reason that limestone is an excellent aquifer is because it has both high porosity and high permeability. Calcite, the dominant mineral in limestone, can dissolve upon contact with water to form extensive interconnected pore spaces, some as large as caverns.

Karst systems are not the only important aquifers on the planet. Sandstone, basalt, and other rocks and sediments, in the right circumstances, can be productive groundwater systems. Naturally, water is released from groundwater systems by draining into lakes, rivers, oceans, or other water bodies via surface or subsurface springs. We also exploit groundwater systems by pumping from wells.

Currently there are a number of problems associated with groundwater use including pollution, saltwater intrusion, and overuse. Some aquifers in arid or semi-arid areas hold fossil water that fell as rain over hundreds of years. The mining of this water in our present era provides distinct challenges for long-term sustainability.

The water cycle is clearly linked to the rock cycle by interacting with the formation of sedimentary rocks. It helps to drive chemical and physical weather and acts to transport rocks. Water bodies are also depositional environments where sedimentary rocks form. As we will see in the next section, the water cycle is also important in chemical cycles.

Figure 2.4 This is the opening to one of the largest caves in the world, Carlsbad Caverns in New Mexico. Caves are found in karst landscapes.

The Carbon Cycle and Global Climate Change

Carbon is one of the most important elements on our planet and is the 15th most abundant element in the Earth's crust. It is present in life forms and it is a key element involved with the regulation of our climate. It is transferred from atmosphere to soil, to rock, to water, and to life in a web of interactions called the carbon cycle.

Carbon is a macronutrient required for basic functions of organisms. Autotrophs like plants collect carbon from the atmosphere for use in metabolic functions. Organisms like humans that take in carbon by consuming other organisms are called heterotrophs. The carbon, once consumed, is used in a variety of metabolic functions.

In rocks, carbon is most commonly present within the sedimentary rocks called limestone or dolomite. These important rocks are important sinks of carbon and store it for long periods of time. They form from the shells, exoskeletons, and fecal matter of marine organisms. The principle mineral in limestone and dolomite is calcium carbonate.

Of course, there are also huge amounts of carbon stored in coal and other fossil fuels. These materials are derived from organisms that lived millions of years ago. Great swampy basins captured the remains of these organisms where, over time, they became the energy resources we use today. As we will see, the burning of these fossil fuels causes serious problems associated with greenhouse gas warming and associated global climate change.

Carbon is also present in soils. It exists as either biologic material such as worms, bacteria, fungi or other organism, or as a stable organic chemical called humus. Humus forms when soil organic matter breaks down with time. Humus is a complex organic material that coats soil grains to stain them dark. It is a very stable chemical that can remain in the soil for decades. It is the material that darkens the soil in some of the world's most productive agricultural regions.

There is also a tremendous amount of carbon stored in the ocean in the form of dissolved carbon in the carbonate ion (CO_3^{-2}). The lithosphere contains, by far, the most carbon on the planet. However, the oceans contain the second largest amount of carbon. Once dissolved in the ocean, it can enter biological systems, deposit on the floor of the ocean in sediments after it combines with a suitable cation (positively charged ion), or stay dissolved for long periods of time. The ocean gets most of its carbon by absorbing carbon dioxide present in the atmosphere.

Global Climate Change and the Carbon Cycle

One of the great challenges of our current era is human caused global climate change, largely due to our disruption of the carbon cycle by adding tremendous amounts of carbon dioxide into the atmosphere over the last several centuries. Carbon dioxide, along with methane, nitrous oxide, water, and other gases, constitute a group of gases that are known as greenhouse gases. These gases have the ability to absorb energy as heat and store it in the atmosphere, thereby warming it. We have added huge amounts of carbon dioxide into the atmosphere through the burning of fossil fuels since the advent of the industrial revolution. We have learned by studying samples of gases trapped in ice that in the 1700s the concentration of carbon dioxide in the atmosphere was approximately 300 ppm (parts per million). Now, the concentration is over 400 ppm, with most of the increase occurring after 1950.

The increase in carbon dioxide is believed to be a driving force in global climate change. Many of these emissions derive from the burning of fossil fuels, particularly petroleum and coal. Over the last few decades, since concerns about global climate change emerged, many have looked at reducing the use of these fossil fuels as a key strategy to slow down the impacts of greenhouse gases on the planet.

China and the United States are the largest polluters of greenhouse gases. Each has high reserves of coal. China is rapidly developing and increasing its greenhouse gas emissions. The United States has been making efforts to reduce greenhouse gas emissions, but has not effectively developed any clear policy to holistically manage the greenhouse gas problem.

Attempts have been made to try to reduce greenhouse gases using treaties similar to the Montreal Protocol. This effort resulted in the Kyoto Protocol, a 1997 agreement that required most of the industrialized nations to reduce greenhouse gas emissions within a particular timeframe. While most of the world ratified it, the United States, the largest greenhouse gas polluter at the time, did not agree to ratify it due to economic concerns and concerns that emerging economies with high emissions, particularly China and India, would not be required to make significant reductions. Canada and Russia also withdrew from the agreement recently.

While many had hoped that the Kyoto Protocol would evolve into a workable international strategy for reducing greenhouse gases, no such agreement currently exists. The Montreal Protocol stands as the only major international air pollution agreement that resulted in effective protection of the environment. A new agreement was reached that extended the Kyoto Protocol agreement to 2020, but there is doubt over its effectiveness.

The atmosphere isn't the only place where there is concern over carbon dioxide pollution. Significantly, it is believed that 30–40% of all anthropogenic sources of carbon end up in a variety of surface waters. The most significant of these is oceans. One of the problems associated with increasing levels of carbon dioxide in our atmosphere is that it leads to a problem known as ocean acidification. When the carbonate ion reacts with water, it can create carbonic acid. As we have seen carbon dioxide pollution increase across the planet, some areas have seen dramatic acidification of ocean waters. When this occurs, it impacts the metabolic processes of some marine organisms. There is evidence that acidified ocean water inhibits shell growth.

The US EPA has started to investigate how to regulate ocean acidification. Some have argued that the EPA has authority to regulate it within the statutory authority of the Clean Water Act. The EPA has stated that ocean acidification should not deviate from 0.2 on the pH scale from natural conditions. Yet some areas have such extremes already and many are struggling with how to develop effective policy to protect the oceans from acidification as a result of widespread carbon dioxide pollution.

China is currently the largest contributor of greenhouse gas pollution. As it has gone through its rapid industrialization and development in the last few decades, pollution has increased significantly. In January 2013, air pollution in Beijing, the capital, reached levels considered hazardous to human health. Local health officials required reduced burning of fossil fuels and people were compelled to stay indoors. The country is spending billions of dollars on alternative fuels and on improving efficiency in order to reduce pollution. However, as many nations of the world reduced or slowed greenhouse gas emissions, China continues to increase their pollution output.

In contrast, the United States saw greenhouse gas reductions over the same time period. Some of this may be due to the broad global economic slowdown. However, the US EPA (Environmental Protection Agency), since the election of President Obama in 2008, has regulated greenhouse gas emissions from many major sources, particularly heavily polluting power plants. The regulations were put in place after Massachusetts and other states sued the EPA for not regulating greenhouse gases. The litigants believed that they were already seeing the damage from global warming and felt that the EPA needed to regulate greenhouse gases as air pollutants. The US Supreme Court agreed and directed the EPA in 2006 to develop strategies for regulating them. This decision is considered the most important directive for establishing greenhouse gas policy in the United States. Since then, the congress has failed to provide more definitive policy and the EPA is moving forward through the executive branch at developing rules.

Yet, there are concerns over the future. The world has seen unusual weather in recent years and many parts of the world are experiencing unusually warm temperatures. People living in low-lying areas are growing more concerned. Some low-lying island nations, including the Maldives, Kiribati, and the Seychelles are worried whether or not they will exist within a few decades. Much of the land in these nations is only 1–2 meters above sea level. Slight changes will make these nations uninhabitable.

One village, the Native Alaskan Village of Kivalina, tried to take matters into their own hands by trying to sue the major energy companies of the world for damages. Kivalina is located on a small barrier island on the northwest coast of Alaska. Here, the sea normally freezes during the spring and fall seasons, protecting the island from the impact of severe storms that come in from the Chukchi Sea. In recent years, the sea ice has not been as persistent during the stormy seasons and large portions of the island have eroded, causing people to relocate their homes.

Kivalina tried to obtain damages from the energy companies for their losses. The litigants believed that the energy companies willfully sold a product that they knew was damaging to the environment and that led to global climate change. They claimed that there was a conspiracy by the companies to hide the impacts of their products. The case was eventually dismissed on a number of grounds. The judge in the case felt that global warming was a political issue and that since we all use and benefit from fossil fuels, it was inappropriate to put the blame of the actions on a product that was difficult to trace to a single user. In the United States, it seems unlikely that the energy companies will be held responsible for any damages caused by greenhouse gas pollution.

The link between greenhouse gas pollution and the carbon cycle provides a clear example how easily humans can modify natural chemical cycles. In the next section, we will see that nutrient cycles are also highly modified and threaten the Earth's fresh water systems.

The Sulfur Cycle

Sulfur is a common element present in rocks. In igneous and metamorphic rock, it is present either as elemental sulfur, or as a sulfide of metals, most commonly pyrite, FeS_2. However, both elemental sulfur and pyrite are easily weathered from rocks, often by microorganisms, after which it enters the atmosphere as hydrogen sulfide gas or surface water as sulfate. Organisms ingest sulfur in a number of ways and it is present in proteins in organisms and thus is a key macronutrient for life.

Yet tremendous amounts of sulfur are present in ocean bodies. Some of it comes to the water as atmospheric deposition from sulfur derived from volcanoes or pollution. The rest comes to the water in runoff from the continents. Once dissolved in the ocean, it can become deposited where it reenters the rock cycle in the form of sedimentary rock as pyrite, gypsum, or anhydrite.

We have greatly modified the sulfur cycle by releasing large amounts of sulfur into the atmosphere as pollution from the burning of fossil fuels or the processing of metallic ores into usable materials. Many of the sources of metals are bound in sulfide minerals. These minerals must be broken down in order to release the usable metals from the unusable sulfide materials. In addition, sulfur is a very common accessory element found in many coal deposits and some petroleum and natural gas reserves. When coal is burned, large amounts of sulfur can be released into the atmosphere.

Upon entry into the atmosphere, the sulfur can combine with atmospheric moisture to create sulfuric acid. This is a natural process that can occur when sulfur naturally enters the atmosphere. However, due to the exceptional amount of sulfur added to the atmosphere every day, the atmosphere has become more acidic since the industrial revolution and there have been distinct changes as a result of this atmospheric modification.

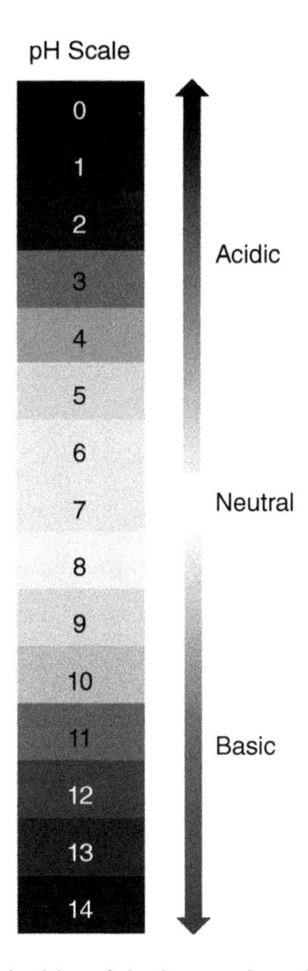

Figure 2.5 The pH scale. Materials that are less than 7 are acidic. Those higher than 7 are basic.

Perhaps the most significant impact of the increase of sulfuric acid in the atmosphere is the formation of acid rain. The pH scale measures the acidity or basicity of materials and ranges from 0–14 (Figure 2.5). Pure water, considered neutral on the pH scale, has a pH of 7. Materials that fall below 7 on the pH scale are acidic and materials that are higher than 7 are basic. Most natural rainwater is acidic and has a pH of around 6. But, it can vary considerably. However, acid rain is any rainwater that has a pH of less than 5.3.

Since the industrial revolution, it has become evident that rain was becoming more acidic. There were many examples of impacts of the acidic rain on the environment. The pH of soil and water bodies decreased as a result of it, thereby changing the ecosystems of some of these environments. Lakes in some of the more vulnerable regions of the world lost their fish populations—the lakes became too acidic to support them. The acid rain can also damage plants, making them more susceptible to diseases. Acidification of soil can cause the replacement of nutrients by the hydrogen ion. Nutrients in these circumstances are leached from the soil, leaving them less productive.

There was also evidence of acid rain in our built environment. Buildings made of limestone, a rock highly vulnerable to solution by acids, were damaged. Likewise, ancient statues made of marble that stood for hundreds of years out of doors in places like Greece and Italy lost their features (Figure 2.6). Tombstones made of limestone or marble became unreadable.

Figure 2.6 The Parthenon in Athens, Greece has been greatly impacted by acid rain and pollution. How are buildings in your area impacted by pollution?

There has been a considerable effort put forward in recent years to reduce acid rain. Many power plants have switched from burning high sulfur content coal to burning low sulfur coal. As luck would have it, the high sulfur coal is more plentiful and less expensive. Because of this, many power plants have developed technologies to reduce sulfur from emissions in order to reduce or eliminate sulfur and other harmful pollutants.

Another approach that has been taken in the United States and other regions of the world is to conduct a cap and trade program for sulfur emissions. These types of programs set regional caps for emissions. Polluters are allowed to trade out credits for cash if they reduce emissions at their plant. Those polluters who are paying are allowed to pollute more. This system provides incentives for reducing pollution as well as funding sources for the development of new technologies. The sulfur cap and trade program has been highly successful. It is believed that it was responsible for a reduction of 47% of sulfur pollution during the first five years it was in operation. A modified cap and trade approach is used to control a variety of pollutants around the world, including carbon dioxide, the main chemical responsible for global climate change.

The Nitrogen and Phosphorus Cycles

Approximately 70% of the atmosphere is made of nitrogen in the form of inert gas, N_2. Nitrogen is released from the atmosphere in two ways. Lightening can break apart nitrogen to turn it into nitrogen oxides. However, most nitrogen is fixed by bacteria in the soil that converts it into ammonium or nitrate for use in plant metabolism. We take in nitrogen by eating plants that have taken in nitrogen via this nitrogen fixing process or by eating animals that have eaten such plants.

Dead plant or animal material, as well as feces and urine, contain nitrogen that is converted to ammonium and nitrates during decomposition. Nitrate is highly soluble. It can therefore easily enter water bodies where it can become a pollutant. Nitrate is converted back into atmospheric nitrogen through bacterial processes. It is also stored in plant and animal remains in the form of coal, peat, and other fossil fuels. When the fossil fuels are burned, nitrogen can be a significant air pollutant that can cause respiratory issues and acid rain.

The most common source of inorganic phosphorus is the mineral apatite. It is a common accessory mineral in many rocks. When it weathers under particular conditions phosphate is released. Interestingly, phosphorus is very unevenly distributed across the planet. In addition, it weathers very slowly. Thus, a great deal of phosphorus is cycled over and over again via organisms through biologic cycling. For example, soils in many tropical areas contain very low amounts of phosphorus. When a tree or plant dies, the phosphorus in the plant is released and taken up again by other plants.

In soil, we often look at the presence or absence of nutrients as limiting factors. When very low levels of a particular nutrient are available, it is a limiting factor for the overall success of a plant. The plant will only do as well as the lowest level of the range of nutrients available. Thus, if a plant is in a soil with very low levels of phosphorus and high amounts of all other nutrients, it still will not thrive because it does not have enough phosphorus to make it a healthy plant. In nature, phosphorus is one of the key limiting nutrients in soil due to its irregular distribution.

As agricultural approaches have improved in recent decades, we have developed new ways to fertilize our farms. The green revolution that started in the 1940s (and that still continues today), produced a wide array of advances to improve crop yields. One of the most important of these advances was the development of synthetic fertilizers (Figure 2.7).

We will discuss sustainability and agriculture in Chapter 7, but it is worth noting here that the development of synthetic fertilizers significantly altered the phosphorus and nitrogen cycles. These two elements, along with potassium, make up the major components of agricultural fertilizers.

In the past, many farmers used manure and other readily available fertilizers to improve their fields. However, since the green revolution, manufactured fertilizers are applied directly onto farm fields to improve yields. The chemical fertilizers are a mix of materials that bring readily soluble nutrients into the soil. This has proved a boon to farmers who have seen dramatic improvements in yields over the last several decades.

However, these fertilizers have proven to be problematic for the environment. Due to their solubility, they easily run off into streams, enter groundwater systems, and impact surface water bodies like lakes, oceans, and bays.

While the green revolution of the last several decades has been great for increasing crop yields and thereby facilitating the expansion of the human population, widespread application of fertilizers creates nutrient pollution.

Figure 2.7 Our modern agriculture uses a great deal of fertilizer to produce highly productive plants like this rhubarb in a garden in Wisconsin.

Fertilizers are soluble in water to allow nutrients to be in solution in water thereby facilitating nutrient uptake in plant roots. Of course, not all of the nutrients will be absorbed by plant roots. Some will run off into surface waters or ground water systems.

When nutrients like nitrogen and phosphorus enter a water body, they change the water chemistry to create conditions that allow phytoplankton to grow unchecked. There are no limiting nutrients for expansive growth of these organisms.

Many ecosystems and water bodies are naturally low nutrient environments. The addition of nutrients into them changes them and allows widespread growth of plant materials like algae or cattail reeds. When these plants die, their decomposition creates hypoxic conditions to form algal blooms. Hypoxia is the absence of dissolved oxygen in water. When hypoxic conditions form, it is difficult for fish and other aquatic animals to survive.

Widespread eutrophication problems were first identified in rural agricultural areas where excess nutrient runoff created difficult conditions in lakes, ponds, and rivers. Since then, eutrophication has been identified in many urban areas where the phenomenon cultural eutrophication was coined.

In many parts of the world, the expansion of eutrophication is associated with the spread of suburban landscapes. We will return to discuss these urban systems in Chapter 12, but one of the hallmarks of suburbia is the lawn.

Up until the 1950s lawns, yards, and gardens were not very common features of the world's urban landscape. However, with the advent of mass-produced homes in the late 1940s, lawns became a much more common feature of the Earth's developed landscape.

With lawns comes the care and maintenance of them. Fertilizers, pesticides, and herbicides were manufactured just for the suburban homeowner. Of course, most homeowners are not scientists well versed on the dangers of over application of these materials. Homeowners, seeking to create beautiful lawns tend to over fertilize using the "more is more" adage. Along with these fertilizers, the building of golf courses, sewage systems, and other nutrient adding systems severely damaged many surface water systems in the late twentieth century.

As we will see, there are ways to reduce the use of nutrients in fertilizers and sewage systems. But, many areas of the world have periods of time where severe hypoxia creates widespread fish kills. These areas are called dead zones.

One of the most significant of these is the Gulf of Mexico Dead Zone near the mouth of the Mississippi River. Here, water draining from extensive agricultural areas of the Mississippi River drainage basin flow into the Gulf. In addition, the Mississippi River receives effluent from dozens of sewage treatment plants throughout its length. Many of the nutrients added as fertilizers in the drainage basin or released in waters from sewage treatment plants end up flowing into the Gulf of Mexico.

The Gulf of Mexico Dead Zone expands and contracts with time, but has been growing over the years. In 2012, it was approximately 17 000 square kilometers in area. While some hypoxic conditions can be present naturally near deltas and estuaries, hypoxic conditions of the size recorded in 2012 are not normal. Dead zones around the world greatly impact natural ecosystems and the fishing industry.

The Gulf of Mexico Dead Zone is just an example of what can happen when nutrients are present as a pollutant. In other cases, hypoxia is not the problem. The excess nutrients attract plants that prefer high nutrient conditions, thereby wiping out natural low nutrient ecosystems that have been in place for millennia. One place where this has occurred is the Florida Everglades where native low-nutrient wetland grasses have been replaced with nutrient loving cattails. Excess nutrients are a problem for ecosystems all over the world, particularly in agricultural and urban settings.

Nitrate Pollution of groundwater

Some nutrients can enter groundwater. The most significant nutrient pollution in groundwater is soluble nitrate. This ion is a very effective nutrient because it is so soluble in water. However, it can easily enter groundwater where it becomes a pollutant of concern for human consumption in drinking water.

When water containing nitrate is consumed, the nitrate can enter the bloodstream where it inhibits oxygenation of the blood. Organs of the body can become oxygen starved and a condition known as methembloinemia may occur. Unfortunately, the young are more susceptible to nitrate poisoning. Blue baby syndrome is an informal name given to babies with nitrate poisoning. According to the US EPA, water containing levels of nitrate over 10 ppm/liter of water are unsafe for drinking.

The rock cycle and the biochemical cycles of carbon and the main nutrients provide a broad understanding of the interconnections of all activities on the planet. The cycles have evolved as the Earth has evolved. If these cycles are modified by humans, significant changes can occur that can be temporally beneficial, but often detrimental to the long-term health of a planet thereby limiting its ability to sustain the diversity of live that we now know.

The following section summarizes the diversity of life and how it is organized by science.

Organisms and Ecosystems

All life on Earth can be classified within a taxonomic system that allows us to better understand the organism and its relationship with others. Originally, this was done within a system called the Linnaean System, which provided a ranking system for naming organism which uses genus and species as a framework for naming and organizing life. With time, the system has become more sophisticated, particularly with the advent of DNA and other biochemical tools for understanding the way organisms work.

It is worth noting that Darwin's evolutionary ideas contribute greatly to the understanding of the organization of life. We look at organisms as evolving from others within a tree of life. Thus, we can see that birds and reptiles are linked through the dinosaurs and that man and whales had common ancestors. This helps us classify life within distinct groupings.

Currently, we recognize that life can be broken into six kingdoms: Bacteria, Protozoa, Chromista, Plantae, Fungi, and Animalia (Figure 2.8).

Ecosystems are assemblages of organisms and their environment. Thus, while we may look at a single plant as a botanist, ecologists and environmental scientists look at

Figure 2.8 This eastern racer snake is part of the Animalia Kingdom, but it is also part of a broader ecosystem.

organisms within their environment. Thus, the concept of ecosystems is key to the field of sustainability.

There are many different kinds of ecosystems on the planet. In many ways, the Earth can be seen as entire ecosystem. This is in line with the thinking of James Lovelock who envisioned the Earth itself as a self-regulating organism (see inset box). However, more commonly we look at individual ecosystems. For a breakdown of the major ecosystems of the world.

The Gaia Hypothesis

The 1960s were an era of ecological enlightenment. Many people started to see the impact of widespread ecological devastation on the planet and became concerned about the overall long-term health of the Earth. At the same time, many scientists started to understand some of the broad bioecological cycles, such as the carbon cycle, that we all know and understand today. At the time, it was a revolutionary idea to think that all corners of the Earth were interconnected. Scientists started to understand the actions in one place can impact places far distant from the action. (Figure 2.9)

Figure 2.9 The Gaia Hypothesis tells us that subtle actions in one part of the world can have great impacts in another. The example that is used is that the fluttering of a butterfly's wings, here in Guyana, could cause a hurricane across the planet. How are your actions impacting other corners of the world far from your home?

At the same time, the ideas of environmental ethics, nicely outlined in an earlier decade by Aldo Leopold, began to have weight within the broader world. We started to become concerned about the impacts of our heavy hand on the environment and we started to question the ethics of our behavior toward the greater good of not only humanity, but also the planet itself.

(Continued)

Within this mix, James Lovelock and his colleague Lynn Margulis developed the Gaia Hypothesis that gained traction in the 1970s. It continues to influence our thinking about planetary scale ecosystems. The hypothesis states that the Earth, and perhaps other planets, are self-regulating systems that evolved as organisms evolved to interact with their environment.

What this means is that the Earth as an entire planetary ecosystem is a product of the ecosystems that inhabit it. For example, the evolution of our atmosphere involved the respiration of oxygen by plants. This allowed animals to emerge onto the land surface. Today, the delicate mix of oxygen and other elements provides a suitable mix for the preservation of life. That is why there is so much concern about the release of greenhouse gases. It has been many tens of thousands of years since our atmosphere has had the carbon dioxide levels we see today.

The Gaia Hypothesis warns us that changes in broad global ecosystems or chemical cycles can throw a tightly regulated planetary system out of balance. Now that we are in a heavily modified era of the planet's history, we really do not fully understand what will happen next. What are the impacts of our significant changes to our planet?

We know from recent research on Mars that this planet once had flowing water and a different atmospheric chemistry. Were organisms involved in this planetary change? To date, no life or evidence of past life, has been found on Mars and it would be inappropriate to suggest that life was involved in changing its planetary ecosystem.

Yet on our planet, we can see the heavy hand of humans in the changing biogeochemical cycles. For example, it is quite likely that early human habitation were in part responsible for the development of the North American and central Eurasian Prairies. As humans evolved hunting skills, they used fire to drive animals to kill zones. These fires destroyed forests and may be one of the reasons that these areas now host vast areas of prairies. The prairies created rich, deep soils, that we have utilized in the American and Eurasian breadbaskets to grown crops like corn and wheat. Many of these soils are now eroded and are losing productivity. As these changes occur, there are modifications of global geochemical cycles that regulate the planet's ecosystems.

The Gaia hypothesis has been particularly useful in understanding the impacts of changing atmospheric chemistry in our modern era. We have significantly altered the atmosphere over the last 150 years. Until recently, there was very little concern about these changes. As we now know, there is great concern about the future of our finely tuned atmospheric systems. Yet, while there is concern, there is limited action. We continue to add compounds like carbon dioxide at unprecedented rates. The Earth will certainly react to our changes. Based on the best estimates of scientists, there will be considerable global warming and significant fluctuations in weather. We are in for some difficult times as the planet reacts to the atmospheric changes. But, this is the key point of the Gaia hypothesis: the planet will react. We cannot think that it is a place that will not change as we alter it. Based on past geological evidence, it is clear that the Earth is a very changeable place.

Ecosystems evolve within particular places, thus they are greatly impacted by the local conditions of climate, geology, soils, and relief. Each place has unique conditions, but areas with similar assemblages of conditions can be mapped and classified as similar. Two broad ecosystem types are aquatic and terrestrial ecosystems.

Aquatic ecosystems evolved under water or in areas that are temporally submerged. They can be divided into marine and freshwater systems. Marine ecosystems can be further subdivided in a vast variety of systems based on highly variable conditions in the sea (beach, tidal, bay, deep water, surface water, etc.). Likewise freshwater systems can be divided into lakes, rivers, ponds, and other large and small water bodies.

Terrestrial ecosystems too are highly variable. They include a vast array including forests, grasslands, alpine ecosystems, deserts, and tundra.

Urban Ecosystems

Expansive urbanization and suburbanization in the last century have greatly modified natural systems. The resultant landscapes are called urban ecosystems. They are highly complex and difficult to assess. They certainly include humans, but they also include a highly unnatural assemblage of other organisms that include rats, squirrels, cockroaches, bedbugs, pigeons, and a vast array of microorganisms. Yet, they can be studied and assessed in similar fashion to natural ecosystems.

We are always trying to influence urban ecosystems to make them better habitats for humans. Whether it is attempting to reduce rats or increase tree canopy cover, urban ecosystems are constantly changing through our direct and indirect actions. Three examples demonstrate the issues of urban ecosystem within a practical standpoint: campus cats, suburban alligators, and invasive bamboo.

Most people like cats. However, in nature, cats are predators. The modern housecat has evolved in partnership with humans. We have depended on them to keep down the population of mice and rats. Yet, many have returned to nature and survive as feral (wild) cats. For some reason, college campuses have become home to many feral cats. Some have speculated that lonely students get them as pets only to release them at the close of the school year. But, for whatever reason, large populations are present on many campuses.

However, campus cats are highly controversial. Many well-meaning university employees feed them and conduct catch, neuter/spay, and release programs to try to humanly keep the population in check. Yet many have advocated for the killing of these animals because they are unnatural within the ecosystem. Indeed, they can cause significant harm to migratory bird populations who happen to land in the area. Should we kill campus feral cats or should we feed them and catch and neuter or spay them?

Another urban ecosystem problem is the Florida alligator. In this case, the organism isn't an invasive species. Instead, the alligator exists naturally within the many surface water bodies and swamps that dot the Florida landscape. Only a few decades ago, the alligator was in danger of extinction. However, in recent years, the alligator has come back strongly.

They are present throughout Florida and have been relatively successful at adapting to urban and suburban regions.

The Florida human population has largely adapted to the increase of the animals within the urban ecosystem. Residents have become comfortable with them around their homes in ponds, lakes, and rivers. Once uncommon, alligators are frequently seen in day-to-day life. For the most part, it is a peaceful co-existence.

Sometimes, however, problems do occur. There are rare alligator/human encounters—often when people invade the territory of the animal or if they happen to get too close to them as they migrate from wetland to wetland (Figure 2.10). Sometimes, strange encounters occur, such as the one pictured here. There are professional alligator trappers who will remove alligators from pools, backyards, or ponds if they become a nuisance or if people become nervous about their proximity.

Sometimes, we change ecosystems in our attempt to make them more beautiful. We have modified ecosystems for millennia as we have tried to improve the assemblages of plants and animals in them. In recent decades, this has taken new forms as we developed the modern concept of urban and suburban

We have all seen beautiful gardens that contain wide varieties of plants from all over the world. Sometimes these plants do so well that they are able to reproduce without any horticultural assistance. They can spread and become aggressive invaders of urban and natural spaces. These exotic species become pests. There have been attempts to ban some exotic plants and animals in some areas in order to limit their expansion.

One such species is the bamboo. Bamboo is a tall grass common to many parts of the world. It is a lovely plant and is featured in some of the most beautiful art produced

Figure 2.10 The American alligator is a nuisance in some areas of Florida where they can wind up in pools and neighborhood ponds in cities and suburbs. Once threatened with extinction, these animals have thrived with sound management.

throughout Asia. When some varieties were introduced to North America they became sensations and were used in landscaping throughout the nineteenth and twentieth centuries. Over the years, they escaped the yards and gardens to spread throughout a vast amount of territory. Currently, some species are banned in some areas of the United States.

Other plants and animals around the world are banned, hunted, removed and in other ways managed to prevent their expansion. Plants are only part of the problem. Animals too cause significant difficulties. Some of the best-known cases in the United States are the expansion of the gypsy moth in the northern portion of the country, the invasion of the zebra mussel in the Great Lakes, and the expansion of the python in the Everglades. The problem with the python has gotten so bad that the state has recruited hunters to take to the swamps to try to kill the snakes. These snakes lay dozens of eggs each year. This has allowed them to expand greatly. They have taken over important predatory niches held by other organisms in the past.

Wetlands and the American Alligator

One of the more interesting examples of animal adaptation to the modern world is the American Alligator. Many once feared this animal. Indeed, it has killed many people in the United States so the visceral fear of the creature is warranted.

The alligator thrives in rivers and wetlands throughout the American southeast and currently it is widely distributed in places like Florida and Louisiana. If you live in these places, you know that you can see them almost anywhere standing water is present throughout the year. They are often associated with well-known rivers of Louisiana, such as the tributaries of the Mississippi River, and Florida's "River of Grass", the Everglades. It is hard to imagine that they almost went extinct in the middle of the twentieth century.

There are two main reasons they were almost destroyed: destruction of their habitat and widespread hunting. Let's take a look at these two issues in some detail.

Wetlands throughout the United States were destroyed to make "improvements" to agricultural land and drainage systems throughout the twentieth century. Wetlands were also drained to eliminate habitat for mosquitos that carried diseases. According to the United States Environmental Protection Agency, the main reasons for the loss of wetlands are drainage, dredging and stream channelization, deposition of fill material, diking and damming, tilling for crop production, levees, logging, mining, construction, runoff, air and water pollution, changing nutrient levels, releases of toxic chemicals, introduction of non-native species, and grazing by domestic animals. Other more natural causes include erosion, land subsidence, sea level rise, droughts, hurricanes, and storms.

While there were certainly wetland losses prior to the 1900s, mechanization allowed us to make significant changes to the natural landscape to divert streams, fill in low places, and create drainage ditches. Over half of the wetlands have been lost in the US. Much of the loss of the wetlands was in low-lying states of the southeastern United States, the prime habitat of the alligator.

(Continued)

Yet the loss of habitat wasn't the only reason that alligators declined. Throughout the twentieth century, suburbanization into the warmer areas of the American South grew as a result of the development of home air conditioning. Millions of people moved from the cold northern regions of the industrial rust belt into the warmer sun belt. The entire Gulf Coast, particularly Florida, grew tremendously during this time. Throughout most of the twentieth century, Florida doubled in population every 20 years. The growth in population put a tremendous stress on the alligator habitat.

At the same time, alligator skin came into fashion. Alligator belts, handbags, and shoes became luxury items that grew in demand. Alligator pelts were worth serious money. Many wild alligators were killed for their skins at the same time their habitat decreased. Plus, may restaurants promoted alligator meat as an exotic food. Hunting put the survival of the alligator in question.

Things have gotten significantly better for the American alligator since its decline in the late 1900s for two main reasons:

1) Alligator management. Hunting was banned for many years to try to stabilize the population. Now, limited hunting permits are given. At the same time, alligator farms were established to provide alligator meat and skin. Now, most of the alligator products available come from animals raised in farms, not in the wild. Plus, there are many high-quality synthetic materials available that look like alligator leather, thereby making alligator skins unnecessary.
2) The rate of wetland destruction has decreased. In the last few decades, the rate of wetlands loss has decreased dramatically. No longer is it easy for us to make widespread changes to the landscape by draining or filling wetlands. Now, special permits are required in the United States and many other parts of the world to make any significant changes to wetland environments. We are still losing too many wetlands, but things are better than in the past.

One other point is worth noting. Alligators and humans have learned to get along with each other. Many people who live in areas with alligator habitat throughout the Gulf Coast regularly see alligators within their neighborhoods. They can be found in backyard ponds, golf course hazards, and any other waterway. They can be found in most major cities and suburbs of the American South. They have almost become a lovable problem. People find them in their swimming pools and on near their walking trails. While attacks do happen, they are rare and often associated with the alligator-mating season.

Gold Rushes

When gold was discovered in California in 1848, many men and women from all over the world traveled to the Sierra Nevada Mountains of California to make it rich. The gold was found in alluvial, or streambed, deposits in valleys in many portions of California's dramatic eastern mountain chain. While many mining companies went west, so too did many individuals.

Just think about the environmental impact of gold mining of sediment in streambeds. Entire fluvial ecosystems were destroyed. Valleys throughout the region were dug up to try to find gold. The sediment was placed in sluices to try to separate the heavy gold from the lighter stream sediments. Today, you can still see the scars in many valleys of California due to this disruption. Plus, the destabilization of the valleys released tons of sediment that was transported down the steep valley streams into the low-lying landscapes of the Central Valley of California. Many low-lying wetlands were destroyed in the process.

Yet the problems didn't just stop there. In order to refine the gold and conduct a separation of it from impurities, the gold needed to go through an amalgamation process using toxic mercury. Most of the miners did not know that the mercury they were using was toxic to the environment and there certainly were no rules about mercury releases into the environment.

Mercury, of course, is an elemental metal that exists in liquid and vapor form. When it was released in the mid-1800s in California, much of it entered the soil and groundwater where it remains today. One can still find streams and groundwater in some areas of the beautiful mountains of the Sierra Nevada Mountain Chain that are contaminated with mercury that was released almost 200 years ago.

Unfortunately, mercury pollution is an all too common problem in gold mining areas—both historic and current. Our modern problem with gold mining is especially problematic because the extremely high price for gold has led to expanded gold mining in many parts of the world.

In remote Guyana and Venezuela, for example, a gold rush is underway that is not all that different from the nineteenth century California gold rush. Miners from all over the world, particularly from South America, have made their way to the llanos region of northern South America where gold is found in alluvial (river) deposits and other sediments and rocks. Once again, mercury is being used to amalgamate the gold and once again long-term pollution problems are occurring to threaten the long-term health of the region.

The miners know that mercury is an environmental pollutant, but the lust and greed for gold overcomes the environmental ethic in the region. Can you think of any other cases where greed or the need for resources is a stronger motivator than environmental ethics?

Understanding the Anthropocene

Geologists have divided the history of the world into distinct time periods that encompass the 4.5 billion year lifespan of the Earth. This chronology is formally called the Geologic Time Scale (Figure 2.11). Most of the history of the world occurred in an eon called the Precambrian. This was a time before the widespread development of life on the planet as we know it. It lasted up until 570 million years ago, nearly 90% of the entire history of the world. Entire long cycles of rock formation, erosion, and deposition occurred during

	Eon	Era	Period		Epoch	Age in millions of years before the present
	Phanerozoic	Cenozoic	Quaternary		Holocene	0.01
					Pleistocene	1.6
			Tertiary	Neogene	Pliocene	5.3
					Miocene	23.7
				Paleogene	Oligocene	36.6
					Eocene	57.8
					Paleocene	66.4
		Mesozoic	Cretaceous			144
			Jurassic			208
			Triassic			215
		Paleozoic	Permian			266
			Pennsylvanian			320
			Mississippian			360
			Devonian			408
			Silurian			438
			Ordovician			505
			Cambrian			570
Precambrian	Proterozic					2500
	Archean					3600
	Hadean					4550

Figure 2.11 The Geologic time scale.

this era. However, with the advent of the Phanerozoic Eon, things began to change rapidly as diverse life evolved on the planet.

During the Paleozoic Era, the Earth experienced widespread expansion of life within the sea and the development of land plants. Following this, giant reptiles and other animals took over the lands during the Mesozoic—the age of the dinosaurs. After a likely meteor impact on Earth, the dinosaurs died off and mammals began to diversify and dominate the Earth during our current Era, the Cenozoic.

As you can see on the Geologic Time Scale, we are able to divide the Mesozoic into smaller periods of time called Epochs. We are able to do this because the time is not all that distant from our own compared with the great spans of time in the Precambrian, Paleozoic, and Mesozoic times. For example, geologists have named the last two Epochs the Holocene and Pleistocene.

The Pleistocene started about 1.6 million years ago with the advance of continental scale glaciers across many portions of the planet. This Ice Age greatly changed the face of the Earth. The major ice sheets melted about 10000 years ago and we are now in an Epoch called the Holocene.

As you have probably deduced, the divisions in geologic time are based upon significant planetary changes like the advent of the Ice Age or the sudden destruction of the dinosaurs as a result of a meteor impact.

If you think about our current time period, we humans are making huge changes on our planet. Some geologists have advocated naming our present time the Anthropocene. If you think about it, this makes some sense. We have significantly altered all of the planetary systems at work on Earth. We have changed atmospheric chemistry, water cycles, nutrient cycles, and even the rock cycle.

Just think about the widespread deforestation that has occurred in the last 200 years throughout the world. Almost all areas of the world have lost the native forests at one point and what is present now are cities, agricultural land, or secondary growth forests that are often monoculture trees for paper or wood harvesting. Deforestation causes widespread sediment loss and changing hydrology. We can point to almost every modern human activity such as agriculture, transportation, and urbanization and see its impact on the planet. That is why we have created the term Anthropocene for this new geologic age.

3

Measuring Sustainability

One of the most important aspects of sustainability efforts is that they must be measured in order to verify success. Over the years, many organizations have advertised that they were "going green" by doing things that actually may not have been the best thing for the environment. These organizations utilize approaches broadly called greenwashing. Greenwashing is embracing the environment without really doing anything positive to improve it. Greenwashing is often used in advertising or in somehow making people feel good about some activity. We all do some degree of greenwashing in our daily lives. For example, we may eat organic, but drive a gas guzzling sports utility vehicle. It is impossible to live without impacting the Earth in some way. However, intentionally deceiving others about environmental efforts is the egregious ethical lapse known as greenwashing.

To avoid greenwashing, efforts have been made to develop clear measurable approaches to improve the environment that are verifiable. Several of these will be reviewed. Each has the following key features:

1) They provide specific goals or benchmarks. A benchmark is a target that can be met by changing behavior. It might be the amount of fuel saved per year or an improvement of food sources. But, the one thing that benchmarks have in common is that they all provide measurable goals.
2) They can be evaluated, or verified by outside parties. In other words, benchmarks and the data that support them, are often publicly available or verified by a trustworthy external party.
3) Benchmarks are comparable within industries or particular activity areas. In other words, it is possible to benchmark a number of different plastic manufacturers, schools, or any other organization against similar organizations.
4) The results of measuring sustainability are evaluated in order to improve them or in order to address emerging technologies or initiatives.

In the following sections, several different benchmarking efforts are reviewed. We will start with very broad benchmarking efforts of the United Nations Millennium Development Goals and Sustainability Development Goals and extend the discussion to national, state, and local examples.

Introduction to Sustainability, Second Edition. Robert Brinkmann.

Companion website: www.wiley.com/go/Brinkmann/IntroductiontoSustainability

The United Nations Millennium Goals

The United Nations Millennium Goals evolved in the late twentieth century from the Brundtland Report (Figure 3.1). This document, briefly discussed in Chapter 1, focused on evaluating the status of sustainability on the planet. The situation was problematic. Many difficulties were assessed, ranging from the state of the world's climate and ecosystems to the overall food supply and economy. The report, in the eyes of many, fell just short of providing a clear pathway for improving the situation in the future.

Yet many worked to find exactly such a pathway. As a result, the United Nations formed a group to develop goals to improve the planet. The efforts took into account not only the situations identified in the Brundtland Report, but also issues with a variety of other development problems that were identified in the last several decades, including diseases like HIV, or the lack of education of women in some regions of the world.

The result of the group's effort was the publication of the Millennium Goals in 2000 after the United Nation's Millennium Summit (Table 3.1). Eight broad goals were established. It is worth reviewing all of them to understand the reach of the effort. Each goal has subgoals with particular targets for measuring and evaluating progress. What is interesting about the United Nations approach is that the world can be measured as a whole. But, individual countries can also be assessed in order to compare, or benchmark, their efforts with other countries around the globe or in their region.

Figure 3.1 The United Nations headquarters in New York City.

Table 3.1 The Millennium Development Goals.

Goal	Focus Area
1	To eradicate extreme poverty and hunger
2	To achieve universal primary education
3	To promote gender equality and empower women
4	To reduce child mortality
5	To improve maternal health
6	To combat HIV/AIDS, malaria, and other diseases
7	To ensure environmental sustainability
8	To develop a global partnership

To provide an example of the types of subgoals that were developed for each goal, it is worth focusing in on Goal 1 which seeks to eradicate extreme poverty and hunger. One of the targets of the goal is to halve the proportion of people living on less than $1 a day. The indicators that the United Nations established for this subgoal included: (1) the number of people living on less than $1 a day and the reduction of poverty. All of the subgoals for each indicator can be found on the United Nation's Website (https://www.un.org/millenniumgoals/)

The main focus of the Millennium Development Goals was enhancing the lives of people in developing countries. They were sunset in 2016 when the new Sustainability Development Goals were put into place. However, while not all areas of the world made progress on all of the goals, the Millennium Development Goals are broadly seen as a successful initiative that made great improvements in many areas (Figure 3.2). However, it was clear that in order to fully engage on environmental sustainability, the developing world needed to be part of the solution.

The United Nations Sustainable Development Goals

The Sustainable Development Goals were established by the United Nations in 2015 with the intent that they will be met by 2030. They evolved, in part, from the Millennium Development Goals. However, it was widely felt at the time that Millennium Development Goals did not have enough focus on global sustainability issues. Developing nations needed to be part of the assessment process since many sustainability challenges the world faces occur due to problems created in the west.

There are 17 Sustainable Development Goals (Table 3.2). As can be seen, they include a number of issues that were part of the Millennium Development Goals (like reducing poverty) but also contain topics that focus on things like biodiversity, clean water, sustainable communities, and climate action (Figure 3.3.). Each of the goals has specific targets that can be assessed. Note that for the sake of space, only two targets are listed for each goal

Figure 3.2 Many areas of the world have made great improvements in poverty reduction. One of the great success stories in poverty reduction is the People's Republic of China. The photo is of the city of Haikou.

in Table 3.2. Most of the goals have several measurable targets. Goal 17, for example, has 19 targets. The goals are very specific. Goal 10, which focuses on reducing inequalities has one target that seeks by 2030 to sustain income growth of the bottom 40% of the population at a rate higher than the national average. Utilizing this number, countries can compare themselves over time and also compare themselves with similar nations. They can also utilize a number like this and look at regional differences within their country. In contrast, goal 6 which involves water and sanitation seeks to achieve by 2030 access to adequate and equitable sanitation and hygiene for all and end open defecation, paying special attention to the needs of women and girls and those in vulnerable situations.

National sustainability planning

Many nations of the world have developed sustainability planning mechanisms for benchmarking their progress. Each country assesses their overall sustainability differently. Most countries measure a number of different categories of indicators that somehow assess environmental, social, and economic sustainability. Some countries utilize the MDGs or SDGs as a baseline of assessment. But most countries that do some form of sustainability assessment create indicators that are unique to that particular place. For example, an agricultural, rural nation like Cambodia focuses heavily on indicators of agriculture and environmental protection. Industrialized nations tend to include indicators such as greenhouse gas (or

Table 3.2 The United Nations Sustainable Development Goals. Note that for each goal, there are several measurable targets (Goal 17, for example, has 19 targets). Only two targets are listed for each goal here. From https://sustainabledevelopment.un.org/sdgs

Goal	Measurable Targets
1. No poverty	1.1 By 2030, eradicate extreme poverty for all people everywhere, currently measured as people living on less than $1.25 a day
	1.2 By 2030, reduce, by at least half, the proportion of men, women and children of all ages living in poverty in all its dimensions according to national definitions
2. Zero hunger	2.1 By 2030, end hunger and ensure access by all people, in particular the poor and people in vulnerable situations, including infants, to safe, nutritious and sufficient food all year round
	2.2 By 2030, end all forms of malnutrition, including achieving, by 2025, the internationally agreed targets on stunting and wasting in children under 5 years of age, and address the nutritional needs of adolescent girls, pregnant and lactating women and older persons
3. Good health and well-being	3.1 By 2030, reduce the global maternal mortality ratio to less than 70 per 100 000 live births
	3.2 By 2030, end preventable deaths of newborns and children under 5 years of age, with all countries aiming to reduce neonatal mortality to at least as low as 12 per 1000 live births and under-5 mortality to at least as low as 25 per 1000 live births
4. Quality education	4.1 By 2030, ensure that all girls and boys complete free, equitable and quality primary and secondary education leading to relevant and effective learning outcomes
	4.2 By 2030, ensure that all girls and boys have access to quality early childhood development, care and pre-primary education so that they are ready for primary education diversity and culture's contribution to sustainable development
5. Gender equality	5.1 End all forms of discrimination against all women and girls everywhere
	5.2 Eliminate all forms of violence against all women and girls in the public and private spheres, including trafficking and sexual and other types of exploitation
6. Clean water and sanitation	6.1 By 2030, achieve universal and equitable access to safe and affordable drinking water for all
	6.2 By 2030, achieve access to adequate and equitable sanitation and hygiene for all and end open defecation, paying special attention to the needs of women and girls and those in vulnerable situations
7. Affordable and clean energy	7.1 By 2030, ensure universal access to affordable, reliable and modern energy services
	7.2 By 2030, increase substantially the share of renewable energy in the global energy mix
8. Decent work and economic growth	8.1 Sustain per capita economic growth in accordance with national circumstances and, in particular, at least 7 per cent gross domestic product growth per annum in the least developed countries
	8.2 Achieve higher levels of economic productivity through diversification, technological upgrading and innovation, including through a focus on high-value added and labor intensive sectors

Table 3.2 (Continued)

Goal	Measurable Targets
9. Industry, innovation and infrastructure	9.1 Develop quality reliable, sustainable and resilient infrastructure, including regional and transborder infrastructure, to support economic development and human well-being, with a focus on affordable and equitable access for all 9.2 Promote inclusive and sustainable industrialization and, by 2030, significantly raise industry's share of employment and gross domestic product, in line with national circumstances, and double its share in least developed countries
10. Reduced inequalities	10.1 By 2030, progressively achieve and sustain income growth of the bottom 40 per cent of the population at a rate higher than the national average 10.2 By 2030, empower and promote the social, economic and political inclusion of all, irrespective of age, sex, disability, race, ethnicity, origin, religion or economic or other status
11. Sustainable cities and communities	11.1 By 2030, ensure access for all to adequate, safe and affordable housing and basic services and upgrade slums 11.2 By 2030, provide access to safe, affordable, accessible and sustainable transport systems for all, improving road safety, notably by expanding public transport, with special attention to the needs of those in vulnerable situations, women, children, persons with disability and older persons
12. Responsible consumption and production	12.1 Implement the 10-year framework of programmes on sustainable consumption and production, all countries taking action, with developed countries taking the lead, taking into account the development and capabilities of developing countries 12.2 By 2030, achieve the sustainable management and efficient use of natural resources
13. Climate action	13.1 Strengthen resilience and adaptive capacity to climate-related hazards and natural disasters in all countries 13.2 Integrate climate change measures into national policies, strategies and planning
14. Life below water	14.1 By 2025, prevent and significantly reduce marine pollution of all kinds, in particular from land-based activities, including marine debris and nutrient pollution 14.2 By 2020, sustainably manage and protect marine and coastal ecosystems to avoid significant adverse impacts, including by strengthening their resilience, and take action for their restoration in order to achieve healthy and productive oceans
15. Life on land	15.1 By 2020, ensure the conservation, restoration and sustainable use of terrestrial and inland freshwater ecosystems and their services, in particular forests, wetlands, mountains, and drylands, in line with obligations under international agreements 15.2 By 2020, promote the implementation of sustainable management of all types of forests, halt deforestation, restore degraded forests and substantially increase afforestation and reforestation globally

Table 3.2 (Continued)

Goal	Measurable Targets
16. Peace, justice and strong institutions	16.1 Significantly reduce all forms of violence and related death rates everywhere 16.2 End abuse, exploitation, trafficking and all forms of violence against and torture of children
17. Partnerships for the goals.	17.1 Strengthen domestic resource mobilization, including through international support to developing countries, to improve domestic capacity for tax and other revenue collection 17.2 Developed countries to implement fully their official development assistance commitments including the commitment by many developed countries to achieve the target of 0.7 per cent of ODA/GNI to developing countries and 0.15 to 0.20 per cent of ODA/GNI to least developed countries; ODA providers are encouraged to consider setting a target to provide at least 0.20 per cent of ODA/GNI to least developed countries

Figure 3.3 Many areas, such as the Mediterranean Island of Serifos shown here, have lost significant biodiversity on land and in water.

other) pollution. What is important to stress is that the indicators are measurable ways to assess how effective a country is within one particular area on which the nation wishes to focus.

Using these indicators, nations can develop national policy that makes sense for the long-term health of the country. For example, if a country has an indicator measuring the number of acres of forested land and the nation wishes to see that number increase, policies can be enacted to increase forest cover.

Each sustainability plan includes many indicators and it is easy to fall down a rabbit hole of examining each one in detail. However, the plans are best assessed by looking holistically at the plan to determine how effective it is at addressing long-term environmental, social, and economic sustainability.

While there are many examples of national plans that we could focus on, two contrasting ones come from Canada and Bhutan. What is interesting about both of the plans is that they are very quantitative in their assessments, but focus on very different themes. The Canadian plan largely addresses broad environmental issues. In contrast, the plan from Bhutan is concerned largely with the overall happiness of their population. Each plan provides a unique viewpoint on sustainability that demonstrates the range of approaches that a country can take to create a more sustainable future for their population.

Canada

Canada is an industrialized country where the population is largely found within a handful of urban centers like Toronto, Quebec, Montreal, Calgary, and Edmonton (Figure 3.4).

Figure 3.4 Montreal, Canada is one of the world's most modern cities. This image shows a typical street in the historic downtown.

Much of the country is wilderness or agricultural land with low population densities. Canada's most recent sustainability plan published in 2019, called *Achieving a Sustainable Future* (http://www.fsds-sfdd.ca/index.html#/en/intro/what-is-fsds), has 13 main themes. Each theme has distinct measurable goals. The themes are listed below:

Effective action on climate change
Modern and resilient infrastructure
Pristine lakes and rivers
Clean drinking water
Safe and healthy communities
Greening government
Clean energy
Sustainably managed lands and forests
Sustainable food
Clean growth
Healthy coasts and oceans
Healthy wildlife populations
Connecting Canadians with nature

In the report, each of the plan goals lists several specific tasks and strategies that can lead to measurable outcomes. For example, the following approaches are listed under the effective action on climate change in the first goal:

- By 2030, reduce Canada's total greenhouse gas emissions by 30% relative to 2005 emissions.
- Zero-emission vehicles will represent 10% of new light-duty vehicle sales by 2035, 30% by 2030, and 100% by 2040.
- Work with provinces and territories to ensure carbon pricing is in place across Canada in a way that meets the federal carbon pricing benchmark.
- Implement regulatory measures to reduce greenhouse gas emissions, including:
  - The phase-out of traditional coal-fired electricity by 2030
  - Reducing methane emissions from the oil and gas sector by 40-45% by 2025
  - Ongoing regulations for light- and heavy-duty vehicles
  - A Clean Fuel Standard to encourage the use of low-carbon fuels in transportation, buildings, and industry
- By 2019, 60% of communities (based on a representative sample of small, medium, and large Canadian municipalities) identify adaptation measures in their plans, strategies, and reports.
- By 2022, at least 100 projects across the country benefit from the Low Carbon Economy Fund and have reduced their emissions

These goals set Canada on track to be one of the most carbon neutral industrialized countries in the developed world. But what is important here is that Canada has listed several key goals within their plan. The United States, Canada's most important neighbor and trading partner, does not have a sustainability plan. See the textbox to see why.

Sustainability planning in the United Kingdom versus the United States

The United Kingdom has gone through several cycles of sustainability planning. For example, in 2013, the United Kingdom published a list of dozens of sustainability indicators to track and measure national sustainability. The categories of assessment indicators are listed below:

- Economic prosperity
- Long-term unemployment
- Poverty
- Knowledge and skills
- Healthy life expectancy
- Social capital
- Social mobility in adulthood
- Housing provision
- Greenhouse gas emissions
- Natural resource use
- Wildlife
- Water use
- Population demographics
- Debt
- Pension provision
- Physical infrastructure
- Research and development
- Environmental goods and services
- Avoidable mortality
- Obesity
- Lifestyles
- Infant health
- Air quality
- Noise
- Fuel poverty
- UK carbon dioxide emissions by sector
- Energy from renewable sources
- Housing energy efficiency
- Waste disposal and recycling
- Land use
- Origins of food consumed in the UK
- Water quality
- Sustainable fisheries
- Priority species and habitats
- UK biodiversity impacts overseas.

This range of sustainability themes provides an excellent context for understanding the range of issues that can be assessed within a highly developed nation. The list includes a number of themes that could be classified as economic, social, and environmental concerns. For each theme, there are quantitative indicators that can be used to benchmark sustainability for the nation. The report was published along with a spreadsheet that provides numerical data on each of the indicators.

Their most recent plan, published in 2019 focuses on how the UK can implement the SDGs (see https://www.gov.uk/government/publications/implementing-the-sustainable-development-goals/implementing-the-sustainable-development-goals--2). For example, for Goal 2 which focuses on ending hunger, achieving food security and improved nutrition, and promoting sustainable agriculture, the UK developed the following goals:

- Lead the world in food and farming with a sustainable model of food production
- Develop a domestic agriculture policy that delivers an Environmental Land Management system based on public money for public good
- Work to promote healthy lives, end malnutrition, and uphold human dignity and promote prosperity. This includes strengthening the Global Financing Facility

The UK's approach is a highly centralized approach that provides clear national goals and strategies for reaching sustainability targets. The plan stands in great contrast to US sustainability goals, or lack thereof. While there are some branches of government that develop targets and guidelines, there is no single office in the United States that coordinates or tracks sustainability indicators. What this means is that it is very difficult for the US to benchmark or conduct comprehensive long-term planning on overall national sustainability.

Some have argued that this approach is not necessarily a bad thing for the United States. The vastness of the country makes it difficult to have a one size fits all approach. It would be difficult, for example, to benchmark sustainability in New Mexico with Michigan. Others, however, argue that the lack of national strategy in the United States leaves states with little guidance on how best to approach their strategic sustainability initiatives.

What do you think? Do you think the UK's approach with a comprehensive highly centralized benchmarking system is better than the US's approach that leaves sustainability management largely up to the states and local governments? How do the UK measurements compare with those of London? Where are there similarities and differences in environmental, social, and economic indicators?

Bhutan

Bhutan takes a very different approach to sustainability planning. Their main focus of measurement is happiness—they are concerned with assessing the overall happiness of the people, while also focusing on broad sustainability goals.

Bhutan's gross national happiness index originated in 1972 when King Jigme Singye Wangchuck sought to find ways to advance his country's economic agenda, while also ensuring that the people of his country maintained contentment with their lives. The index measures seven main categories:

- Economic wellness
- Environmental wellness
- Physical wellness
- Mental wellness
- Workplace wellness
- Social wellness
- Political wellness.

Each of these indicators has quantitative and/or qualitative ways of measuring wellness within the framework of national happiness. This index focuses on issues very different from those measured in the Canadian sustainability plan. The focus on economic, physical, mental, workplace, social, and political wellness along with environmental wellness bring it closer to the broad assessment of sustainability as defined within the environmental, social, and economic framework discussed in Chapter 1.

This unique approach to sustainability has garnered much attention and many countries all over the world are finding ways to add assessments of happiness into their broad sustainability indices. For example, France and Venezuela have found ways to incorporate happiness into their national indices. Canada has also created the Canadian Index of Well-being that brings together eight measurable "domains" that include: community vitality, democratic engagement, education, environment, healthy populations, leisure and culture, living standards, and time use. There is not one right way to measure a nation's happiness and Bhutan and Canada have found ways to measure and assess happiness that make sense for their own unique cultures.

Some have been critical of the happiness measurement approach. Even in Bhutan, where the happiness measurement system originated, the index is losing its luster. Many in the country have criticized the index as being too focused on things that do not solve the nation's key problems. The current prime minister of that country feels that it is time to turn the page on the index and focus on more tangible measurements like gross domestic product.

Regardless of how the index is evaluated today in Bhutan, there is no doubt that the development of this index was a significant contribution to the way our world thought about assessing progress and development by putting the happiness of people above more standard ways of measuring national success.

US sustainability planning

The United States does not have a single sustainability plan for the nation. This might surprise some of you, since the United States has some of the strongest environmental rules in the world. However, there are two important reasons why the nation doesn't have sustainability planning: (1) politics and the federal systems; and (2) organization of the executive branch of the government. Let's take a look at each of these issues.

Politics. The United States has been politically divided for the last two decades—and this division is bitter. Many trying to achieve consensus and agreement between the two sides are often branded as traitors to their parties. As a result, few important initiatives have come through congress. For example, early in President Obama's first term, the

US congress tried to pass a modest climate change bill advocated by the President. However, even though his party had control of congress, they were unable to get the votes needed to ensure its passage. Plus, the idea of national sustainability planning goes against the grain of more populist candidates on the right who look at most types of national goal setting or strategies with disfavor. For this reason, most of the regional sustainability planning is taking place at the state or local level (see discussions of New York City in this chapter for an example). Congress has largely been silent on sustainability.

Organization of the Executive Branch. In the United States system of government, the executive branch is charged with carrying out the laws passed by congress and the overall business of the country. Whoever holds the presidency has tremendous power through his or her appointees, to direct the tone and direction that different offices will take. Within the Cabinet of the President, there are 15 executive officers as well as the Vice President and the Attorney General. The head of the Environmental Protection Agency (EPA) is not an official member of the Cabinet, but has Cabinet-level authority. Most of these offices in some way deal with sustainability issues and the President can use executive authority to ensure that sustainability initiatives are addressed. For example, in the Obama years, the EPA and the Department of Energy focused heavily on greenhouse gas issues and strived to move the country forward on solving some vexing greenhouse gas problems. The United States Military is surprisingly one of the leading governmental organizations on sustainability initiatives. Most agencies have developed sustainability initiatives for their organizations and strive to advance sustainability in the nation in measurable ways.

However, it must be noted that these presidential actions are not always long-standing. The election of a new president can sweep away the initiatives of the old. For example, sustainability was not a high priority of the presidency of George W. Bush. However, when President Obama took office, he started a number of executive initiatives. When he left office, many of the initiatives disappeared under the Trump administration

Overall, it is clear that at the current time, the United States does not have a clear or coherent sustainability policy compared to the other examples of Canada and Bhutan presented here. Most of the action in the US is at the state and local level.

Regional sustainability planning

Very often, states are able to do more comprehensive sustainability benchmarking than nations, largely because they can address regional issues more effectively than the national government. Just think about large countries like Russia, China, Brazil, or the United States. Each of these places has significant differences that make a "one size fits all" approach to sustainability inappropriate. Just think of the vastness of China. Parts of it are remote desert-like places, while others are more tropical and urban. Each region of the country has its own problems and challenges. While large national goals are often set in these places, regional goals might be more appropriate.

There are many examples of state- or regional-level sustainability planning. For example, the Chinese Provence of Hainan, a special economic zone in the tropical region of China in the South China Sea, has done extensive sustainability planning and assessment. They have identified clear goals and plans to try to develop the island into a more sustainable place, while promoting economic development around tourism (see box).

Sustainability planning and tourism in Hainan

Many call our current century the Chinese Century The growth of the Chinese economy and their expanding international influence have transformed the country from an agrarian inward-looking society to one that is industrialized and increasingly international in scope.

Yet this growth has had a significant cost to the environment and to Chinese society in general. Most of us have heard of or seen first-hand the terrible air pollution in Beijing or heard of the massive displacement of population as a result of the construction of the Three Gorges Dam. However, China is working hard to try to correct the environmental problems of the past, while finding ways to promote economic development. The province of Hainan is a great example of the new approach.

Hainan is a large (roughly 34 000 square kilometers) island just off the south shore of China in the South China Sea near Vietnam. It has a decidedly tropical environment. Indeed, it is China's only truly tropical state. As such, it provides the country with the majority of its tropical crops and a significant portion of its winter fruits and vegetables (Figure 3.5).

Hainan is ethnically different from the rest of China. A number of different ethnic groups are present on the island that have lived in relatively simple ways for hundreds of years. However, with the advent of modern China, many Han Chinese from mainland China have been attracted to Hainan for the pleasant weather. Today, Hainan is a highly diverse society with two main cities (Haikou and Sanya) and extensive areas of tropical forests and agricultural fields.

Several leaders of China attempted a number of development schemes to try to diversify the economy to expand beyond agriculture. These efforts largely focused on developing industrial facilities. None of these were entirely effective, largely because of Hainan's relatively remote location and lack of supporting industries and natural resources. As occurred with many failed industrial schemes around the world, there are problems with left over pollution from the past.

However, since Hainan was made a Special Economic Zone in 1988, many of the region's leaders have worked to advance tourism as a main economic development strategy This makes a great deal of sense since this is China's only tropical region, and the province was perfect for tourism development, particularly taking advantage of tourists seeking a respite from China's and Russia's brutal winter seasons. In addition, the strategic location of Hainan in the South China Sea has drawn greater attention to the region. Here, China, Vietnam, and the Philippines have made competing claims to resources, particularly oil and fisheries. There has been growing concern about the risk of international conflict in the region.

Figure 3.5 Tourism in Hainan, China is hot now. It is one of China's main winter tourist destinations. The Nanshan temple shown here is one of the main tourist attractions on the island.

> Since 1988, a number of important tourism attractions have been developed. These include some of the world's largest golf courses, several theme parks, luxurious beach resorts, and extensive yacht basins. In many ways, the vacation spots of Florida and Hawaii served as models for the kinds of amenities that tourists might want. If you travel to Hainan today, you will find it a place very reminiscent of tropical beach communities around the world– but with a Chinese flair.
>
> What has made tourism such a big success are the abundant natural assets of the island. The beautiful beaches, tropical rainforests, spectacular mountains, and warm weather all contribute to drawing people to the place. Yet might these be at risk?
>
> Some are concerned over the impacts of tourism on the environment. For example, if sufficient sewage systems are not in place, the crowds of tourists might hurt the very assets that draw them to the island in the first place. Thus, the provincial government of Hainan has constructed a statewide sustainability plan that sets very specific goals for the future of the region. A strong component of the plan is environmental protection and expansion of infrastructure (like sewers) to ensure that the environmental impact of tourism is limited.

In County Durham in England (relatively equivalent to a state government in the United States), several themes emerge within their environment and planning efforts. They include animal welfare, archaeology, building control, calibration and testing, conservation, environmental health, land management, land drainage and stormwater, landscape and agriculture, pollution, waste management, street maintenance and cleaning, climate change, and civic pride. The county seeks to infuse sustainability within all of their planning efforts.

While these initiatives are not always measurable, the county does provide very specific guidelines for the region in order to ensure that appropriate sustainability measures are being taken.

In the United States, the State of California, one of the largest and most influential states in that country, has set distinct goals for state government. The former Governor of the State, Jerry Brown, released an executive order that called for all new or renovated state buildings larger than 10 000 square feet to achieve the US Green Building Council's LEED Silver rating and to incorporate green energy such as solar or wind (Figure 3.6). In addition, he directed state agencies to reduce greenhouse gas emissions by 10% by 2015 and 20% by 2020 as measured against a 2010 baseline. Related to this, he is required a reduction of grid-based energy purchase of 20% by 2018, as compared to a 2003 baseline.

Brown also addressed water resources; he required 10% reductions of water usage by 2015 and 20% reductions by 2020, as measured against a 2010 baseline. These goals were largely met. They were important because they set guidelines that were also followed by many local governments and businesses within the state. Since that time, the new governor, Gavin Newsome, working with the state's Office of Sustainability, has set new guidelines. For example, California recently made changes to its very strong purchasing guidelines to only allow the purchase of vehicles from manufacturers that support stricter fuel efficiency standards.

Figure 3.6 Wind farms are becoming more common across the world. This one is in France. Where is the closest one to you?

The three examples, Hainan Province, County Durham, and the State of California demonstrate different ways that state governments influence sustainability at the regional level. In the case of Hainan, sustainability is part of an overall economic development strategy. In County Durham, sustainability guidelines are established for those working with the local government. In California, the state sets rules for its own behavior, thereby setting the stage for local governments to follow. As we will see in Chapter 12, regional planning efforts around sustainability can transform regions into more sustainable places.

Local sustainability measurement

Many large, medium, and small cities, towns, and villages are involved with sustainability planning in their local communities. They bring together a number of stakeholders (see box) to develop appropriate strategies for ensuring a sustainable future for its citizens.

There are several organizations that assist with local sustainability planning. Perhaps the best-known organization is ICLEI for Local Governments. This group formed in 1990 as a result of efforts by the United Nations to develop strong local strategies for sustainable development strategies in urban areas, as outlined in the Rio Conventions and Agenda 21. It is now independent of the UN. There are now over 1200 communities engaged with sustainability initiatives with ICLEI in 84 countries.

ICLEI works with communities directly by providing sustainability best practice guidance and a number of tools that sustainability managers can use as they develop programs. Their work generally follows their "Five Milestone" agenda that includes:

1) *Milestone 1.* Develop a baseline. This can be any type of sustainability assessment of the community and may be in the form of a sustainability assessment and/or greenhouse gas inventory.
2) *Milestone 2.* Set goals. Based on the baseline, what do you hope to achieve? Do you want to reduce greenhouse gases? If so, by how much? Do you want to reduce the number of pollutants? There are many goals that should be crafted while achieving this milestone.
3) *Milestone 3.* Develop a plan. In order to achieve your goals, you need to create a roadmap as to how you are going to get them accomplished. This roadmap is the plan.
4) *Milestone 4.* Implement the plan. After you have developed the plan, you need to work with stakeholders to accomplish its goals.
5) *Milestone 5.* Evaluate the progress. Did you achieve your goals? Why or why not? What could you have done better? How could the plan be improved?

Stakeholder input

One of the more important aspects of any modern planning operation is working with stakeholders to get input and advice as to what should happen on any given project. For example, if you are developing a community sustainability plan, you don't want to draft the plan first and drop it on the community as a *fait acompli*. Instead, you need to engage your community of stakeholders in order to come to appropriate decisions about what should be included in a sustainability plan. In the past, some planners used

(Continued)

a top down approach in which projects would be presented to a community with limited input. Community members had little say in the nature of the project or how it would be developed. Great examples of these types of projects are the urban renewal projects of the mid-twentieth century that created highways in the inner cities of American urban centers.

This top down approach to planning tends to be driven by a small group of decision-makers. For a number of reasons, such schemes are no longer in favor and most projects go forward with input from a variety of stakeholders so that the flow of information is from the top down and from the bottom up.

But who is a stakeholder? A stakeholder could be any person or organization that may be impacted by a project or plan. Usually they include community members, governments, and a number of non-government organizations that could include non-profits (schools, churches, charity organizations, credit unions), for profit businesses and banks, and labor and trade organizations. The goal of any stakeholder work is to get as many people together talking about the project as possible.

Just think about how these various stakeholders might provide input to a sustainability plan.

Community Members. These are the people who will be impacted in their day-to-day lives by any new project. If the plan calls for new guidelines on energy consumption, solar energy car charging stations, or any new project, these are the people who will have to pay for them through their taxes. They may also need to change their behavior.

Governments. Governments are often responsible for implementing any plan. The elected officials have to feel confident that the plan is the right thing for their community and something the voters want. Government staff have to feel confident that they can implement the plan. They can provide valuable input if a plan is too far reaching or inappropriate for the community. It is also useful to have regional, state, or national governmental experts aware of your efforts and they should be invited to any discussions on local sustainability planning.

Non-Profit Organizations. Non-profit organizations often help to achieve the results of a plan. For example, if a plan includes greater education on sustainability issues, it is important to get schools and faith-based organizations on board. There are thousands of different kinds of non-profits engaged with sustainability issues. These organizations range from those strictly concerned about the environment to those that focus on economic equity. All of these kinds of groups in a community should be invited to participate in any sustainability planning effort. Credit unions are also non-profit organizations and they can assist with local financing, particularly for smaller home-based projects such as home solar energy financing.

For-Profit Organizations. For-profit organizations are also involved with many projects associated with sustainability. If your plan, for instance, focuses on trying to create a denser, walkable downtown, you will need to get input from experts in real estate and property law. Banks are also important partners in the financing of large projects.

Labor and Trade Organizations. Labor unions and trade organizations are crucial to have at the table for discussions on any new shift in labor that might occur as a result of the outcome of the plan. For example, if you have a goal of creating more energy efficient homes via home renovations, labor needs to be available for the project. Specific training may be required.

Stakeholder engagement takes a great deal of preparation and planning in order to get the main parties in one room to work on a sustainability plan. However, the outcome of such efforts is well worth the effort.

Green local governments in Florida

While ICLEI works with local governments all over the world, many nations and regional organizations provide sustainability benchmarking tools to assist local communities in developing strategies. One of the most successful and widely used strategies is the Green Local Governments benchmarking tool developed by the Florida Green Building Coalition (FGBC).

The FGBC's main mission is to support green building initiatives throughout Florida. However, it also developed the Florida Green Local Governments Program. This initiative focuses squarely on the activities of local governments.

The benchmarking for this system is done using a spreadsheet approach to capture all activities undertaken by local governments. Points are earned for good practices in a number of categories. They are rated as follows:

Platinum: > 70% of maximum points earned
Gold: 51–70% of maximum points earned
Silver: 31–50% of maximum points earned
Bronze: 21–30% of maximum point earned.

The categories where a community can earn points are listed in Table 3.3. Each category has many ways to earn points. Only one way is listed in the table. There is also a special category for innovation. Communities may have a unique project, such as a green energy project, for which they could earn points. For a complete list, see the FGBC Green Local Governments Website here: http:// www.floridagreenbuilding.org/local-governments

To date, nearly 50 local governments in communities of all sizes across Florida have participated in the certification process. Large cities like Miami and Tampa have been certified, as have rural and urban counties. Many small communities, such as the tourist community of St. Pete Beach, have been certified as well.

The cost for local government certification ranges from $3000 to $6000 and depends on the size of the population (Table 3.4). In recent years, some communities have let their certification lapse. However, new communities are joining the program each year. One of the most recent communities to get recertified is the small Gulf Coast community of Dunedin which earned a Platinum rating.

The Florida Green Building Coalition's efforts to develop a regional green rating system for local governments is unique in that it provides a way for communities of any size to find

Table 3.3 Major indicators used to assess the greenness of local governments in Florida, as developed by the Florida Green Building Coalition.

Area of government	Example
Administration	Incorporate an environmental mission statement within the mission statement
Agriculture/Extension Service	Offer incentives to create organic farms, or sustainable/water efficient agriculture
Building and Development	Enact and enforce a Florida friendly landscaping ordinance for new construction and implement enforcement
Economic Development/Tourism	Track amount of tourism that takes part in eco-related activities
Emergency Management/Public Safety	Fire departments review training operations and conserve water where appropriate
Energy Efficiency, Conservation, and Supply	Enable customers to track and analyze energy usage via the Internet
Housing and Human Services	Affordable housing constructed by local governments mandated green
Human Resources	New employee orientation includes general city/county commitment to environment
Information services	Enact policy so all computer electronic equipment purchased has conservation features
Natural Resources Management	Enact automobile emissions regulations for vehicles registered in the community
Parks and Recreation	Implement energy efficient lighting and controls for outdoor courts, parks, and playfields
Planning and Zoning	Develop a system of sustainable community indicators related to local government planning
Ports and Marinas	Host boater education classes, educational signs, and materials
Property Appraiser/Tax Collector	Tax incentives for lands qualifying as historic, high water recharge, greenbelt, etc.
Public Transportation	Implement and enforce carpool or express bus lanes and provide an express bus to the suburbs
Public Works and Engineering	Maintain a green fleet program for department or entire local government
School Board	Involve students on green projects within the school
Solid Waste	Community wide hazardous waste collection
Water and Wastewater	Adopt policies to encourage alternative onsite wastewater and water reuse technologies and approaches

Table 3.4 The costs to participate in the Florida Green Building Coalition's Green Local Governments program.

Size of Population	Fee
Population < 20 000	$3000
Population 20 001–100 000	$4000
Population 100 001–200 000	$5000
Population > 200 000	$6000

Figure 3.7 Many Florida communities benchmark their sustainability initiatives against each other. Governments encourage homeowners to improve homes and landscaping to achieve regional goals.

ways to contribute to regional sustainability efforts and to be rewarded for their efforts. It is one of the most comprehensive local government rating schemes. It is worth spending some time online looking at the system to get acquainted with the range of issues where local governments can make a contribution to sustainability initiatives. Are there rating systems for local governments where you live (Figure 3.7)?

Greening sports: baseball

Think of the impact of any major event that you might go to (Figure 3.8). It could be a concert, a festival, or a sporting event like a baseball game. Tens of thousands of people show up to one place for a few hours of activity. But think of everything that goes into making sure that the event goes off as planned:

- Space needs to be built to hold that many people
- Roads, parking lots, and other transportation options must accommodate the travel needs of the attendees
- Food and beverages must be provided
- The facility must have power and lighting
- Advertising and community outreach about the event must be completed
- Bathrooms and sewer lines must be constructed to accommodate the significant sewage waste produced by crowds
- Waste management of paper, food waste, and other materials must be coordinated
- Water must be provided for bathrooms, showers, catering operations, and grounds
- The grounds of the facility must be landscaped
- The stars of the show, whether a sports team or superstar singers, must be transported to and from the venue

When considering all of the things involved with putting on these spectacles, there are many ways to coordinate a sustainability initiative. That is exactly what many teams are doing. They are looking at ways to measure their impact on the environment. Indeed, this initiative has been supported broadly by Major League Baseball in the United States. In 2005, the organization created a partnership with the National Resources Defense Council to promote greening of the professional sport among all the teams.

Figure 3.8 It might be a surprise to some, but professional sports focuses heavily on sustainability in their operations. What is your sports team doing? This photo is of CITI Field, home of the New York Mets. Photo by Josh Grossman.

Each team has completed different initiatives based in part on their unique geography and fan base. For example, some of the teams (for example, Washington Nationals and Target Field) have built LEED (Leadership in Energy and Environmental Design (see Chapter 8)) certified stadiums that are among the greenest of sports venues in the world based on criteria established by the United States Green Building Council. The San Francisco Giants play in a field that is LEED rated for existing buildings.

Other teams have focused on green energy by reducing energy used by lighting or by building solar panels. The Houston Astros and the Boston Red Sox use solar energy to power some of their operations. The Minnesota Twins help to transform their waste to energy. Still others have focused on adding organic and local food options for visitors. The Boston Red Sox use only hormone-free meats from animals that are pasture grazed.

Transportation is also an important issue for baseball. Many teams have worked to locate their parks in downtown areas, where it is easy for someone to attend a game by taking mass transit. The Tampa Bay Rays, with their downtown St. Petersburg, Florida, location, offer free parking for car poolers and the Florida Marlins reserve spaces for hybrid vehicles in their stadium parking lot in Miami.

As noted in the first chapter, sustainability is not just about the environmental aspects of an operation. Economic and social aspects must be considered as well. Major League Baseball does a tremendous amount of community work in most of their markets. For example, the Philadelphia Phillies offer a number of environmental outreach programs. The Arizona Diamondbacks focus efforts on environmental education in an outdoor pavilion.

Most of Major League Baseball's efforts have made a measurable impact in trying to lessen the environmental footprint of professional sports in their home communities.

Take a look at a professional sports team in your area. What types of initiatives do they have for trying to positively influence their local community?

Specific community plans

While the Florida Green Building Coalition and ICLEI for local governments provides guidance on sustainability planning, some local governments have developed their own unique approaches to local sustainability planning. In this section, we will review two important plans that have influenced the way communities address sustainability in government.

PlaNYC

New York City is one of the most densely populated cities on the planet. However, by some measures, it is also one of the greenest cities that we could ever create. It might seem strange to think of a place covered in concrete and with very little green space as a sustainable community. Many have complained about the pollution, traffic, and overall frenetic pace of the city. Indeed, the magazine *Travel and Leisure* rated it the United States' most dirty city due to all the litter and garbage present in the street. However, by many measures, particularly indices of energy, water, and transportation, New York City is one of the most efficient urban areas in the world. However, in 2007, a new effort to make it even greener was launched.

The initiative, called PlaNYC, is the brainchild of New York's former mayor, the entrepreneur, Michael Bloomberg, owner of the financial media company, Bloomberg L.P. Bloomberg is one of the leading voices in the United States on issues of climate change. He has long advocated for greenhouse gas reduction policies and for broader improvements in environmental protection. He has used his platform as a business leader and mayor to advocate for a number of issues in the United States, such as gun control regulation, LGBTQ rights, and immigration reform. However, his work on climate change and sustainability planning may be his largest legacy.

While many parts of the plan are not controversial, such as repairing infrastructure and conserving resources, many were critical of the plan's expectation to greatly increase the population of densely crowded New York. Yet, New York remains one of the main centers for international immigration in the world and it is wise to plan for increased population (Figure 3.9).

The Statue of Liberty welcomes visitors; Frank Sinatra says "If you can make it here, you can make it anywhere"; and Jay Z says "New York, concrete jungle where dreams are made of, There's nothing you can't do, Now you're in New York."

The expansion of New York is evident everywhere in the city. For example, in Midtown Manhattan, not too far from the famous Carnegie Hall, a new residential building was just

Figure 3.9 New York City is encouraging many sustainability initiatives including modifying older buildings and building new ones to green standards.

Table 3.5 Major New York city sustainability themes. Selected specific indicators are listed. For a full review of all of the goals in the plan, see http://www.nyc.gov/html/planyc/html/home/home.shtml.

PlaNYC goal	Measurable indicator
Housing and Neighborhoods	Create homes for almost a million more New Yorkers while making housing and neighborhoods more affordable and sustainable
Parks and Public Space	Ensure all New Yorkers live within a 10-minute walk of a park
Brownfields	Clean up all contaminated land in New York City
Waterways	Improve the quality of waterways to increase opportunities for recreation and restore coastal ecosystems
Water Supply	Ensure high quality and reliability of the water supply system
Transportation	Expand sustainable transportation choices and ensure the reliability and high quality of the transportation network
Energy	Reduce energy consumption and make energy systems cleaner and more reliable
Air quality	Achieve the cleanest air quality of any big US city
Solid waste	Divert 75% of solid waste from landfills
Climate change	Reduce greenhouse gas emissions by over 30%

constructed as the tallest building in the city— a 1350 foot tower that is only 43 feet wide at the top! It is taller than the new World Trade Center building if you do not count the antenna. The apartments will go for millions of dollars.

While the millionaires have taken over Manhattan, middle- and low-income workers have been displaced to the surrounding boroughs of the city. Nevertheless, many new building requirements in PlaNYC required that all housing be built to greener standards than they were in the past.

Housing was not the only component of PlaNYC. Specific targets are set for almost every component of the city's governance. While the Florida Green Building Coalition rates communities based on a salad bar of options, PlaNYC was very specific about creating measurable goals that can be met within a specific time period. The goals are listed in Table 3.5. While the plan has transitioned into a new plan (ONENYC), it is worth understanding this important plan since it is really the first of its kind to be developed.

As you can see, the ten goals listed are all quantitative and measurable. Each provided a way to assess how the city is doing at achieving its goals each year. The goals created a pathway by which decisions were made and results assessed as to whether or not appropriate decisions were made toward the goals.

London and sustainability

Another city that has made tremendous progress in promoting sustainability is London. The city has been working on sustainability initiatives in a concentrated way under the auspices of the London Sustainable Development Commission (LSDC) since 2002. Over the years,

Table 3.6 Environmental indicators used by the London Sustainable Development Commission. Note that progress was made on most indicators since 2012. Positively, there was no deterioration associated with any of the indicators.

Environmental indicator	How measured	Improvements since 2012?
Carbon dioxide emissions (scope 1 and 2)	Total scope 1 and 2 emissions for London	Yes
Carbon dioxide emissions (scope 3)	Total scope 3 emissions for London	Insufficient information
NO_x emissions	Tonnes of emissions in London	Yes
Particulate matter 2.5 micrometers and smaller	Tonnes of emissions in London	Yes
Particulate matter 10 micrometers and smaller	Tonnes of emissions in London	Yes
Recycling	Percentage of household waste recycled or composted in London	No change
Waste	London's performance against the greenhouse gas Emissions Performance standards	Yes
Flood risk, tidal and fluvial	Properties at risk of tidal and fluvial flooding	Insufficient information
Flood risk, surface water	Properties at risk of surface water flooding	Insufficient information
Water consumption	Per capita consumption	Yes

they have developed a series of indices that can be measured and benchmarked in order to better address sustainability issues within the city.

The indicators developed by the LSDC are divided into three broad categories: Environmental Indicators, Social Indicators, and Economic Indicators. The most recent report that was produced in 2017 compares indicators with measurements from the last report, which was published in 2008.

The Environmental Indicators are listed in Table 3.6. There has been positive movement on all but four of the eleven environmental indicators. None of the indicators have had any deterioration. It is clear that the city, for the most part, is making progress at making their city more sustainable within the realm of the environment.

The social indicators are listed in Table 3.7. One of the indicators, education, has had deterioration. Most of the remaining indicators have seen improvement.

The economic indicators are listed in Table 3.8. These indicators have not done as well as the environmental and social indicators. This may, in part, be due to the global economic meltdown that occurred after 2008 and the challenges with Brexit. However, only three indicators (Gross value added, Employment, and Child poverty) saw significant improvements.

Table 3.7 Social indicators used by the London Sustainable Development Commission. Note that progress was made on most of the indicators since 2012. One (education) saw deterioration.

Social indicator	How measured	Improvements since 2012?
Healthy life expectancy	Healthy life expectancy at birth for men and women	Yes for men No change for women
Education	Proportion of pupils obtaining at least 5 GCSE passes at A–C or equivalent including English and Math	No
Travel	Share of journey stages in London made by a sustainable mode	Yes
Crime	Recorded crime	Yes
Decent housing	Percent of decent housing stock in London	Yes
Happiness	Self-reported levels of happiness	Yes
Satisfaction with London	Percentage of Londoners satisfied with the capital as a place to live	Insufficient information
Volunteering	Participation in formal or informal volunteering over previous 12 months	Yes
Social integration	Proportion of people who think their local area is a place where people from different backgrounds get on well together	No change

Overall, progress has been made in London in each of the three indicator groups (environmental, social, and economic). Indicators are used to measure progress year by year. The environmental indicators demonstrate that there is the most success in this area. However, there are remaining challenges in social and economic areas of sustainability that need to be addressed.

Measuring sustainability indicators in London provides a way to evaluate how the city is making progress on sustainability over time. By adding new indicators, the scheme stays current to the needs of the community, while continuing to measure long-term success. Finding areas where little progress is being made or where there is a deterioration of conditions provides evidence for the need for investment or the development of new policy in key areas. Finding success is also important in that it provides a way of understanding what techniques are useful in making improvements.

Take a look at the 2017 report from the London Sustainable Development Commission here: https://www.london.gov.uk/sites/default/files/lsdc_-_qol_2017_summary.pdf for more details. Why do you think they were successful in some areas of sustainability, but less successful in others? What kinds of strategies can you think of to try to make improvements in some of the areas where there was little change or deterioration?

Take a look at your own local community. Do they have ways of benchmarking or assessing sustainability or green initiatives on an annual basis? What kinds of measurable indicators might be useful in your own community?

Table 3.8 Economic indicators used by the London Sustainable Development Commission. Note that progress was made only on three of the eleven indicators since 2012 (Gross value added, Employment, and Child poverty).

Economic indicator	How measured	Improvements since 2012?
Gross value added	Gross value added per head in London	Yes
Employment	Employment rate	Yes
Business survival	Survival of London business after one year of trading	No significant change
Human capital	Full human capital per head	No significant change
Innovation (products)	Proportion of firms reporting introducing product innovations	No
Innovation (processes)	Proportion of firms reporting introducing process innovations	No
Income inequality	Disposable income differentials in London	No significant change
Child poverty	Children living in households below 60% median income	Yes
Fuel poverty	Proportion of fuel poor households in London	No
Housing affordability	Ratio of lower quartile house prices to lower quartile earnings	No
London living wage	Proportion of people earning less than London Living Wage per hour in London	No

Small towns and sustainability

The ideas represented by the New York and London sustainability plans represent a number of great ideas that can be used by communities of all different sizes. Many medium to large cities have many of the same issues represented by the indicators used by New York and London. Yet small towns and rural communities have different types of sustainability issues. They also have a number of problems including the fact that they often have: (1) a very narrow economic base, (2) experience a loss of population as young people move to the city, (3) undergo environmental degradation due to poor agricultural management or mining activities, (4) are confronted with poor infrastructure such as sewage, electrical, roads, etc., and (5) have poor income potential due to limited job and educational opportunities. For many communities, the idea of sustainability is far from their top priority due to the myriad of challenges they face. Nevertheless, some rural communities have come to understand that some of the activities in their environments are highly unsustainable and are seeking to transform their rural communities into models of sustainability.

Think about rural sustainability in your own region and country. What types of sustainability issues are associated with agriculture, mining, or fishing? What are the main environmental, social, and economic challenges in these areas? What organizations are working on rural sustainability?

Business sustainability

One of the challenging things in our world of global economic interconnectedness, and widespread and expanding consumer culture is how to achieve sustainability, while also encouraging economic growth and development. Many organizations have tried to achieve this balance by finding ways to address sustainability within a business model. Instead of focusing in on the three "E"s of economics, equity, and environment, they focus on the three "P"s of people, planet, and profits. This will be discussed in much more detail in upcoming chapters (Figure 3.10).

The concept of people, planet, and profits suggests that there are ways to advance sustainability, while making a profit for individuals or shareholders. The concept has been embraced by many in the business community. Some of the readers of this book might be surprised to learn that companies such as Wal-Mart and IKEA have offices of sustainability and that these large companies annually turn in a sustainability report.

These companies set distinct goals to try to improve various aspects of their business. There are about as many approaches to business sustainability as there are businesses. Yet most focus on some aspect of supply, manufacture, shipping, or personnel. Take, for example, one of the first major businesses to fully embrace the concepts of sustainability, Interface Inc. This company makes a number of different types of carpet products, including industrial carpeting for high-use areas such as airports or hotels.

The president of the company, Ray Anderson (1934–2011), wasn't particularly interested in sustainability issues when he took over the leadership of the firm. However, he started to realize the impact of his supply chain on the environment. He started to see that the choices

Figure 3.10 How can businesses that you use benchmark their operations and products to become more sustainable? The Lakefront Brewery in Milwaukee has created an organic beer to help their organization become more sustainable.

made by the company over the years had a distinct negative impact on the environment and on other cultures around the world.

As a result of this epiphany, Mr. Anderson did a complete review of the company's operations to try to limit the impact on the environment. When he did this, he realized that the decisions made to improve the environment also improved the profit of the company. It made good business sense to go green. It also made the business itself more sustainable by ensuring that they did not deplete resources that were important to the manufacturing of their products.

After realizing the benefit of moving his business in the direction of sustainability, Mr. Anderson spent the rest of his life promoting business sustainability with other business leaders around the world. Today, his efforts have paid off in many ways. Most of the major global companies try to address sustainability issues in some way. While there is a long way to go, there is distinct progress. We'll review several case studies in an upcoming chapter.

However, think about businesses you know well. Perhaps you work in a restaurant part time or maybe you have a small lawn care business or a babysitting service. How might you make your workplace greener? What has already been done? How could the efforts be measured and tracked? In order for our planet to move forward into a more sustainable future, all organizations, no matter how small, must think about sustainability and try to find ways to reduce their overall footprint. Yet what can we do personally?

If you are taking this class and reading this book, you probably have some personal commitment to sustainability principles. Perhaps you are a true environmental "tree hugger" like me, or perhaps you just like the idea of environmental protection. Or, maybe you realize that there are new business opportunities in the emerging "green economy" and want to learn more about it. Your motivations and feelings about the environment are not all that important from a quantitative and scientific perspective. What matters is your measurable impact on the planet. You could have the greenest outlook, but have a much bigger impact than someone who thinks global warming is a hoax. It is easy to point the finger at companies that promote their sustainability initiatives without looking at our own impact.

For the record, I have a huge carbon footprint. I am writing this on a ferry in the Aegean Sea as I travel to visit a friend on the island of Serifos. I flew to Athens from New York with London as a stopover to catch this ferry. Plus, I travel on average in the United States via plane for business or personal reasons seven times a year. Air travel is among the biggest emitters of carbon on the per person basis, so I probably have a bigger carbon footprint than most people on the planet who do not regularly travel by plane (note, I purchase carbon offsets for my air travel). As I noted earlier, there are many different ways of viewing sustainability. In my case, I need to try to compare my impact with others like me, not against others very unlike me. In other words. The cultural context of sustainability matters. Just think about your overall planetary impact. How would your impact on the planet differ in other parts of the country where you live? What about in other parts of the world?

There are many ways to measure and assess personal sustainability. As I stated above, a carbon footprint is a very common approach. Table 3.9 provides a method by which you can measure your carbon footprint.

Yet carbon is not the only thing that can be measured to assess personal sustainability. We can calculate our water budget, energy use, food impacts, and many others. Once we have measured these, we can set targets to make reductions. As noted several times in this

Table 3.9 An example of how to develop a measurable indicator for assessing one's personal carbon footprint. This example focuses on reducing carbon as a result of air travel.

Problem	I have a high carbon impact on the planet as a result of air travel
Goal	Reduce my impact on the planet from air travel
Methodological approach	Use carbon credit purchase options for air travel
How achieved	Purchase of carbon credits
How reported	Personal blog

chapter, it is all about measuring impacts, setting goals, finding ways to achieve goals, and reporting on the outcomes.

I stated earlier that I fly approximately seven times a year. I have made it my personal goal to find ways to mitigate the impact of the carbon reductions by purchasing carbon credits for all travel in the coming years. Many travel websites offer the purchase of carbon credits, which makes it very easy to achieve my goal. I will report on my goal using my blog, On the Brink, as a way to keep me honest. Think about what goals you can set to reduce your impact on the planet? How will you measure them and report them?

4

Energy

We all require energy to exist in our modern world. This book was written on a laptop computer, sent electronically to my publisher, printed or otherwise disseminated, and you used energy in some way to complete the purchase of the book. This text, therefore, has a distinct carbon footprint. Almost everything we do uses energy in some way. As much as we try to reduce our energy consumption or try live as simply as we can, we all use energy. In this chapter, we'll examine different forms of dirty and green energy sources and discover some innovative energy programs in some parts of the world. We'll see that the world is moving more and more into using clean sources of energy and leaving dirty energy sources behind.

World Energy Production and Consumption

Energy resources are not distributed evenly across the planet. In fact, just a few nations have extensive energy resources and most of the world must import energy in order to thrive. A myriad of problems have evolved in recent years as a result of this imbalance.

The top ten energy producers are shown in Figure 4.1. China, the United States, Russia, Saudi Arabia, and India are all major producers of energy. Each of these countries has abundant resources, such as coal, oil, or natural gas. Many of them also have large populations that require large amounts of energy.

In contrast, the top ten energy consumers are shown in Figure 4.2. China, the US, India, Russia, Canada, Brazil, and Iran are both high energy producers and consumers. However, Japan, South Korea, and Germany are all low energy producers, but high consumers. As we will see, each country has developed a unique strategy to confront their lack of energy resources.

Another way of looking at energy is by assessing energy intensity. Energy intensity is a way of calculating the energy efficiency of national economies. For example, if a nation uses a great deal of energy to produce low value for the national economy, it would be considered a relatively high energy intensive economy. Nations have been seeking to develop low energy intensive operations to improve the energy efficiency of their economies.

A list of the least energy intensive economies (energy efficient) are shown in Figure 4.3. Energy efficiency is calculated as a kilogram oil equivalent at 2005 purchasing power parities. What is interesting about this list is that it only contains one nation (Germany) that is

Introduction to Sustainability, Second Edition. Robert Brinkmann.

Companion website: www.wiley.com/go/Brinkmann/IntroductiontoSustainability

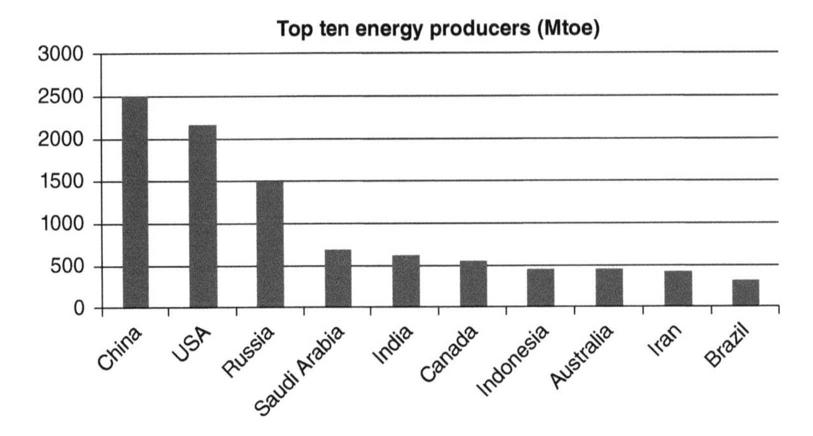

Figure 4.1 Top world energy producers in 2018. The figure includes all sources of energy. (Source of data: http://yearbook.enerdata.net)

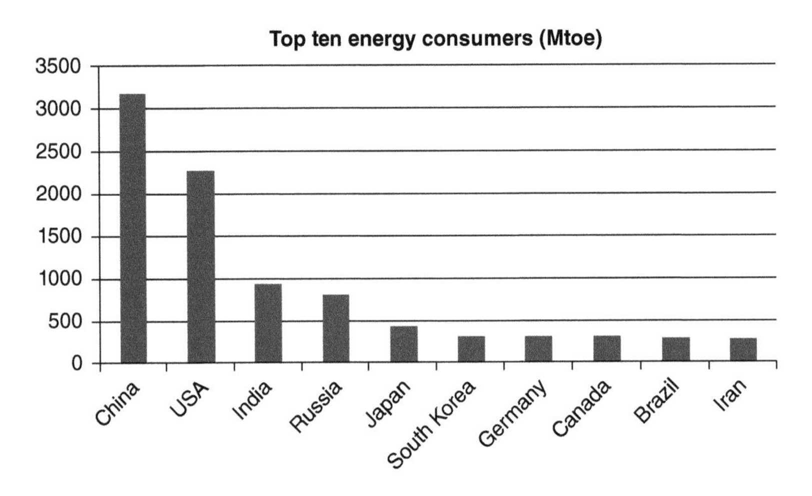

Figure 4.2 Top world energy consumers in 2018. (Source of data: http://yearbook.enerdata.net)

one of the top ten energy consumers. What this means is that the economies of most of the major energy producing nations are relatively inefficient. The major energy producers are not present on the list, suggesting that energy producers use their resources inefficiently.

By far, most of the energy that is utilized on the planet comes from traditional fossil fuel sources (Figure 4.4). Coal, oil, and natural gas account for over 80% of all energy consumed on the planet. Nuclear energy makes up nearly 6% of the world's total energy use, and biofuels and wastes another 10%. Hydroelectric sources make up approximately 2.3% and alternative energy sources, such as solar and wind, account for only 0.9% of global energy use.

Now that we have reviewed the global energy situation, let us now move on to discuss specific energy sources. We will do this by examining traditional "dirty" energy, green or clean energy resources, and nuclear energy. I put nuclear in a category alone because there is currently tremendous debate as to whether or not it is a clean or green energy source.

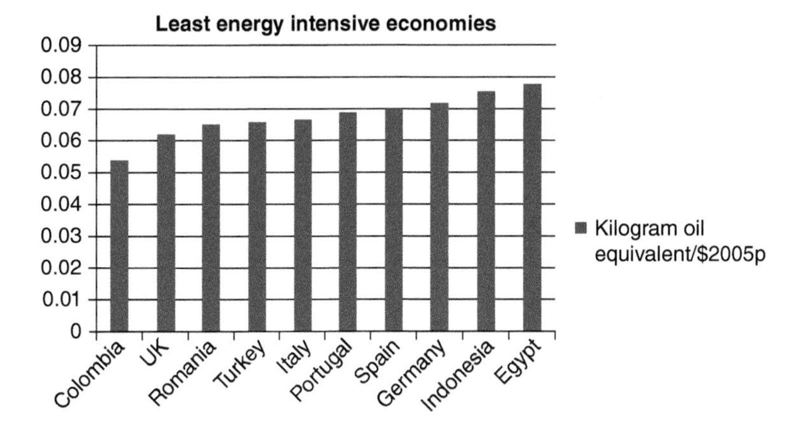

Figure 4.3 Least energy intensive economies in the world. (Source of data: http://yearbook.energdata.net)

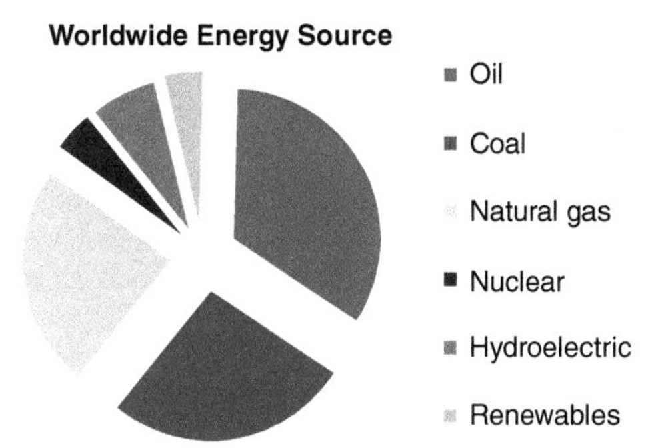

Figure 4.4 Global energy use by source. (Source: https://www.bp.com/content/dam/bp/business-sites/en/global/corporate/pdfs/energy-economics/statistical-review/bp-stats-review-2019-full-report.pdf)

Traditional or "Dirty" Energy Resources

Many of the traditional sources of energy are considered dirty energy because they produce significant pollution, particularly carbon dioxide, the main gas behind global climate change. But there are many different types of dirty energy, including oil, natural gas, coal, and tar sands. Each of these will be reviewed. However, they account for approximately 80% of the world's energy production. While the world is moving rapidly toward the development of alternative energy sources, it is important to understand the nature of these traditional energy resources in order to assess how to make them more sustainable in the future. In addition, overall energy use in the world is increasing and renewables are not developing as rapidly as the traditional sources.

Oil

Oil, or petroleum, is a natural liquid that is found in some sedimentary rocks. The liquid formed as a byproduct of the decomposition of animals and plants in vast ancient swampy basins. In these settings, layer after layer of organic material was deposited for millions of years to produce vast hydrocarbon reserves that include coal, natural gas, and oil.

Scientists have found that oil is found mainly in rocks of the Mesozoic Era. This time period lasted from 252 to 66 million years ago. This vast time span is commonly referred to as the Age of the Dinosaurs. Large dinosaurs roamed large areas of the planet and much of the Earth's surface contained extensive cover of vegetation. When the plants and animals died, their remains were preserved if they ended up in water. With time, the deposits of organic matter transformed into oil, natural gas, and coal.

Oil is a highly versatile product. It is relatively easy to extract in wells on land or on platforms in the ocean. It can be made into a wide variety of products, including gasoline and other fuels, plastics, and pharmaceuticals. It is also very easy to distribute through pipelines, by trains and trucks, and by car. It is relatively abundant and readily accessible as a commercial product all over the world.

As noted above, the location of energy resources is not evenly distributed across the planet. That is largely because the Mesozoic Era basins where the organic matter collected occurred in a few key areas of the planet. Thus, once remote, isolated, and unpopulated Saudi Arabia becomes important due to its ancient geologic history. And Japan, with its lack of oil reserves due to its volcanic landscape, becomes dependent upon outside sources of energy.

Over the last several decades, hundreds of oil fields have been exploited throughout the world. New oil fields are being brought on line each year. Nevertheless, there is concern as to whether or not the world has achieved peak oil (see box).

Several countries produce most of the world's oil (Figure 4.5). The United States, Saudi Arabia, and Russia are the main crude oil producers, with Canada, Iraq, Iran, China, the United Arab Emirates, Kuwait, and Brazil following. Contrary to the perception that oil resources are decreasing, production in some of these countries is increasing. For example, the United States saw a 12% increase in crude oil production between 2017 and 2018.

Crude oil must be processed to create other oil-based products, the building blocks of the modern carbon economy. Many crude oil processing plants, or refineries, are found

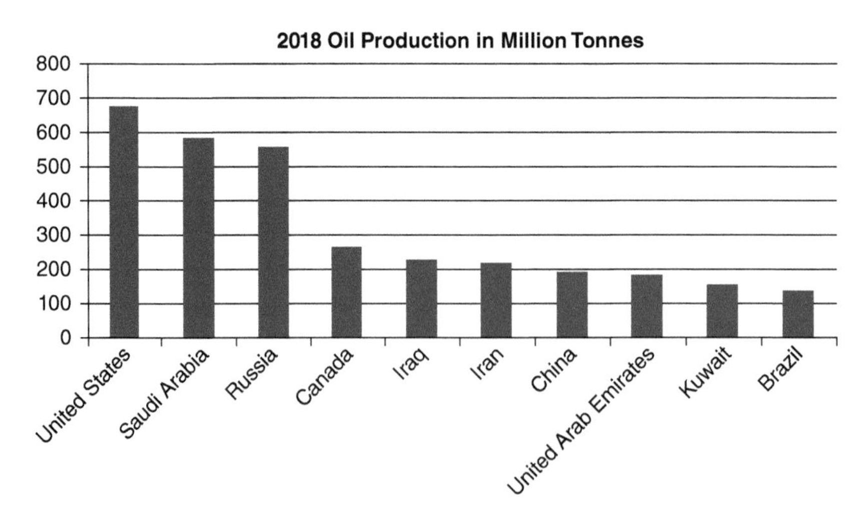

Figure 4.5 Top ten nations for petroleum production. (Source of data: http://yearbook.energdata.net)

throughout the world. These plants transform crude oil into highly refined chemicals like gasoline and plastics.

Many nations without oil reserves have built oil refineries in order to process cheap crude oil into value-added products. For example, oil-poor Japan has dozens of refineries that process crude oil into gasoline and other petroleum products.

There have been concerns about the use of gasoline in recent decades. One issue relates to political and international challenges. For example, the oil crisis of the 1970s presaged the problems associated with the first Gulf War when Iraq invaded Kuwait in 1990. The Falklands War between Argentina and Great Britain was largely thought to be driven by access to oil. Even today, the South China Sea basin is a flashpoint due to conflict between China, Vietnam, the Philippines, and other nations over oil present in the area.

Of course, there are also concerns over pollution. Significant attention has been given to greenhouse gas pollution associated with fossil fuel production. I will address this in detail in an upcoming section. However, there are a wide variety of other chemicals released when oil is burned besides carbon dioxide. Nitrogen and sulfur oxides, as well as impurities and particulate matter can enter the atmosphere when oil products are burned. Many countries have severe air pollution problems as a result of the burning of fossil fuels. Some countries, such as the United States, have developed strict laws about the release of these types of pollutants and have set guidelines for manufacturers.

There is also concern over pollution caused by the extraction and transport of oil. On the night of April 22, 2010, the Deepwater Horizon oilrig exploded in the Gulf of Mexico with 126 people on board; 11 people were killed. The explosion set off a massive oil leak that lasted 87 days. When it was done, the leak ended up being the largest accidental oil spill in history.

Oil spills don't just happen at the source of oil. They can also occur along pipelines and in transit. Thousands of miles of pipelines traverse the world bringing oil from wells to refineries. In March, 2013, a pipeline carrying oil from Canada to the Gulf of Mexico was breached in Mayflower, Arkansas near Littlerock. Approximately 5000–7000 barrels of oil were spilled in that event in a quiet suburban neighborhood. However, perhaps the most notorious oil spill is the Exxon Valdez spill that occurred in March, 1989.

The Exxon Valdez was a tanker that was transporting 55 million gallons of oil from Alaska to Long Beach, California, of which approximately half leaked out of the ship when it hit a reef in Prince William Sound. Eventually, over 1300 miles of coastline and 11 000 square miles of ocean floor were impacted. The spill caused extensive ecological damage in one of the world's most pristine landscapes.

Oil spills are difficult to clean up. The crude oil is very sticky and can cling to animals, thereby making it difficult for them to fly or swim. The oil can also pollute soil and water, making them unsuitable for natural ecological habitat for plants and animals. The oil also can persist for long periods of time. Oil from the Exxon Valdez oil spill continues to wash ashore.

Oil shale and tar sands

While most oil is easily extracted from porous rocks deep under the surface of the Earth, there are rock and sediment deposits that contain oil that can be extracted on the surface—oil shale and tar sands. Oil shale is a sedimentary rock that contains very fine-grained clays and silts intermeshed with oil in very small pores. The pores are so small that it is difficult to get the oil out using conventional pumping techniques. It is estimated that there are 3 trillion barrels of oil in oil shale. The United States, Russia, and Brazil have the largest reserves.

Oil is extracted from the rock by heating it. Thus, the extraction of the oil requires significant energy use. Once heated, the vapors are collected from the rock. Huge amounts of remaining shale are left behind as a waste product.

Tar sands are another source of energy. They are found in sandstone that contains high amounts of semi-solid oil, called bitumen. Most of the tar sand deposits are in Canada, Russia, and Kazakhstan and it is believed that the deposits hold 2 trillion barrels of oil. The extraction of oil from the tar sands is complex. High volumes of hot water are used to extract the oil from the rock and sand. The processing, therefore, uses considerable amounts of energy to heat the water in the extraction process. There are significant waste problems associated with the sands left behind and the water that was used to extract the oil.

Oil shale and tar sand oil extraction are problematic for a number of reasons. First, each requires significant energy to extract the oil. Second, the high energy use significantly increases the greenhouse gas impact of these fuels. Third, the extraction of oil from tar sands requires a significant amount of water. As we will see in Chapter 6, water is a resource that is growing in scarcity these days and its use in energy extraction is questionable. Finally, each of these energy sources leaves behind vast areas of waste in their production.

Nevertheless, the high cost of energy has made the extraction of energy from these unusual sources of energy profitable. It is likely that the importance of these energy sources will continue to grow. Those of us concerned with sustainability should work to limit the impacts of these fuels on the environment.

Peak oil

Many around the world are concerned that the world has achieved peak oil, or the maximum rate of oil extraction after which production is expected to decline. In fact, people have been arguing for years that we have been at peak oil, but oil production continues apace in many parts of the world and new technologies to extract oil have aided this.

Nevertheless, oil is a finite resource and it will run out at some point. Demand for oil is also increasing, especially in the developing world. Will we have enough oil for the future?

Some predict that as supplies diminish and oil prices rise, the world will move aggressively to develop alternative sources of energy without causing any interruption in the current lifestyle. Others are more pessimistic and argue that significant societal changes will occur due to the movement away from cheap oil.

What kinds of changes would you have to make if oil was unavailable to you? How do you think your school or family would react to the lack of oil?

Figure 4.6 Natural gas is among the most common fuels in use around the world.

Natural gas

Natural gas, mainly consisting of methane, is a fossil fuel often found associated with oil or coal deposits (Figure 4.6). In the past, it was considered a waste product of oil extraction and was burned. However, with improved transportation and storage technologies, natural gas has become a highly valuable product and a significant component of the world's energy supply. It is worth noting that natural gas can also be derived from landfills and as a byproduct of biomass production. It is considered one of the cleanest sources of fossil fuels, although it can be a significant greenhouse gas contributor when released unused.

There is a huge amount of natural gas reserves (approximately 7000 trillion cubic feet) in the world. North America is the largest producer of natural gas and accounts for nearly 25% of all natural gas production (Figure 4.7). The United States and Russia are the largest national producers. Iran, Canada, Qatar, China, Norway, Australia, Saudi Arabia, and Algeria each produce less than a quarter of the natural gas of either the United States or Russia.

Natural gas is considered a dangerous fuel and it was not deemed a viable global energy commodity for many years. It is a colorless and odorless gas and it can cause serious damage if large quantities of it are exposed to flame. However, the development of natural gas resources has been aided by the development of modern transportation systems and new power plant technologies. In addition, natural gas can be converted to a liquid for ease of transit. Natural gas pipelines traverse many areas of the world. Many of the pipelines lead directly to power plants where electricity is directly produced using conventional steam generation processes.

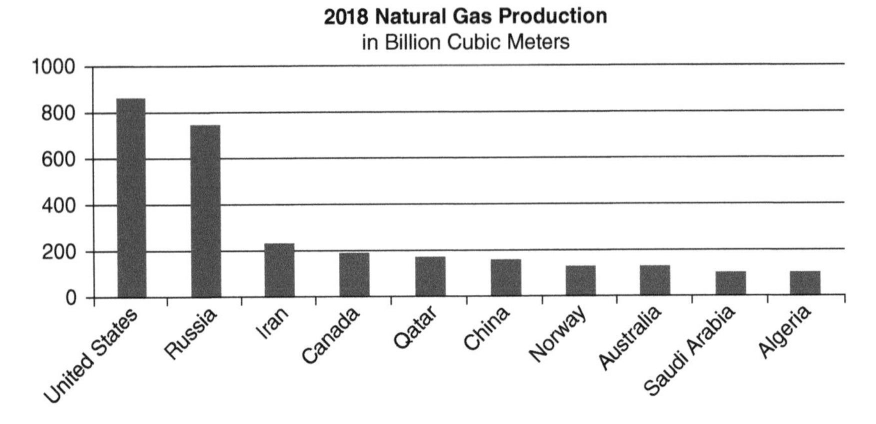

Figure 4.7 Top ten producers of natural gas in 2018. (Source of data: http://yearbook.energdata.net)

There are two broad concerns associated with natural gas production: greenhouse gas pollution and hydraulic fracturing (fracking). Each will be discussed below.

Methane is a greenhouse gas. It helps to trap energy in the atmosphere that leads to atmospheric warming. While I will discuss greenhouse gas and global climate change in detail in another section, it is worth highlighting that methane has 25 times the impact on atmospheric warming as an equivalent amount of carbon dioxide. While we are concerned with carbon dioxide emissions from the burning of fossil fuels, we should be 25 times as concerned about leaking of methane in the extraction and transit of natural gas.

Another problem associated with natural gas production is hydraulic fracturing, or fracking. Fracking is a process by which natural gas is extracted from small pores in rocks deep underground. Fracking is done when conventional extraction techniques are unable to draw out natural gas trapped within pores in rocks.

Using specialized drilling equipment, fluids are pumped underground into the rocks at high pressure to break or fracture the rock. When this happens, the natural gas is released for extraction. Using these processes, many new natural gas fields have been opened around the world.

However, the process is highly controversial. The fracturing process can release methane into groundwater where it can enter wells. In addition, the fluids used in the high-pressure process to break the rocks have been of some concern. Many of the gas companies have not released the variety of chemicals that are utilized in the fracking process. There is concern that these chemicals will enter groundwater and ecological systems.

Some communities have banned the process of hydraulic fracturing extraction of natural gas, while others have embraced it. The state of New York has banned the process. In contrast, the state of North Dakota has embraced hydraulic fracturing. This has led to a huge boom in oil and natural gas in the sparsely populated state. Many areas of the world are looking at hydraulic fracturing as a form of local economic development, especially in rural, sparsely populated areas where there is limited concern about environmental impacts. Supporters of the process state that environmental concerns have been exaggerated. Only time will tell if fracking detractors will be proven unfortunately correct.

Coal

Coal is a carbon-based sedimentary rock made up of organic material that has turned into rock after considerable time. Most of us are familiar with peat, a thick mat of organic material that forms in wetland environments. In some wetlands, water conditions prevent the decomposition of organic matter after it dies and thick layers of undecomposed organic matter, or peat, can form.

There are many different types of coal that can form from peat, depending on what happens to the peat over time. If it has not undergone extensive heat and pressure, it becomes lignite coal, a soft low-carbon coal that is responsible for about 8% of the coal produced in the United States, one of the largest coal producers in the world. Subbituminous coal is somewhat harder than lignite coal and has a carbon content that of approximately 40%. It is the source of 40% of the US coal production and is used largely for coal electricity plants. Bituminous coal is the most commonly used coal in the US and has a much higher carbon content than the other sources. It is used not only in power production, but also in steel and iron production. Anthracite, which is the least common form of coal, is a very hard form of coal that contains up to 97% carbon.

Like oil and natural gas, coal is largely found in rocks that formed in the Mesozoic. Vast wetlands covered a number of different portions of the planet and these are now extensive coalfields. These coalfields are not distributed evenly across the planet and some regions of the world have extensive deposits of coal while others have limited coal resources.

China is by far the largest coal producer in the world (Figure 4.8). As of 2018, it is responsible for over half of the world's coal production and most other producers are dwarfed by the magnitude of China's output. China is also the world's largest user of coal. In fact, it uses four times the amount of coal used by the United States. See text box for information on China's air pollution problems caused by its strong reliance on coal.

Coal is used largely for the production of electricity in coal-burning power plants. Some coal is used in the production of iron and other metals. In addition, some is liquefied or turned into a gas to use as a fuel. There are some other chemical and manufacturing uses of coal. In recent years, coal has become an important component in water purification operations.

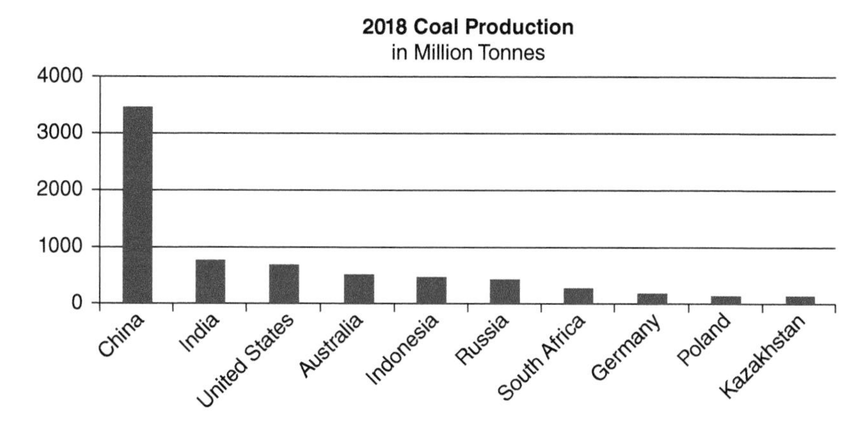

Figure 4.8 Coal production in the world in 2018. (Source of data: http://yearbook.energdata.net)

China's air pollution woes

I remember flying into Beijing airport for the first time in 2012 and looking for the city from my airplane window. The pollution was so bad that I didn't see the ground until just about the time we landed. Clearly Beijing has some serious air pollution problems. However, there is no doubt that China has made great strides in cleaning up its air. Plus, Beijing's unique geography adds to the air pollution woes. It receives a great deal of particulate matter and dust from western provinces and the air around Beijing can get trapped during some meteorological conditions.

When China was awarded the right to host the 2008 summer Olympic Games in 2001, there was immediate concern about the status of the air quality in Beijing. Many parts of the country, particularly Beijing, are known for severe air pollution and there was concern over the safety of athletes and tourists who would attend the games. However, China immediately began planning how to clean the air for the Olympics.

Some factories that were responsible for major air pollution problems were phased out, closed, or moved. New environmental controls were implemented in others. As the Olympics approached, even more efforts were made to lower pollution by reducing the numbers of cars on the streets, improving mass transit, and decreasing working hours to limit the potential for pollution.

Overall, the plan worked. The Olympics were enjoyed in relatively clean air. However, many of the changes were temporary and China's air pollution woes have not lessened. It has been reported that air pollution cuts life expectancy in polluted northern China by up to 5.5 years (http://money.cnn.com/2013/07/09/news/china-air-pollution/index.html). Many seek to get out of Beijing to find jobs in less polluted areas of the province.

China has been getting very serious about trying to solve its environmental problems, but this is a challenge. China burns half of all the world's coal used every year. Unless significant changes are made, China's air pollution woes are likely to continue.

Coal mining

Coal is a sedimentary rock that forms within layers, or strata, of rock. Most of the time, this rock can be mined across a single layer of rock. When it is not at the surface, shafts are dug in the rock to produce accesses to mine the rock out of the subsurface. Coal mining is one of the most dangerous occupations in the world due to problems with roof collapses and exposure to poisonous gases that are sometimes associated with the coal beds. In addition, sudden fires can break out in the subsurface. Some of these have caused loss of life. In some rare instances the fires can burn for years.

Extensive subsurface mining is usually conducted using a process called roof and pillar mining. Coal beds are followed through the subsurface and extracted leaving behind pillars to hold up the roof. Vast empty rooms are left behind in the process. With time, the ceilings of the rooms can collapse to cause damage to the surface.

Another form of mining is open pit or surface mining. There are a number of different ways that surface mining can be conducted. Simply stated, the rocks above coal deposits are removed to expose the valuable energy resource. This is done with very large heavy equipment. Once exposed, the equipment is used to excavate the coal for shipment. This process is clearly safer for miners than subsurface coal mining.

When surface coal mining is done at a large scale, it is very controversial. One of the most controversial forms of surface coal mining is sometimes called mountaintop removal. In some areas of the world, large landforms or mountains are made in large part out of coal. Thus, the entire mountain is excavated, leaving behind a changed landscape largely unfamiliar to local residents. In addition, hydrology and ecosystems (as well as other natural systems) are heavily altered.

Some places have banned mountaintop removal in order to protect the local environment. However, it continues apace in many areas of the world. The large quantities of coal removed using this process makes it highly efficient and profitable. Yet, the environmental consequences are substantial. A great deal of environmental activism on the topic of mountaintop removal has emerged in recent years that has brought attention to environmental and social problems associated with coal mining.

Pollution from coal

Of course, one of the larger concerns associated with the use of coal is pollution. The burning of coal is by far the largest contributor to anthropogenic carbon dioxide in the atmosphere, one of the major greenhouse gases associated with global climate change. A wide range of other gases is released into the atmosphere as well, including nitrogen and sulfur oxides. The burning of coal also releases a variety of metallic pollutants, including mercury and arsenic. These pollutants have very complex environmental cycles, but can enter ecosystems, where they do great damage. They can also enter the food chain.

Sulfur dioxide is a pollutant that has gotten special attention in recent decades because it contributes to acid rain. When sulfur oxides are released into the atmosphere during the burning of sulfur-rich coal, they react with water to produce sulfuric acid. The acid combines with atmospheric moisture and can fall with precipitation. There are many well-documented cases of acidification of rain downwind from power plants and factories that burn sulfur-rich coal.

Over the years, technology has been developed to reduce sulfur dioxide from power plant emissions. In addition, low-sulfur coal has become the fuel of choice among power plants. The impacts of acid rain are many. It has been found to be responsible for acidification of lakes and other water bodies to the point that they do not support indigenous animals and plants. Acid rain also causes problems when it falls directly onto plants; many forests have been damaged as a result.

Waste from burning coal, called fly ash and bottom ash, is also problematic. As noted above, the carbon percentage in coal varies considerably. The remaining materials are often left behind after the coal is burned in the form of ash. This ash is fine grained and often contains hazardous materials like heavy metals. If dumped in the open air, water could easily flow through the ash like water flowing through coffee grounds and extract metals and other materials that can enter the groundwater.

In many parts of the world, the disposal of fly ash is heavily regulated. Some have found useful ways of recycling this material by turning it into paving or building material. However, huge amounts of fly ash must be disposed of every day.

Green energy

Biomass

Biomass is a significant source of energy in many parts of the world. Energy from biomass comes in many different forms, including the burning of firewood or manure, the burning of garbage, and the conversion of biomass to liquid or gas fuel. Each is discussed below.

Biomass: wood, manure, peat, and other organic sources

Wood and manure are common fuel sources in areas that do not have access to standard electricity. Wood is also used as a primary or supplemental heat source in cold climates. Many rural areas of the world, particularly in lesser-developed regions, rely on burning wood or other biomass such as manure for heating or cooking.

Firewood is a renewable resource. However, if collection outpaces regeneration, problems will ensue. Ecosystems can decline and long-standing ecosystems degrade. In hilly or mountainous areas, soil erosion can cause significant devastation.

The collection of firewood is problematic in some areas, particularly semi-arid regions where tree growth is slow. In some areas of the Sahel of Africa, a semi-arid belt bordering the Sahara Desert, extensive firewood collection has led to the expansion of the desert in a complex anthropogenically driven process called desertification. If firewood collection proceeds more rapidly than regeneration, the desert can expand and claim former semi-arid regions, thereby disrupting the ecosystem for generations.

Some tropical areas have also seen significant deforestation from both agricultural expansion and the collection of firewood. It is important to note that many people are working to solve the deforestation problems caused by the collection of firewood. New fuel sources are being delivered to communities reliant on firewood. Plus, reforestation efforts are underway in many areas. Please see text box for a review of reforestation efforts in Haiti.

Haiti and its biomass problem

Haiti is a country in the Caribbean on the western half of Hispaniola. It is home to nearly 10 million people on its nearly 28 000 square kilometers of land. It is the poorest and one of the most densely populated countries in the western hemisphere. Haiti was originally an agricultural colony of France. During this time, much of Haiti was deforested to make way for plantation-style agriculture. The ecology of the island has not fully recovered from the devastation wrought during this era. Plus, aggressive timber extraction in the twentieth century made matters worse. Since 1923, Haiti has gone from 65% tree cover to less than 5%.

Haiti established independence from France in 1804 and today some areas of Haiti are thriving, even though the country has been troubled by political instability. However, most areas of Haiti remain in extreme poverty and many do not have access to basic services, including electricity. Thus, many poor Haitians rely on wood as a fuel source, thereby exacerbating the colonial legacy of deforestation.

The colonial plantation agricultural system and more modern deforestation caused significant soil erosion from Haiti's mountainous and hilly landscape. Thus, the natural soil ecosystem is severely degraded and many areas cannot support the type of vegetation that once existed on the lush tropical landscape. The collection of firewood furthers the problem.

The government and many non-profit groups are striving to solve these problems in Haiti today by establishing aggressive tree planting campaigns. For example, The Lambi Fund (www.lambifund.org) is involved with promoting reforestation using trees that can aid in economic development. They have been promoting the planting of coffee and fruit trees. Other organizations like the Timberland Company, a manufacturer of shoes, have been involved (see http://community.timberland.com/Haiti). Timberland planted over 6.5 million trees between 2010 and 2017 and plans to plant millions more.

However, planting trees only solves part of the problem. Ecosystems must be repaired. Among other organizations, the Audubon Society of Haiti has focused on promoting ecological restoration of the country (http://audubonhaiti.org/program/conservation/). While much needs to be done in Haiti, there is no doubt that there are a number of public and private sector organizations working to make improvements.

Burning of garbage: waste-to-energy

The burning of garbage has emerged as a significant energy source in some areas. The process, more commonly called waste-to-energy, burns garbage in a power plant to create electricity.

There are many sources of garbage in the world. Many cultures have become major consumers of global products. The consumer culture that has been created thrives on the manufacture and purchase of inexpensive products that do not have a long lifespan. Products are built to last a few years before they are out of fashion. A great example of this is our current obsession with phones. A new model comes out every year with slight improvements, thereby making older models out of date and out of fashion.

This drive for the new is sometimes called "fast fashion." We are driven for the new, the improved, and the exciting. We throw out the old and used. This has led to huge amounts of waste that must be managed by landfills and dumps throughout the world. We talk more about garbage and waste in Chapter 10, but one way that we have found to deal with all of this garbage is to burn it for energy.

In the past, waste-to-energy facilities had few environmental controls and were notorious polluters. Garbage contains a myriad of waste products that are harmful to health and the environment when burned. Because of this, the plants had bad reputations and were unwelcome in communities. However, most waste-to-energy incinerators have strict controls that require them to remove harmful chemicals.

When the garbage is burned, ash remains as a byproduct. Metals are recovered from this material and the remaining ash must be sent to a landfill. The ash must be handled in particular landfills because it can be a significant pollution source if water entering ground or surface water percolates through it. However, burning of garbage reduces landfill volume by up to 95%.

Today, there are hundreds of waste-to-energy facilities all over the world. They operate by heating water in a boiler to produce steam that drives turbines to produce electricity.

Many modern plants are co-generation plants, meaning that they produce electricity as well as useful heat as a byproduct of the electrical generation.

Many involved with the sustainability movement have criticized the growth of the waste-to-energy facilities. They note that they burn resources that could have been recycled. They also suggest that waste-to-energy reduces a community's desire to recycle.

Understanding waste-to-energy

Waste is a problem all over the world and we constantly struggle with how best to manage waste in our culture. The best options are to reduce, reuse, and recycle. But in many places, options for reducing, reusing, and recycling are limited. The reality is that many areas of the world have severe problems with waste and waste management. We just produce too much waste for most places to handle. In Long Island, New York, for example, there are no landfills available to take the waste of millions of people. Instead, the waste is either transported great distances to landfills far from Long Island, or the waste is burned in a waste-to-energy facility.

Waste-to-energy is the process by which garbage is burned and turned into electrical energy. The heat from the burning garbage creates steam, which turns turbines that create the electricity. The company that runs several plants where I live is Covanta and they are the largest operator of waste-to-energy facilities in the US. Many do not realize it, but waste-to-energy is one of the greenest options out there for dealing with garbage (except for recycling and reusing it). Putting it in a landfill creates methane gas that has a significantly higher climate impact than carbon dioxide.

The largest plant on Long Island employs 84 people, most of whom are in the operations side of the plant. They burn everything that comes into the plant and collect metals or other materials from the ash. All of the material that enters the site is scanned for radiation. Waste-to-energy plants typically have emissions-monitoring systems and use LN (low nitrogen) technology to collect nitrogen oxides. According to Covanta, the best waste-to-energy plants running continuously for a year have fewer emissions than ten diesel cars running continuously for the same time period. The largest facility on Long Island provides enough electricity for 75 000 homes.

To stress again, the best options for garbage are to reduce, reuse, and recycle it. But, turning your remaining waste-to-energy is the best current option for dealing with garbage in many parts of the world. One of the challenging issues with waste-to-energy facilities is the use of the remaining ash. In some cases it is landfilled, while in other cases it is turned into usable products such as grit in cement.

Conversion of biomass to liquid or gas fuel

One other source of biomass fuel is agricultural crops such as sugar and corn. Sugars from these crops are converted into bioethanol via fermentation. The ethanol is an alcohol-based fuel and burns differently from traditional fossil fuels. However, it can be mixed with gasoline as an additive or it can be burned directly in various types of fuel cells.

The United States is the largest ethanol producer in the world, much of it derived from corn. It produces approximately 15 million gallons per year, roughly accounting for

10% of the volume of gasoline. Brazil, the second largest producer of ethanol, produces approximately 7.2 million gallons of ethanol per year, derived largely from corn and sugar cane. Together, the United States and Brazil produce nearly 90% of world production. Unlike the US, Brazil's ethanol production makes up about half of its gasoline usage.

Ethanol is one of the more controversial green energy sources, in part because it takes food to produce energy. Thus, land that could be used to feed people is instead used for other purposes. As the world's population has increased in recent years, there have been ethical questions raised as to whether or not it is appropriate to utilize agricultural land to produce ethanol.

Ethanol is also controversial because it takes a number of different resources to produce crops that create energy. Water, fertilizer, pesticides, and herbicides must be utilized. In addition, creating ethanol utilizes a tremendous amount of energy.

The high use of energy and other resources has led many to doubt the sustainability of using crops to produce ethanol. Many have suggested finding other biological sources for the production of energy. A variety of crops show promise. For example, some have suggested using native shrubs or grasses instead of standard agricultural crops to produce methanol.

In addition, scientists have been able to derive fuels from oils produced by algae. This too requires energy and nutrients. Others have developed forms of energy production from a variety of agricultural waste products or human or animal wastes. There is no doubt that many are working to derive less energy intensive forms of biofuels in laboratories. These biofuels will be with us for many decades and it is likely that improvements will be made to reduce their impact on the planet.

Wind energy

One area of green energy that has seen tremendous growth in the last several years is wind energy. Certainly we all know that wind energy has been used for centuries in a number of different capacities. However, it is only recently that it has been widely used to produce electricity. Today, expansive fields of windmills, called wind farms, can be seen all over the world.

Before reviewing these large sources of electrical energy, it is worth noting that small-scale wind energy is growing in popularity. Individuals or organizations can purchase or build small wind turbines to produce electricity for their own needs. In many places, the electricity that is produced can be placed into the grid where utility companies purchase it. In addition, many remote areas rely on wind energy as the principal source of energy.

Overall, wind energy produces 4.5% of the world's electricity—nearly doubling from just a few years ago. China and the US are the leaders in world wind energy production with over 50% of the world's total. Germany produces about 10% of the world's total and Spain, the India, and the United Kingdom each produce about 5%. However, this total energy production doesn't really tell the full story. The United States and China consume huge amounts of electricity. So even though they are producing large amounts of electricity from wind, it is a relatively small percentage of their overall energy usage. In contrast, Denmark, which has significantly lower energy needs, gets approximately 42% of its electricity from wind.

Electrical wind generation is increasing rapidly at approximately 25% per year. However, not all areas are suitable for wind production. Those that are, will often be found in coastal regions or interiors of continents where wind speeds are suitable for the production of wind energy.

Interior portions of continents have highly variable winds and may not be entirely suitable for electrical power generation. Wind variability makes it difficult to rely on wind power as a steady source of energy; it must be supplemented with other energy sources. Offshore winds are much more regular and thus offshore areas are more suitable for the development of wind farms for reliable energy production. However, with proper energy planning, wind can be integrated into a broad energy plan for a region.

In places where wind energy produces more energy than needed or during times of high winds, energy can be stored by pumping water into tanks or reservoirs for release through electricity generating turbines. The water can be released during periods of low winds or during times of high demand. This process essentially stabilizes wind power as a constant and reliable source of energy. Batteries can also be used to store the energy.

The development of wind farms has increased greatly in the last few years as costs for electricity have increased and as costs for the production of windmills has declined. Most of the expense of developing a windmill is during the initial planning and construction phase. There are minimal maintenance and upkeep costs. However, many wind farms are located far from electricity users. In some cases, it must be transmitted great distances.

Part of the reason that wind farms are not closer to population centers, is that many do not want them in their neighborhoods. Some find the windmills unsightly and there is concern over their impact on property values. The windmills are not particularly loud, but they do produce a constant whooshing sound when they are in operation. Some have even suggested that the sound impacts farm animals that graze in the vicinity of wind farms.

There is also concern over the impact of wind farms on birds and bats. The wind turbines are immense: they can be over 100 m tall, about the height of the Statue of Liberty. There is no doubt that some wind turbines cause the death of birds and bats. However, there is considerable disagreement over the total impact of the turbines on these animals. Raptors, particularly the brown eagle, seem to be the most affected birds. Nevertheless, wind energy is considered one of the greenest forms of energy available to our modern society.

How safe are windmills for wildlife and people?

One of the biggest critiques of wind energy is that the windmills are responsible for the deaths of birds and bats. There is no doubt that if windmills are sited on major migratory bird routes that they can cause significant problems. Windmills can also harm bats if they are located near their roosting sites.

Yet how significant is this damage? Statistics demonstrate that feral cats kill far more birds than windmills. In recent years, some have suggested that the regular whooshing sound from the windmill blades makes grazing animals and wildlife nervous. However, there is little research demonstrating any long-term problems with the sound of the windmills.

Some dramatic windmill accidents have been filmed that show dramatic breakdown of the windmill turbine and tower. Such accidents can occur when braking systems in the windmill fail and the blades spin at high speeds to cause overall instability of the windmill structure. When such events happen, debris can jettison at high speeds across a wide area to do damage to structures or nearby animals or humans. However, such accidents are extremely rare.

While care should be taken when siting windmills to ensure the least amount of damage possible to migrating birds or bats, it is clear that in most cases, windmills are not a major ecological problem. Environmental impacts occur with any project. Windmills definitely have an environmental impact. The question is whether or not the impact from windmills is more palatable than the impact from developing other energy sources.

Solar energy

Solar energy is energy that derives from the Sun for useful purposes. There are two broad ways that solar energy is utilized: passive and active systems (Figure 4.9). Each will be discussed below.

Passive solar energy

Passive solar systems use the heat of the Sun to supply energy to homes directly or seek to limit the impact of the Sun's rays during hot months. This is done by constructing and

Figure 4.9 This solar panel helps to power a farm in Vermont.

orienting a home to address the unique needs of the climate. In cooler climates, for example, a building will be designed to collect solar energy in the form of heat through windows. Buildings are oriented to capture Sunlight and heat. Today, many windows are designed to transfer solar heat to buildings when needed, by using special glazing technologies.

In warmer climates, buildings are designed to limit the impact of the Sun's rays. Many indigenous buildings in arid and semi-arid climates utilize thick mud brick walls that warm during the day, but keep heat away from the interior of the building. At night the walls radiate heat, thereby heating the inside when it is cool. In our modern era, building materials have been developed to reflect the Sun's rays to limit solar heating. Shading of windows is employed, and windows are oriented to avoid the impact of the Sun's heat.

Passive solar energy use is often built into many building code requirements throughout the world. The use of this relatively simple technology has saved the world considerable energy.

Active solar energy

Active solar energy is what most people think of when they think of solar energy. It uses solar panels to collect the Sun's rays. The panels convert the Sun's energy to either heat or electricity.

When panels are used to create heat, they are often used for heating water for household use. In these cases, the Sun's rays heat water and glycol (to prevent freezing) within tubes in the solar panels. The warm liquids are carried to water tanks where a heat transfer operation is undertaken through a coil system within the tank. Due to the unreliability of the Sun's rays, the tanks typically have electrical back-up to ensure a steady supply of hot water. Nevertheless, these hot water systems can save considerable energy. In places with a reliable supply of solar energy, some have taken their entire hot water systems off the electrical grid.

In the last decade, much more significant attention has been given to the development of photovoltaic solar energy. A photovoltaic cell converts basic solar energy into electrical energy using converters. In the early stages of photovoltaic development, solar cells were highly inefficient and required significant investment that made them relatively cost prohibitive. Today, photovoltaic energy is approaching the costs of conventional energy, particularly with incentives that are given by many governments to advance the use of them.

Photovoltaic solar panels can be installed on homes or businesses to reduce the overall cost of conventional energy. In recent years, there has been considerable attention given to concentrating the number of cells on large buildings, parking lots, or vacant land to create managed large energy operations. There are also some large photovoltaic power plants that have been built in recent years. These are called solar farms and they are growing in importance throughout the world.

China, the United States, and Japan are the largest producers of solar energy in the world. Cloudy Germany is the fourth largest producer. Other large producers include India, Italy, the United Kingdom, Australia, France, and South Korea It must be noted, however, that many small countries have moved strongly into solar energy, particularly in rural areas and rapidly developing regions far from main power grids. Kenya, for example, is one of the largest per capita producers of solar energy in the world.

The development of solar energy is aided by the advent of high-tech solar tracking devices that can automatically subtly change the position of solar collectors throughout the day in order to take advantage of the best angle to collect the Sun's radiation. These tracking systems utilize energy collected by the solar panels, so there is no net energy cost to this improved efficiency. However, because these panels require electronics and equipment that moves the panels automatically, there are some inherent maintenance costs.

Concentrated solar power

One of the other significant innovations in solar energy is the development of concentrated solar power stations. These systems utilize arrays of mirrors to concentrate solar energy into a beam of light that is utilized to produce heat. The heat is used to create steam that is used to produce electricity. The development of these concentrated solar power plants is advancing rapidly. The amount of power generated by these plants is roughly doubling every year. Most of these plants are located in Spain and the United States.

Critiques of solar power

There have been a number of critiques of solar power. Some of the large solar farms and concentrated solar power stations utilize considerable amounts of land, thereby limiting other productive uses. Some of the photovoltaic cells utilize rare elements that can cause environmental problems. There are also concerns over the long-term supply of some of the components of photovoltaic cells. The major manufacturer of solar energy equipment is China and some have called into question the overall quality of some of these products.

Nuclear energy

Some classify nuclear energy as a green energy source and others consider it a dirty energy source. I have opted to discuss it separate from both green energy and traditional dirty energy sources due to its unique properties.

There is no doubt that many look to nuclear energy as the energy of the future. However, concerns over the safety of nuclear power plants and the long-term problems of nuclear waste call this assertion into question.

Nuclear power plants typically operate through the splitting of uranium-235 via nuclear fission. This releases heat that can be used to generate steam that turns turbines to create electricity. In this way, nuclear power plants are similar to other types of power plants such as coal-burning systems. However, the fuel is rather different from traditional power plant systems.

Uranium-235 is an isotope of uranium that is utilized in most nuclear power plants. Due to the toxic nature of uranium, its mining and handling can prove difficult. One of the clear challenges of the use of uranium and other nuclear materials is that it has a very long half-life, about 704 million years.

Nuclear power produces roughly 10% of the world's electricity in over 400 nuclear reactors. The largest producer of nuclear energy is the United States. Roughly 20% of the electricity in that country comes from nuclear energy. Although it doesn't produce as much nuclear energy as the US, France gets 70% of its energy from nuclear power plants. Nuclear energy use has declined in recent years over safety concerns. However, some nations, particularly China, are building new power plants. Thus, it is likely that nuclear energy production will increase in the coming decades.

Many have looked to nuclear energy as a way to provide energy for areas with few energy resources like coal, oil, natural gas, or wind. Indeed, some countries, like France and Japan, have invested heavily in the development of nuclear energy as a significant source of fuel. Others also argue that nuclear power is a suitable substitute for carbon emitting energy sources. Indeed, some experts believe that rapid development of nuclear energy as a replacement for dirty energy sources is the only way to reduce greenhouse gases in order to prevent global climate change. Indeed, Andrew Yang, a contender in the US 2020 presidential election, highlighted the need to grow nuclear energy to reduce greenhouse gases in our atmosphere.

However, others point to the dangers of nuclear power plants and associated radioactive waste as reasons why we should not develop nuclear power. While most nuclear power plants operate safely, there are examples of widespread devastation as a result of accidents in power plants. The most famous of them are the accidents at Three Mile Island plant in the United States, the Chernobyl plant in what was then the USSR, and the Fukushima plant in Japan. Each case will be briefly reviewed.

The Three Mile Island accident occurred in 1979 near Harrisburg, Pennsylvania in the United States. The cause was a malfunctioning valve, compounded by the poor reaction of the operators. Radioactive gases and waters were released in the accident, but there was no known health or environmental impact. However, this accident raised awareness of some of the dangers of nuclear power plants and the public, particularly in the United States, became concerned about the implications of using nuclear technology to produce energy.

In comparison, the 1986 Chernobyl nuclear power plant disaster was much more significant. It occurred due to a power surge during normal testing of the plant. The surge caused steam explosions in the reactor and the nuclear core of the plant was exposed to air. When this happened, the core caught fire and a series of explosions ensued, causing some of the nuclear fuel to eject from the building. During these explosions, massive amounts of radiation were released. Overall, dozens of employees and emergency workers were killed in the accident. More than 100 000 people were evacuated from the region and radiation spread over wide areas of Europe. Even today, there is an exclusion zone 30 kilometers around the doomed plant. The area will not be safe for settlement for 20 000 years.

The Fukushima power disaster, more formally called the Fukushima Daiishi disaster, occurred after an earthquake and tsunami in 2011. Nearly 20 000 people were killed in the natural disaster, although, thankfully, there were no deaths associated with the immediate aftermath of the nuclear power plant disaster. Interestingly several nuclear power plants are located near major fault zones or in areas that could be impacted by tsunamis or other flooding events. However, the Fukushima plant is the first one that experienced significant damage as a result of a natural disaster.

When the earthquake occurred, the power plant shut down automatically in order to prevent any significant problems. However, when the tsunami hit, the waters flooded the generators, causing them to fail. When this happened, the nuclear reactors did not receive the normal supply of cooling water and they began to overheat. A number of explosions occurred that caused the leakage of radioactive elements into the atmosphere. Radioactive water also was released into the ocean.

Unlike the Chernobyl disaster, large areas were not impacted. However, areas around the plant are still off limits to most and some areas remain evacuated as of this writing. While no one was killed in the initial release of radioactive materials, there is concern over the long-term health of people exposed to the radiation and the long-term environmental impacts of the release of radioactive elements into the environment. Fish have been caught with very high levels of radiation offshore and some crops have been found to be contaminated. Plus, there are long-term waste and clean up problems that are rather daunting.

These cases illustrate the concerns that some people have about the future of nuclear energy. To many, nuclear power plants are disasters waiting to happen. Given the broad problems associated with Chernobyl and the unknown environmental consequences of the Fukushima radioactive releases, these concerns are entirely understandable.

There is also concern over what to do with radioactive waste that is produced in power plants. While most scientists suggest deep underground storage of nuclear waste in stable geologic settings, most places do not welcome the waste due to concerns over contamination. Plus, it must be noted that for considerable amounts of the waste, the long half-lives of radioactive decay make the waste a problem for thousands of years. There are very few governments that have been stable for that long and there are concerns over what happens to the waste as time progresses. In addition, there is concern over the waste being turned into a dirty bomb that could contaminate major cities. A nuclear explosion is a horrific event that kills many people at once. However, a dirty bomb would spread nuclear waste across vast areas, making them unusable for long periods of time.

In the United States the problem with waste is compounded by the fact that there is no set national repository of high-level nuclear waste. For many years, a place called Yucca Mountain was considered the best choice for the location of a national nuclear waste management facility. The site is located in the state of Nevada about 160 kilometers from Las Vegas, near the California border. Work progressed on the site for many years after the US Congress approved it in 2002. However, due to strong local opposition, the funding for the site's development was removed. Today, there are no approved repositories for the storage of high-level nuclear waste.

There was also concern about the transport of the waste to the facility. Yucca Mountain is in a remote portion of the United States far from most sources of nuclear waste. The shipment of this waste raises safety concerns, no matter how it is shipped. Many communities were concerned about the waste travelling through on roadways or rail lines. Although relatively uncommon, accidents on roads and rail do occur.

Since there are no places to take dangerous nuclear waste, it has to be stored on site. Thus, at dozens of nuclear power plants in the United States, significant amounts of nuclear waste are stored in cooling ponds. While this material is inventoried and guarded, there are concerns over the wisdom of storing large amounts of nuclear waste at dispersed sites all over the United States. Accidental releases could occur and some of the waste could be stolen.

Given all of these issues, nuclear energy has been on the decline in some areas of the world. While some areas are aggressively pursuing nuclear energy, many areas of the world are having second thoughts about the development of their nuclear facilities.

Yet it is important to note that there are some emerging nuclear technologies that hold promise. Improvements in safety and reduction of waste products are particularly important innovations. Plus, one of the most interesting innovations is the development of small-scale nuclear power plants that could be utilized to power small towns or distinct urban neighborhoods. These systems are compact and much safer than many current power plants.

Nuclear energy remains an important, although controversial, energy source in today's world.

Other innovations

While all of the previous discussion has focused on fuels, it is worth spending some time to discuss other innovations that help with energy reduction. Specifically, this section will review energy efficiency innovations, smart grid technology, and small scale, off the grid efforts.

Energy efficiency

When electrical innovations were first created, they were highly inefficient. Many basic products, like the light bulb, used tremendous amounts of energy. Today, we are using a number of innovative ways to improve energy efficiency of basic products. There are thousands of innovations, from light bulbs to automobiles. A few will be highlighted.

The basic light bulb that has been used for decades is called an incandescent light bulb. It works by running an electrical current through a wire filament. Unfortunately, these light bulbs are highly inefficient and utilize only about 5% of energy in producing light. Much of the rest is lost in heat.

Over the years, several innovations have been developed to improve the efficiency of light bulbs, including the invention of fluorescent light bulbs (including compact fluorescent bulbs) and LED lights (Figure 4.10). Fluorescent light bulbs utilize a mercury vapor that excites a fluorescent coating on the bulb to produce light. The bulbs are much more efficient than traditional light bulbs and they last much longer. They have been used for decades in tube fixtures in large buildings and institutional settings for years. It has only been recently that they have been put into production for home use in the development of compact fluorescent bulbs.

The implementation of compact fluorescent bulbs does have some controversy. There is concern over the disposal of these bulbs because of the amount of mercury they contain in the tube. They need special handling in order to recover the mercury. However, it is believed that a significant number of these bulbs enter the normal waste stream where the mercury can be released into the environment.

Our modern computers demonstrate another innovation: the automatic shutoff. When things are not in use for some period of time, they can be shut off automatically to reduce energy consumption. A typical desktop computer uses about $1.00 worth of electricity if it is left on overnight. However, when it goes to sleep, it uses an almost unmeasurable amount of energy to keep it ready to turn on when needed. Many modern electrical applications utilize this technology and it saves considerable energy.

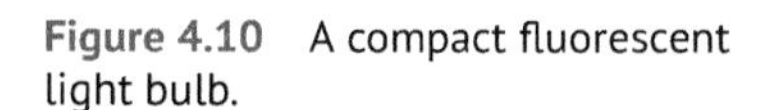

Figure 4.10 A compact fluorescent light bulb.

Smart grid technology is another source of energy efficiency. The traditional electrical grid consists of a series of wires that radiate outward from power plants. If a major power line went down, it would impact the entire grid. The smart grid system in contrast breaks this grid down into smaller components so that it is much more reliable and resilient to storms or other power line interruptions. In addition, it utilizes modern electronic communication systems to adjust the electrical load on the power lines. It will understand where there is demand and will communicate with electrical equipment in homes or offices to reduce loads if there are extreme demands on the line.

Smart grids also allow for time of day metering. The highest demands for electricity are during the day. Thus, this is the most expensive electrical production period. Using smart grid technology and two-way communications, users can reduce their electrical consumption in order to reduce their costs.

In addition, there is more interest in distributed energy systems. These are small electrical generating facilities that can be added to the electrical grid. Some of these might be rather small, in the form of a household solar system, or larger, in the form of an institution-based solar array on top of buildings. The smart grid can easily accommodate these additional energy sources into the broader system.

The development of smart grids has progressed rapidly in the United States and many other areas of the world are adopting its implementation. It saves considerable amounts of energy and makes the overall energy grid much more reliable. When Superstorm Sandy hit New York in 2012, many areas of Long Island were without power for over ten days.

A smart grid would have prevented many of the outages as they were due to the downing of key power lines. With a smart grid, the energy is much more distributed across the grid and is less reliant on single major lines.

There is of course a wide-range of energy efficiency improvements that have been made in the last several decades, ranging from improvements in engines and motors to improvements in energy transmission. Some of these improvements, such as energy management systems that are used to manage large buildings and innovations in energy consumption in transportation applications, I'll return to in other chapters. However, it is worth noting that these innovations have saved considerable amounts of energy and will likely save considerable energy in the future. Thus, while societies continue to develop alternative energy sources, it is important to also focus in on improvements in energy efficiency.

Living off the grid

Do you think you could live off the energy grid? Some people have given up living with regular power systems and are living off the grid. Some have rejected all energy systems and live with just wood or bottled natural gas for heating and cooking. Others have gone high-tech and have installed solar and wind energy systems for electrical systems.

Just take a look at Dancing Rabbit Farms (see www.dancingrabbit.org). This is an intentional community in Missouri in the United States. Here, dozens have come together to live simply in a small collective agricultural settlement. Members have their own chores such as gardening, cooking, teaching, etc. while also living much more simply than the rest of us. They do have limited energy use, but have opted to live without television, regular lighting, and the hi-tech gadgets that many of us have.

The idea of *intentional communities* is not new. People have been choosing to live with like-minded people for years. The idea of leaving your family or your birth residence for a new place with individuals who want to live like you probably goes back to the earliest of times. Even the idea of simple living separate from the technological world goes back hundreds of years. Many have rejected the growth of technology to form new ways of living. In the United States, the Amish are an example of a large group of people who have rejected conventional life ways. They live in tightly knit agricultural communities without most modern conveniences such as cars and televisions. Their conventions are based, in part, on Biblical teaching.

Dancing Rabbit Farm is certainly a new intentional community that incorporates many of the ideas of the anti-technology communities like the Amish that came before them. However, this modern community is rooted in our current concerns about issues such as global climate change, consumerism, and quality food. Dancing Rabbit is just one example of a modern intentional community rooted in some of the ideals of sustainability. What intentional communities exist in your area? Could you live in such a community? Why or why not?

5

Global Climate Change and Greenhouse Gas Management

One of the most significant challenges the world is facing today is global climate change. There are a small number of people who feel that it is not real, but the vast majority of scientists believe that it is a serious problem for the future of our planet. In this chapter, we will look at how greenhouse gas pollution is the major driver of global climate change, review greenhouse gas pollution, examine ways in which greenhouse gases can be managed, and finally discuss adaptation strategies that have been adopted due to the overall failure to date of our society to confront the challenges of global climate change in a serious and consistent way.

Over the years, we have made great strides in managing a number of environmental pollutants. For some reason, we have not been able to come to terms with managing greenhouse gases. Unless serious attention is given to solving this problem, we will be facing a significantly changed environment in our lifetimes.

The end of nature?

In 1989, Bill McKibben published a book called, *The End of Nature,* which noted that nature, which for millennia was immune to the broad impacts of human activity, was changing all around us (Figure 5.1). McKibben provided abundant evidence for the heavy hand of man and its impact on the planet. The book especially highlighted the impacts of global climate change on our world. This book came out decades ago, yet we still have not found ways to stop the emissions associated with global climate change. In fact, they have increased. So, while many scientists and environmental activists have been sounding the alarm, many have sought to deny the problem or deceive the world into thinking the problem doesn't exist.

While we can find the impacts of global climate change and other environmental problems all around us, there is still hope for the future. We have confronted other serious environmental problems in the past and solved them. We can still address the serious issues of global climate change. The world may not be the same and we may be at the end of nature as we know it. However, we have to recognize that we have to not only try to solve the problem of global climate change, but also to seek ways to become more resilient in the future. We may be entering an era of great climate uncertainty. How we react to this new form of nature may decide the fate of generations to come.

Introduction to Sustainability, Second Edition. Robert Brinkmann.

Companion website: www.wiley.com/go/Brinkmann/IntroductiontoSustainability

Figure 5.1 This is Manhasset Bay, an inlet off of Long Island sound. The landscape here is highly altered as a result of human activity. How many kinds of alteration can you see in the land, water, and air? What kinds of alteration can you see from your bedroom window?

As we will see, unless we significantly reduce greenhouse gases, we will see a warmer world. This added heat will cause oceans to rise, change weather patterns, and disrupt long-held agricultural traditions. It will change ecosystems and ocean currents. Added carbon dioxide also fundamentally changes the chemistry of the oceans, thereby adding stress to already polluted environments. How we act in the next several years to curtail greenhouse gas pollution will determine the fate of the planet.

The science of global climate change: The greenhouse effect

In order to understand why so many people are concerned about global climate change, it is important to understand the basics of the greenhouse effect. Imagine that you have entered a very hot car that has been sitting in the sun. The air in the car is much warmer inside than outside. The same thing happens in greenhouses. Horticulturalists take advantage of the greenhouse effect to keep their greenhouses warm and toasty for plants when it is too cold outside for their survival. But how does the greenhouse effect happen?

The greenhouse effect occurs because certain gases in the atmosphere have the capability of absorbing infrared energy, or heat. These gases are called greenhouse gases because of their special properties. When they are in abundance, they can trap more heat than when there are fewer of them. Thus, the amount of greenhouse gases in a place determines how much heat can be absorbed.

The inside of a car or the inside of a greenhouse are both confined spaces that contain gases, but our planetary atmosphere is another space that can absorb heat in just the same way. In fact, our planet is just the perfect temperature for us because our atmosphere contained greenhouse gases and it was able to absorb heat to allow life to occur. However, if there are too many or too few greenhouse gases, planetary cooling or heating can occur. Greenhouse gases help to regulate our temperature. Too many gases and we become a super hot, lifeless planet. Too few and we become a cold, dead planet. Earth has it just right for us (Figure 5.2).

There are several major greenhouse gases. The most significant of the greenhouse gases are water vapor, carbon dioxide, and methane. Others are nitrous oxide, fluorinated gases, and surface ozone. The second group is relatively insignificant compared to the other greenhouse gases, so we will focus our attention on the big three: water vapor, carbon dioxide, and methane.

Figure 5.2 This is Paynes Prairie, Florida, an extensive wetland that covers 21 000 acres. This ecosystem evolved over thousands of years. How will it survive in a rapidly changing climate?

Water vapor

Water vapor is highly variable on the planet. It is most abundant in warm, moist tropical areas, and least abundant in deserts and in polar regions, where the air is too cold to hold much water vapor. Atmospheric water is a very important vehicle for transferring heat energy from one place to another. For example, warm tropical air can move north, taking with it the heat trapped in the water vapor. When storms occur where the warm air meets the cold air, the heat energy is released, thereby completing a transfer of heat from tropical areas to cooler areas closer to the poles.

Thus, we see that variations in water vapor significantly impact our weather and climate. But water vapor is not significantly altered by human activity. The only thing that can modify the amount of water vapor available to drive global climate change is by modifying overall global temperature. Warm air can hold more water. Certainly, if the planet continues to warm we will see more water vapor in the atmosphere, but water is not considered a significant driver of global climate change because it is generally constant globally, even though there are local variations by day, season, and year.

So, while water vapor is an important greenhouse gas, it is not one that we are concerned with in our conversations about global climate change.

Carbon dioxide

Carbon dioxide is a naturally and unnaturally occurring gas in our atmosphere, where it is the fourth most abundant gas and accounts for approximately 0.04% of its naturally occurring gases (the others by abundance are nitrogen (78%), oxygen (21%) and argon (0.9%)).

There are a number of natural sources of carbon dioxide in the environment. They include volcanic eruptions, decaying organic matter, and respiration of organisms. As noted in Chapter 2, the Earth's carbon cycle eventually transfers these naturally occurring forms of carbon dioxide into carbon sinks, like fossiliferous limestone or reefs. Thus, it is believed that the Earth is in balance with the naturally occurring forms of carbon and the sinks where it is deposited. Certainly there have been variations in atmospheric carbon dioxide prior to the development of humans (see text box on Ancient carbon), but we have ranged from 180–300 ppm over the last 700 000 years (see text box on understanding parts per million).

Understanding parts per million and measuring small amounts of materials

When we measure very small amounts of things in materials, we have to have a way to express the value. Over the years, scientists have developed a number of ways to express these small concentrations. The most common way we do this is by parts per million, which is usually abbreviated to ppm. Some other ways to express small concentrations are shown below, using the concentration of carbon in the environment.

396 ppm carbon dioxide
0.0396% carbon dioxide
396 milligrams per liter

The measurement of small amounts of materials in the atmosphere or any other media has improved significantly in the last decade and we can now measure very small concentrations of materials with great accuracy. For example, the most well-known carbon dioxide observatory in the world in Mauna Loa, Hawaii, uses a non-dispersive infrared analyzer that is calibrated several times a day. Such instruments are highly sensitive and must be managed carefully to ensure accuracy.

Ancient carbon

The natural carbon dioxide content of the atmosphere is constantly changing. The geologic record of our planet tells us that carbon dioxide levels were significantly higher in the past and they were significantly lower. However, the very high levels were at times prior to the development of human life. Indeed, plants were a major force in reducing levels of carbon dioxide in the atmosphere and helped to transform our planet into an oxygenated environment.

More recently, ice ages have had an impact on carbon in the atmosphere. When ice covers large areas of the Earth, carbon dioxide in the atmosphere increases, since there is less plant activity. Temperatures increase with the increased carbon dioxide, thereby causing the ice to melt. Thus, carbon dioxide ice age cycles have been detected that demonstrate patterned rises and falls in carbon dioxide with glacial cycles.

It must be stressed, however, that the atmospheric concentration of carbon dioxide that we are currently experiencing has not been seen on our planet for millions of years. Throughout the ice age, indeed throughout the last few million years, the parts per million of carbon dioxide was less than half that of what we see today.

Scientists understand the concentrations of carbon in the ice age by carefully collecting air samples that are trapped in bubbles in glacial ice. Over the years, many scientists have collected cores from large continental glacial ice sheets in Antarctica and Greenland. An extensive record has been produced that clearly documents how the atmosphere regularly changes with time.

Since the industrial revolution, we have been releasing tremendous amounts of carbon into the atmosphere through the burning of fossil fuels. At the start of the industrial revolution, we had roughly 280 ppm of carbon dioxide in the atmosphere. Today we are seeing levels in the atmosphere of 414 ppm and rising. This is the highest it has been in the last 700 000 years.

There are many sources of anthropogenic, or man-made, carbon dioxide in the environment. They are largely from the burning of fossil fuels, particularly coal, oil, and gasoline, and natural gas (Figure 5.3). Globally, power generation produces about 25% of all anthropogenic carbon dioxide, although this varies considerably by country. In the United States, electrical power generation is responsible for about 30% of its carbon dioxide emissions.

One of the challenges we are facing from the production of so much carbon dioxide is that some of it is entering the oceans, causing acidification of the water. In some areas, the pH of the water has dropped to the point that it is causing limitations for economic fisheries (see Chapter 3 for greater discussion of the carbon cycle).

Figure 5.3 We produce huge amounts of greenhouse gases through our driving choices. In this photo, what alternatives exist for using a car?

The release of carbon dioxide varies from place to place. The top emitters by weight in order are China, the United States, the European Union, India, Russia, Japan, Brazil, Germany, Indonesia Canada, Mexico, and Iran. However, if one looks closer at the data, one can break it down into the top emitters per person. Using this approach, the top emitters are in order: Kuwait, Brunei, Niue, Qatar, Belize, Oman, Bahrain, Australia, and the United Arab Emirates.

This comparison tells us that there are places that are big emitters, but that there are other places where individuals have large impacts due to their overall national energy use. For example the United States and China together produce over 40% of the world's carbon dioxide emissions. However, Kuwait has nearly three times the per capita emissions of the United States.

Methane

Methane is the third major greenhouse gas and the second most important of the anthropogenic sources. It is also the main component of natural gas. It is 25 times more potent a greenhouse gas than carbon dioxide, which has caused scientists to look more closely at the sources of methane in the environment in recent years.

There are many anthropogenic causes of methane in the environment. By far, the largest source of anthropogenic methane is livestock. Gases released via digestive processes of cattle produces approximately 16% of atmospheric methane. Thus, many have looked to the agricultural sector to find ways to reduce the amount of methane produced by animals.

Decomposition of waste deposited in landfills also contributes significantly. In the past, gas wells were drilled into landfills to burn it off to reduce releases of methane into the environment. Large releases have caused explosions. Today, many of these wells produce methane for municipal power production. The agricultural sector is trying to find ways to capture animal produced methane for power production in a similar way.

Methane is also released during the production and transport of natural gas. The growth of fracking has led to a widespread increase in methane release. In recent years, there has been growing concern over leaking of methane in pipelines.

In recent years, there has been concern over the release of methane in periglacial environments. Here, organic matter that has been frozen for centuries is slowly thawing. As it thaws, it decomposes and releases methane in the process. What is unique about this release is that the organic matter has been building up in these places for hundreds of years. In most parts of the world, organic matter decomposes regularly and there is not a large buildup of the stuff. However, as the world's periglacial environments warm, the ancient carbon is being released in one giant pulse as organic matter decomposes and transforms it into methane.

Periglacial environments

Periglacial environments are places with permafrost in the subsoil. Areas of permafrost are found in places that rest between ice covered landscapes like the northern Arctic and more temperate zones without permafrost, like we see in southern Canada. These places are incredibly sensitive to environmental change since their natural conditions are so impacted by subtle changes in temperature. A few degrees cooler and the landscape would be covered with ice. A few degrees warmer and the permafrost would melt to turn into a temperate landscape and ecosystem.

Areas with permafrost have distinctly unique ecosystems. During the cold parts of the year, the upper layers of soil are frozen and frequently covered with snow and ice. However, during the warmer months, the upper layer of soil thaws above a permanently frozen layer. Moisture in the upper layer cannot drain through the ice, so permafrost areas become a vast muddy wetland in the summertime. Places with permafrost are notorious for summer mosquitos and flies that thrive in the moist landscape.

Because permafrost areas have such poor drainage, organic matter accumulates in the soil as muck and peat. Hundreds of seasons of plant growth are stored in the wet soils of the periglacial landscapes. If the permafrost melts, the soils can drain and the organic matter can quickly decompose to form methane and carbon dioxide.

There have been many examples of other problems that occur in periglacial environments as a result of human action. For example, some have documented problems with polar bear habitats in the Arctic Ocean area. Plus, melting of permafrost has impacted roads and towns. Roads and buildings will become unstable as the permafrost disappears.

Sinks of carbon

There are many types of carbon sinks that can be used to try to manage greenhouse gases.

Forests

Plants and animals store tremendous amounts of carbon (Figure 5.4). However, forests have the greatest potential to trap carbon for long periods of time. Unfortunately, we have cut down many of the old forests of the world, thereby releasing carbon in the process. However, there have been many efforts in the last several years to sequester carbon in trees by replanting forests.

Reefs

Reefs are made up of animals that create carbonate exoskeletons by pulling carbon from the ocean. After the animal dies, calcium carbonate remains. Much of the limestone and dolomite in the world formed in reefs and thus we have huge amounts of carbon stored in geologic deposits on all continents. However, there are many who believe that developing carbonate reefs will assist with carbon sequestration. Yet, we have evidence that many of the natural reefs in the world are in danger. They have been damaged by environmental pollution and harmed by boats. There are questions whether or not important reefs, like

Figure 5.4 One of the benefits of urban forests, such as this one in Temple Terrace, Florida, is that it stores carbon. Many communities are looking to enhance and protect urban forests in order to mitigate their carbon emissions.

the Great Barrier Reef and the reefs of the Caribbean will survive into the future. Thus, unless we find ways to protect existing reefs, it is unclear if we will ever be able to use reefs as a way to store carbon.

At the same time, it is worth noting that there are many man-made reefs that have been built in the last several decades to promote biodiversity of the oceans and to provide nursery grounds for fish. These reefs do have the potential to store carbon if they can be maintained as viable ecosystems with healthy reef-building organisms.

The IPCC and evidence for climate change, and the future of our planet

In 1988, as international concern about climate change began to grow, the United Nations established an organization called the Intergovernmental Panel on Climate Change. This group monitors the major research that is coming out on the issue of climate change and provides international policy recommendations to try to deal with the issue. This group consists of some of the best climate scientists in the world. The IPCC produce what are considered the most comprehensive reports on the state of climate change. Thousands of researchers are involved with writing and reviewing them. To date, they have completed five summary reports, the most recent of which was published in 2013. They have also published a number of thematic reports that can be seen here https://www.ipcc.ch/reports/

What is most important about the latest report is that the group concluded that the evidence of climate change, especially planetary warming, is unequivocal, and that humans are the most likely cause of the change. For full access to the report, you can find it here: http://www.ipcc.ch/report/ar5/wg1/

The IPCC reached a number of conclusions from the research they reviewed:

- The last few decades were the warmest decades since 1850
- The period between 1983 and 2012 was the warmest 30 years in the last 1400 years
- The upper ocean is warming
- The oceans hold 90% of the excess energy produced on our planet over the last 30 years
- The major ice sheets in Antarctica and Greenland are shrinking
- The ice cover in the Arctic is decreasing
- The extent of spring snow cover in the northern hemisphere is decreasing
- The rate of global sea level rise is increasing
- Since 1901, the global sea level rose 0.19 meters
- Concentrations of major greenhouse gases (carbon dioxide, methane, and nitrous oxides) are at levels not seen for at least 800 000 years
- Carbon dioxide has increased by 40% since the start of the industrial revolution—mainly from fossil fuel use and land use changes
- The ocean has absorbed 30% of increased carbon dioxide. This has caused ocean acidification
- There is more energy in the atmosphere due to radiative forcing
- Based on highly reliable climate models, continued emissions of greenhouse gases will cause greater changes to the climate system and cause greater planetary warming

- In order to limit climate change, significant decreases in greenhouse gas emissions must take place
- Failure to make significant change can lead to increases over this century of 1.5–2 degrees Celsius (note: some believe this will be higher)
- The modifications of the climate in the future will continue to cause changes in global water systems
- The oceans will continue to warm and heat will enter deeper into the oceans thereby influencing established ocean currents
- Glaciers throughout the world will continue to shrink or will disappear
- Spring snow cover will continue to decrease with warming temperatures
- Sea level will continue to rise, thereby putting more people at risk from inundation during high tides or tropical storms
- Ocean acidification will increase
- Because of the magnitude of the changes under way, climate systems will continue to change, even if all anthropogenic greenhouse gases are eliminated.

Clearly, the IPCC report does not sugar coat our problem. We are in a time of fundamental climate change. While we should do what we can to limit future warming by reducing or eliminating greenhouse gas emissions, the changes we have started in the climate are so severe that we will have to learn to adapt and become more resilient to environmental changes in the future.

Indeed, in many parts of the world, the conversation has changed from trying to prevent global climate change by reducing greenhouse gases to trying to adapt to changing climate. Communities are focusing on how to become resilient in a time of widespread global change. While many are trying to find ways to reduce greenhouse gases, many others are trying to find ways to deal with a changing planet.

What is so problematic about climate change is that while we can predict that we are warming, we cannot accurately predict exactly what will happen from place to place. Regional models have concluded that some places on the Earth will get wetter or drier, or hotter or colder. However, at the practical level, each place will experience climate change in its own unique way. Plus, because we experience weather as a day-to-day phenomenon, we may not fully "feel" the impact of climate change on a daily basis. Indeed, as I am writing this, a snowstorm is about to hit Long Island and it is hard to imagine that our planet is warming. Yet at the same time, I know that my friends in California are experiencing a dangerously dry season and that my friends in Australia are having a warm season with many fires.

Ocean acidification

Because the oceans have been absorbing so much carbon dioxide due to its increase in the atmosphere, they have become more acidic in some areas. This occurs because some of the carbon dioxide transforms into carbonic acid.

Over the last century, the pH of the oceans has gone from 8.2 to 8.1. It is predicted that by 2100 it will be at 7.8. While this may seem like a small change, it must be noted that the ecosystems of the ocean are very susceptible to pH changes. Many of the life forms have calcareous shells that dissolve under acidic conditions. Scientists have already noted changes

Figure 5.5 Many people work in the shellfish industry and many of us like the shellfish we can find in markets and restaurants. Ocean acidification is likely to impact the shellfish industry throughout the world.

in the life cycles of some of these organisms as a result of acidification. Those involved with shell fishing activities for their livelihood are very concerned about the long-term sustainability of the shellfish industry. In fact, in the United States, the EPA is evaluating how best to manage ocean acidification and is developing guidelines for management of waters that are impaired due to decreases in pH.

The concern over ocean acidification extends beyond the shellfish industry (Figure 5.5). The ocean is teeming with life that evolved within a narrow pH range. Considering the damage that is occurring in oceans due to pollution and overfishing, ocean acidification may significantly damage some unique environments beyond repair. Indeed, some habitats, such as reefs and tidal pools can be destroyed due to acidification.

This issue, of course, is very difficult to repair. It is not as if the ocean can take a giant antacid pill to alleviate its problems. Only through the broad decrease of carbon dioxide in the atmosphere will this problem be solved. Until then, it is likely that we will see greater ocean acidification.

Phenological changes

Another important issue with global climate change is that it leads to phenological changes. Phenology is the study of the timing of the life cycle events of plants and animals. We are most familiar with these cycles when we see particular flowers bloom or birds return to nest in our region. Scientists have been studying the timing of these kinds of events for years and have found distinct patterns to the behavior of many common organisms. In some cases, the study of these events has been taking place for centuries. For example, hunters in many parts of Europe have been observing the timing of the life cycle of deer and other prey, and agriculturalists in China have been keen observers of rice growth.

Phenology studies have been helped recently by citizen scientists who share records on animal and plant behavior in their localities. In the United States, citizens can join the USA National Phenology Network (see https://www.usanpn.org) to report data, but most areas of the world have some type of citizen reporting network.

What researchers around the world are observing from these phenological studies is that there is a significant change underway in the natural world. Organisms are reacting in unusual ways to the planetary changes.

Of course, some of these changes might be temporary or last just a few years. We all understand that trees may blossom early one year compared with other years. However, by long-term monitoring, we are able to better understand trends in biological behavior as a result of climate change.

Conducting greenhouse gas inventories

Greenhouse gas inventories are completed to better understand current greenhouse gas emissions and how best to reduce them. According to the US EPA, there are seven steps to conducting a greenhouse gas inventory (Table 5.1). While these steps were designed for inventories conducted by state and local governments, they really are suitable for any organization seeking to create a greenhouse gas inventory.

Table 5.1 Steps to complete a greenhouse gas inventory as defined by the EPA.

Step	Description
Set boundaries	Define an inventory's physical, organizational, and operational boundaries.
Define scope	Decide which emissions source and/or activity categories and subcategories should be included in the inventory as well as which specific greenhouse gases.
Choose a quantification approach	Depending on the data available and the purposes of the inventory, choose to take a top-down, bottom-up, or hybrid approach to data collection.
Set a baseline	When choosing a baseline year to provide a benchmark to compare progress going forward, consider whether: (1) data for the year are available, (2) the chosen year is representative, and (3) the baseline is coordinated to the extent possible with baseline years used in other inventories.
Engage stakeholders	Bring stakeholders into the inventory development process early on to provide valuable input on establishing a baseline; help build public acceptance of policies to address climate change; and provide data, information on data resources, and personnel resources or outreach assistance.
Procure certification	Consider a third-party review and certification of the methods and underlying data in an inventory to assure that the inventory is high quality and that it is complete, consistent, and transparent. Certification may be required for participation in some greenhouse gas registries.

Step 1 Setting boundaries

Before collecting any data, it is important to figure out the nature of what it is that you are measuring. What is the geographic extent of the space? If you are a large corporation, are you measuring just your central operations within the corporate office or are you measuring all of your subsidiaries? If you are a university, are you measuring all campus properties? Some campuses have complex land holdings with property not just on the main campus, but also at external farms, research facilities, hospitals, or housing. Understanding the fundamental spatial limits to a greenhouse gas operation is an important defining step.

So too is evaluating the organizational or operational edges of activities that you are measuring. Many organizations have complex operations that influence greenhouse gas emissions. These may include travel, shipping, and commuting. For example, if you are a company that produces materials for shipping, how much of the emissions from shipping do you include? Does it include shipping after the order delivery if it is delivered to a large distributor? What if you make a product that releases greenhouse gases? Do you include the greenhouse gases that are produced over the lifetime of the product? Do you only include those portions of the operations that are under your financial control? What if some of the operations are managed by two or more different organizations? What if your property is leased? What portions of the emissions are your responsibility? As you can see, the answers to these questions are complex.

Think about your own life. If you were to calculate your personal greenhouse gas emissions, would you include the emissions associated with your computer use on campus? Or the emissions associated with heating and cooling your residence hall? Defining parameters like this at the front end of the greenhouse gas emissions inventory process is crucial to the success of the activity.

Step 2 Defining scope

Once the boundaries are set, it is also important to set the scope of the inventory by deciding the kinds of emissions and gases that are being measured. For example, in the agricultural sector, it may be important to focus heavily on fertilizer use and fuels for farm equipment. However, in settings like a university, it may be more important to look at heating, cooling, and electricity use. Thus, the scope is defined, in part, by the types and sources of greenhouse gases that are produced by the organization.

While it would be useful to complete a comprehensive greenhouse inventory of all sources of emissions, that is sometimes very difficult. For example, in some complex organizations, it is very difficult to assess greenhouse gas impacts from the travel of employees. We tend not to keep records related to business travel that are useful to greenhouse gas inventories. For example, if you were to be reimbursed for air travel for a business trip, you would turn in your receipt for the airline ticket and the rental car. The costs are recorded—not the miles travelled by air or car. Thus, creating a scope of the greenhouse gas inventory helps to define the workflow for the inventory and sets limits on the nature of the inventory based on the available data.

It is important to note that many organizations use well-established greenhouse gas inventory protocols that clearly define the scopes of emissions to be measured and the kinds of data required to measure the emissions. Therefore, in many cases, the scope of the emissions analysis is pre-defined. There are several inventory tools that can easily be found online to assist in creating a greenhouse gas inventory. For example, the American College and University Presidents Climate Commitment has a spreadsheet driven tool that can be accessed on their website that is especially useful for college campuses. Greenhouse gas inventory tools have been developed not only for universities, but for a number of different types of organizations, including governments and complex businesses. Greenhouse gas emissions scopes will be discussed in more detail below.

Step 3 Choosing a quantitative approach

There are three broad ways to organize data when conducting greenhouse gas inventories: top-down, bottom-up, and hybrid. A top down approach uses data collected at a high organizational level. For example, an organization may keep track of total energy purchased by the entire company. This allows one to collect data at the highest level without having to gather it from individual departments. However, in many cases, bottom-up approaches are more appropriate if data are not available at the macro level.

In thinking about an organization like a business with multiple departments such as purchasing, accounting, shipping, and sales, each department may have data to contribute to a greenhouse gas inventory. When one looks at an organization as complex as a city, state, or country, it becomes even more complicated. Some data may be best approached from a top-down perspective, while other data might be best gathered from the bottom-up. Such complex data approaches are considered hybrid approaches. While any of the three approaches will produce a greenhouse gas inventory, it is important to record all of the steps used in the methodology so that the inventory can be replicated year to year and that it can be compared with other organizations.

Step 4 Setting a baseline year

Perhaps the most important step in any greenhouse gas inventory is setting a baseline year for comparative purposes. Often, technicians utilize the first year that data are available for conducting a complete inventory for a baseline year. Organizations do not normally hang on to data that will be useful in completing a greenhouse gas inventory. Think about your own data collection for your own energy use. Do you have records that go back a decade on travel? Electricity consumption? Waste?

Plus, we may measure these things differently over time. Maybe a university measured waste produced by weight one year and another year by the number of containers tipped into garbage trucks. How can you compare these data? You cannot of course. Thus, often, the baseline year is the year of the first inventory that is completed.

One of the things that happens once an organization completes its first greenhouse gas inventory, is that it begins to get better about record keeping on data important to greenhouse gas inventories. So, while the first inventory might be difficult for everyone involved, it is much easier in subsequent years.

Yet it is important to consider how your inventory is going to be used prior to selecting a baseline. If you are trying to compare with inventories that are done every five years, it would be worth putting in the effort to dig up the necessary data to match your inventory with the time of the last inventory of other comparative organizations.

Step 5 Engaging stakeholders

Engaging stakeholders in the inventory process is crucial in any organization. One must work across an organization or community to obtain data and produce results. However, engaging stakeholders also provides an opportunity to educate your organization about the importance of greenhouse gas reduction. It allows you to create a conversation on how best to develop strategies to reduce the impact of your organization. Engaging stakeholders also brings in community expertise to try to better conduct the inventory and develop long-term strategies.

Step 6 Procuring certification

Once an inventory is complete, organizations sometimes seek to obtain a third-party certification of the inventory to verify that appropriate procedures were followed and that the data are sound. While this step may not be necessary for all organizations, it is often used in situations where there is a need for certification. This is done, for example, to participate in climate registries that evaluate greenhouse gas emissions of some organizational sectors like particular businesses or governments.

Greenhouse gas equivalents used in greenhouse gas accounting

The Kyoto Protocol recognized six greenhouse gases that are of special concern to global climate change:

Carbon dioxide (CO_2)
Methane (CH_4)
Nitrous oxide (N_2O)
Hydrofluorocarbons (HFCs)
Perfluorocarbons (PFCs)
Sulfur hexafluoride (SF_6).

Each of these greenhouse gases absorbs heat differently. Thus, they each have different heat absorbing potentials. In order to measure the impact of greenhouse gases on the planet, scientists had to develop a system to compare the overall impact of all greenhouse gases on the environment. That is why they developed a system of greenhouse gas equivalents that allows comparisons of the real impact of anthropogenic releases of greenhouse gases.

The basis for measuring greenhouse gas equivalents is carbon dioxide. This makes sense since carbon dioxide is the most dominant greenhouse gas in the environment. The heating potential of other greenhouse gases is compared to carbon dioxide in this system. Thus, regardless of the type of greenhouse gas emitted, it can be converted into carbon dioxide equivalents in order to assess the overall environmental impact of the emissions. This ability to account and compare greenhouse gases from all emissions from place to place is a basic building block of greenhouse gas management.

Using this comparison, methane is 25 times more powerful as a greenhouse gas than carbon dioxide and nitrous oxide is 298 times more powerful. When conducting inventories, the amount of these chemicals is converted to their CO_2 equivalents or CO_2e. Thus, the release of 1 tonne of methane is equal to 25 CO_2e tonnes of greenhouse gases. So, one of the most important steps in any greenhouse gas inventory is the conversion of the different kinds of emissions to CO_2 equivalents.

Greenhouse gas emission scopes

When conducting a greenhouse gas emissions inventory, there are three scopes of emissions that are used to organize the data:

Scope 1 emissions. These are the direct emissions produced by the organization. They include emissions from stationary sources, such as power plants or steam production, as well as mobile sources that are transportation sources from vehicles owned by the organization. Emissions from manufacturing or other processes are also considered Scope 1 emissions. These processes may be things like chemical production or mining. Fugitive emissions that are released unintentionally or as a result of normal wear and tear are also included in this category. They include leaks of CFCs from refrigerant lines and methane from solid waste landfills.

Scope 2 emissions (Figure 5.6). These are the emissions associated with purchased electricity, steam, heating, or cooling. The organization does not directly produce the greenhouse gases, but is responsible for their release as a result of the purchase of the energy. This is often the most important contributor to an organization's greenhouse gas inventory, since most do not have their own power source.

Scope 3 emissions. These are all other emissions. This category includes, among other things, employee travel (on vehicles not owned by the organization: business travel and commuting) and waste. Some established inventories make the reporting of Scope 3 emissions voluntary, in part, because they are so difficult to calculate.

Figure 5.6 Just because you purchase electricity from somewhere else doesn't mean you are not responsible for including it in your greenhouse gas inventory.

De minimis emissions

As noted earlier, there are some sources of greenhouse gases that are difficult to measure and that do not add up to a significant amount of CO_2e. Greenhouse gases that fall into this category are considered "de mimimis" if they account for less than 5% of the total amount of emissions. By most greenhouse gas measuring procedures, these emissions do not need to be included in the inventory. While efforts should be made to try to compute all greenhouse gas emissions, sometimes it is not worth the effort if small sources are difficult to assess.

Computing greenhouse gas credits

The measurement of Scope 1, 2, and 3 emissions provides a measure of the total greenhouse gas production for an organization for a single year (assuming the year is the time of the assessment —most are conducted for a single year). However, some organizations have purchased greenhouse gas credits that mitigate their overall greenhouse gas emissions. The credits are purchased from organizations that work on projects to reduce greenhouse gases.

Climate action plans

Climate action plans are written plans that guide an organization to reduce greenhouse gases. Of course, the first step in producing a climate action plan is to conduct a greenhouse gas inventory that details the amounts and sources of greenhouse gases produced by an organization. Upon its completion, the inventory guides decision-makers in developing appropriate strategies for the reduction of greenhouse gases.

The climate action plan details all of the goals and strategies involved in reducing greenhouse gases. The goals are often divided into short-term, medium-term, and long-term goals (Figure 5.7). The short-term goals are those that can be met very quickly. For example, it might be easy to quickly modify light bulb policy in an organization to switch to energy saving LED lights. Medium-term goals might include things that will take longer to accomplish, say over a time span of 5–10 years. These might include things like converting vehicle fleets to electric vehicles as the current fleet ages. Or, it may include the renovation of buildings to meet energy efficient standards. Long-term goals are those goals that could be met over a longer time frame of perhaps 10–50 years. These goals might include becoming climate neutral or converting landscaping to native vegetation to avoid fertilization.

The goals that are developed in a climate action plan must address the kinds of emissions that are identified in the greenhouse gas inventory. For example, if the bulk of the greenhouse gases are produced from electricity use, the climate action plan should focus on this area of greenhouse gas production. While goals should be set for reduction of all sources, the greatest attention should be given to the largest sources.

Climate action plans should include realistic and measurable goals. While it might sound good for an organization to strive to be carbon neutral in five years, such a goal is usually impossible to achieve. Thus, the goals must be set with care to ensure success.

Once the goals are established, a listing of policy or strategy changes should be devised to achieve the goals. For example, a goal to become carbon neutral may include a mix

Figure 5.7 Part of a community action plan might be to reduce car and truck traffic in the central portion of the city. This makes room for pedestrians and other activities.

of strategies that could include purchasing carbon credits, developing renewable energy sources, and improving energy efficiency. The timeframe for developing the strategies must also be part of this process.

Climate action plans should be conducted in partnership with key stakeholders. Within the context of climate change, stakeholders should include not only those seeking to reduce greenhouse gases, but also the organizations responsible for producing them. By bringing everyone together, realistic plans can be made that will have significant outcomes. If climate action plans are made without the full support of greenhouse gas producers, they are likely to fail.

The most successful climate action plans address multiple realistic goals at the same time. The State of California climate action plan (see: http://www.climatechange.ca.gov/climate_action_team/reports/2006report/2006-0 4-0 3_FINAL_CAT_REPORT.PDF) has a mix of strategies that address short- and long-term goals in a number of areas that include transportation, energy production, and a range of other issues. It is a comprehensive approach to addressing long-term greenhouse gas reduction.

Climate action plans are produced by a number of different organizations. They can be produced by national, regional, or local governments. Schools and other types of public institutions often develop them. Businesses, particularly those with large greenhouse gas emissions, often develop plans.

Do the governments (local, regional, and national) in your area have climate action plans? Does your school have a plan to reduce greenhouse gases? Take a look at major businesses in your region. Do they have climate action plans? You can find these plans via simple internet searches. If you cannot find a climate action plan in your region, take a look at some other climate action plans you find on the internet. What are the components of the plans? What goals do they set? What strategies are developed to meet the goals? How do the plans vary from each other? Do they set short-term, medium-term, and long-term goals? Think about your own greenhouse gas production. What type of personal climate action plan can you develop to reduce your greenhouse gas emissions? What are your goals? What strategies will you take to achieve your goals? How are your goals divided into short-term, medium-term, and long-term strategies?

Is nuclear energy the answer?

Some have suggested that the only way to solve our global climate problem is to significantly reduce our fossil fuel use immediately. But how could we do this? Green energy sources like wind, solar, or geothermal are not at the point that they can be scaled up to immediately take over the role of carbon-based energy like oil, coal, and natural gas. Many environmentalists argue that nuclear energy is the only solution that will allow us to maintain our dependence on energy while significantly reducing carbon.

Yet, others argue that nuclear energy is not a wise choice due to problems with waste and the threats of terrorism. Given the problems we've seen with the Fukushima, Chernobyl, and Three Mile Island nuclear power plants, are you willing to risk the use of nuclear energy in order to quickly reduce the addition of carbon into the atmosphere?

Many believe that nuclear energy is the only answer to the carbon problem, or at least a temporary answer to the problem until we ratchet up our production of renewable energy. However, can we put the nuclear genie back in the bottle if we greatly expand its use? Once it is developed, will it be easier to rely on nuclear than to develop solar, wind, or geothermal energy sources?

Michael Mann, hockey stick hero?

Michael Mann is a well-known climatologist from Penn State University who found himself at the center of considerable controversy over the famous hockey stick graph he published 1989 that demonstrated that the climate was changing considerably in our modern era compared to the past 1000 years (Figure 5.8). The graph was used by many to demonstrate the significance of greenhouse gas pollution.

Because Mann's work was used to bolster the research need for greater understanding of global climate change, many conspiracy theorists attacked him and accused him of various inappropriate activities that ranged from making up data to misinterpreting the results to meet a political agenda. Mann has defended himself against all of these allegations and none of them stuck. His emails have been hacked and he has been sued in court.

There isn't any real question in the scientific community as to whether or not climate change is real or if there is a link between greenhouse gas pollution and global climate change. In fact, most of the major energy companies understand that global climate change is a real and present danger to our long-term sustainability on

(Continued)

Figure 5.8 Your author (left) with Michael Mann.

this planet. Yet, there are still a few political corners of the world where climate change is considered a hoax. For whatever reason, Michael Mann is still the man that climate change denialists go after when they need a climate change scientist to beat up in the public arena.

He has taken the heat for the entire scientific community that patiently collects data, writes papers, and develops deeper knowledge about climate change. To them, Mann is considered a hero for standing up for the truth of what their studies are showing them.

What is interesting about Mann's case is that it demonstrates how individuals involved in the environmental science and policy field can become part of the public debate about sustainability—whether or not they wish to be part of it.

Can you think of any other scientists that have found themselves in a public debate on scientific issues? Why do you think the climate change debate has become so political in the United States and other countries of the world? Can you think of any scientific heroes that are working hard to inform the public about environmental problems at the local, national, or international level?

Believing in climate change

I live in the New York area that was so terribly impacted by Superstorm Sandy in 2012. When it hit the region, there was quite a bit of discussion around global climate change and the storm. The mayor of New York at the time was Michael Bloomberg. When he endorsed President Obama for the 2012 presidency days after the storm, he noted that climate change and Superstorm Sandy influenced his decision, since Obama was the only candidate that seriously addressed issues of climate change in his platform.

I was without power for several days after the storm passed. During that time, I listened to a hand-crank radio to try to get updates on conditions in the area. One of the stations I listened to was a sports station that switched from sports to taking calls from around the area to get updates on what was happening in the New York region. One caller mentioned Bloomberg's statement about climate change and the election. The host countered by saying he did not "believe" in climate change. He said he didn't know enough about it and was not comfortable with the idea that climate is changing.

I thought it was an honest answer. I, and many climate change advocates, have many friends and family members who are climate change skeptics. However, it always seems odd to me to find someone who doesn't believe that our climate is changing as a result of human activity. There are thousands of scientific papers that document how the climate is changing and how it is impacting our planet. The idea that one does not "believe" in climate change is akin to not believing in chemotherapy or wireless internet.

We trust our scientists and engineers to take care of our bodies and our technology. However, a relatively large (but thankfully shrinking) number of people are hesitant to believe the scientific community on global climate change. I find this so strange given the great deal of data on the subject. Even one of the most vocal climate change denialists of the past, Richard Muller, has now gotten on the climate change bandwagon.

To me, not "believing" in climate change infantilizes science and puts climate change on par with the tooth fairy. It would be much more honest to claim that one doesn't know enough about the topic or that one doesn't care if it is happening or not.

We do not know if Superstorm Sandy occurred as a direct result of global climate change or if it was a random event. However, we do know that storms like Sandy are very unusual events caused by the steering of high- and low-pressure systems that were located in unusual locations for that time of year. Plus, the changing global sea temperatures and Arctic conditions have changed the overall background environment of the atmosphere in recent decades. While Sandy could have been a random event entirely unassociated with global climate change, some scientists suggest otherwise. Regardless, the New York region and many other parts of the world are highly vulnerable to rising sea levels, tropical storms, and other changes associated with global climate change. It is important to have a real discussion about vulnerability in these areas, whether one believes in climate change or not.

Of course, politics gets in the way of much of what we do around the world. Politicians are agenda motivated and are usually in office for a few years. Getting elected, following a party agenda, and keeping constituents happy is their main agenda. If working on climate change issues is outside of a regional priority, it is difficult for some politicians to support climate change initiatives. That is why it is important that citizens stay educated on the issues facing our world on a myriad of sustainability issues.

(Continued)

Glaciers and climate change

Glaciers are bodies of ice that last year to year and that move due to the action of their weight and the forces of gravity. There are two main kinds of glaciers: continental ice sheets like the ones that cover Antarctica and Greenland, and alpine or mountain glaciers like the ones found in the Alps, Rocky Mountains, Andes Mountains, and other highlands around the world. Glaciers exist in one of three states: equilibrium (where the rate of melting is equal to the rate of accumulation of new ice), growth (where the rate of accumulation of new ice is greater than the rate of melting), and retreat (where the rate of melting is greater than the rate of accumulation of new ice). Glaciers react to subtle changes in climate. They will grow when there is greater moisture and cool temperatures, and they will retreat when there is an absence of moisture and warm temperatures. In the last several decades, scientists have been keeping extensive records of glacial ice change and have noted that over all, most glaciers of the world are decreasing in volume and mass. While there are some years when glacial ice thickens due to cooler temperatures or a wetter winter, the overall evidence points to the fact that the ice sheets are melting. For more information on glacial change, see http://www.epa.gov/climatechange/science/indicators/snow-ice/glaciers.html

Because many of the glaciers are very thick (the Antarctic Ice Sheet is 4000 m thick), they contain layers of ice that can be quite old. Scientists have taken cores from some of the thickest areas of Greenland and Antarctica glaciers and found that the ice dates back to over 125 000 years in Greenland and 800 000 years in Antarctica. The ice cores can also be used to assess past climates because bubbles of ancient air get trapped in the ice. The chemistry of this air can be tested to see how it varies from our own atmosphere. The ice also tells us how temperature and climate varied over the time.

One of the most intriguing discoveries in recent years was the noted 5000 year old "iceman" that was found emerging from a melting glacier in the Alps near the border of Switzerland and Italy. The ancient man was considered an amazing archaeological find because the body contained numerous clues about the prehistoric culture of the region. The man was covered with numerous tattoos and had a number of health problems including elevated levels of arsenic, some body abnormalities (missing ribs), and injuries. He died from a head wound. However, what is most intriguing for those of us interested in climate change is not the messages he sends us about our past. Unfortunately, he tells us that glaciers are melting to the point that long-hidden materials are now emerging from the melting ice. While it is fascinating to learn from these new finds, it is a warning to us about the challenges we are likely to face in the future as glaciers continue to lose mass and disappear.

Caves and climate change

Caves are underground voids that form when soluble rock dissolves underground. Most caves form in limestone as water flows through joints or cracks in the rock. Caves form very slowly and range in size from just a few meters in length to many kilometers. The longest cave in the world, Mammoth Cave in Kentucky in the United States, is over 650 kilometers long.

Caves often contain speleothems. These features, some of which are called stalactites and stalagmites, are deposits that form after the formation of the cave. Speleothems consist of minerals that are deposited when mineral rich waters flow through the rock to enter the cavern. With time, these deposits grow into distinct formations (stalactites, stalagmites, flowstone, soda straws, and a variety of other features). These formations represent some of the most continuous record of climate change on the continents since they grow like tree rings, but for much longer periods of time than the life cycle of trees.

The reason that these speleothem deposits are so helpful in puzzling out climate change is that their chemistry subtly changes with climate shifts. They can record warmer and cooler conditions as well as wetter and drier conditions. Because they contain carbon, radiocarbon dating can be done of individual layers in the deposits to precisely record exact timing of these changes.

Along with speleothem deposits, cave scientists also use sedimentary deposits found within caves to assess climate change. Sediment, often high in organic matter, washes into caves during rainy seasons. They contain a wide variety of information on climate that is useful in puzzling out long-term changes that impacted cave sedimentation. Plus, the sediments also contain pollen and other biological evidence to give a sense of local ecological conditions during distinct periods of time.

Scientists have also found the study of ice caves useful in puzzling out subtle changes in climate. Ice caves are natural rock caves that contain ice throughout the year. Many of these caves are found in cooler mountainous regions of the world. During the winter, ice accumulates in the cave when temperatures are low. However, during the summer months, some melting does occur. However, the cave is in one of three conditions: (1) at an ice equilibrium where the ice melt is equal to the ice growth, (2) at a condition of ice loss where the melting is greater than ice accumulation, or (3) at a condition of ice gain where the accumulation of ice is greater than melting. In many ways, these ice caves react very much like underground glaciers in that they are constantly reacting to changes in temperature and moisture. As a result of this, they store tremendous information about past climates.

Taken together, the speleothems, cave sediments, and ice caves are among the best sources of evidence of climate change on the continents. Of course, caves also contain fossils, like cave bear bones and teeth, that tell us a great deal about the kinds of organisms that lived within the caves.

Religion and climate change

In recent years, many religious organizations have recognized the importance of faith-based organizations in encouraging conversations around ethical issues of climate change. These groups have seen the impacts of climate change on people and environments and recognize that there are important ethical dilemmas inherent in dealing with greenhouse gas reduction and the impacts of climate change. They feel that it is important to address climate change within their churches, synagogues and mosques because global climate change is about the protection of creation. Of course, issues of social justice and fairness often are paramount in these concerns.

Evangelical Environmental Network

One of the oldest campaigns by any religious organization is the "What Would Jesus Drive?" movement that was started by the Evangelical Environmental Network in 2002 to promote ethical issues in transportation, specifically those that relate to global warming, oil dependence from unstable regions, and health. By asking members of churches, "What would Jesus drive?" they provide an opening for asking parishioners to consider the ethics associated with their personal transportation options. It also is an entry to discuss the global dimensions of climate change. The major emitters are from wealthy areas of the world and many in poorer countries are impacted without the resources to easily adapt.

The campaign expands on the well-known WWJD (What Would Jesus Do?) campaign that asks Christians to consider the ethical dimensions of their personal choices. The Evangelical Environmental Network clearly urges people to recognize that global climate change is a moral issue that individuals must confront in the decision-making in their everyday lives.

Young Evangelicals for Climate Action

An outgrowth of the Evangelical Environmental Network is the organization, Young Evangelicals for Climate Action (Figure 5.9 here). This group formed in 2012 to help to organize young Christian evangelicals to take action on climate change. They do this by trying to educate and organize other young people by creating chapters of the organization on college campuses, by educating senior leaders in the evangelical movement, and by holding politicians accountable for their policies and deeds.

Catholic Climate Covenant

Since 1990, when Pope John Paul II spoke clearly on the modern problems of pollution and climate change, the Catholic Church has been advocating for action on reducing the impacts of climate change. They have framed it as an environmental justice issue. One of the outgrowths of the discussion on climate change and environmental justice is an organization called the Catholic Climate Covenant which formed in 2006 in the United States as the main expression of Catholic activism on the issue of climate change.

The group uses The St. Francis Pledge as a main way that they educate Catholics about the issue of climate change. Those taking the pledge promise to (direct quote: http://catholicclimatecovenant.org/the-st-francis-pledge/):

- Pray and reflect on the duty to care for God's Creation and protect the poor and vulnerable.
- Learn about and educate others on the causes and moral dimensions of climate change.
- Assess how we—as individuals and in our families, parishes and other affiliations—contribute to climate change by our own energy use, consumption, waste, etc.

Figure 5.9 This is a group of students from Young Evangelicals for Climate Action that visited my campus when Hofstra University hosted the last presidential debates.

- Act to change our choices and behaviors to reduce the ways we contribute to climate change.
- Advocate for Catholic principles and priorities in climate change discussions and decisions, especially as they impact those who are poor and vulnerable.

Each pope, since John Paul II, has written forcefully on climate change and urged people to consider how the most vulnerable among us are impacted by the actions of the wealthy.

Jewish Climate Change Campaign

There are a number of initiatives involving sustainability within the Jewish tradition, in part because of the sustainability issues associated with Israel. However, the Jewish Climate Change Campaign, which started in 2009, is working to mobilize the Jewish community to work toward meeting the challenges presented to our society as a result of climate change.

The Jewish Climate Change Campaign was initiated by Hazon, a non-profit group that is working "… to create a healthier and more sustainable Jewish community, and a healthier and more sustainable world for all." (http://hazon.org/about/overview/). This organization works with individuals and groups by providing transformative experiences, by offering thought leadership, and by developing capacity building among individuals and groups.

The International Muslim Conference on Climate Change

Many have written about the teachings of Islam in understanding how to confront climate change. Several important meetings have been held to develop unifying statements on the issue. In 2010, the International Muslim Conference on Climate Change was held in Bagor, Indonesia to try to bring together a wide variety of stakeholders to discuss an international Islamic response to climate change. The meeting was an outgrowth of the Muslim Seven Year Plan for Climate Action that was agreed upon in 2009 by a number of international organizations in Istanbul. The outgrowth of the Bagor initiative was to urge that the Organization of the Islamic Conference (which included 57 nations) convene a special council to lead the way on climate change that would address climate change policy that would fit Islamic values.

Buddhist Declaration on Climate Change

Buddhists often have strong connections to nature and environment is a key element of Buddhist imagery. Some of the key teachings in Buddhism involve ethical conduct and acting in a non-harmful way. Thus, many Buddhists look at climate change as throwing the Earth out of balance due to harmful actions of man. As a result, some Buddhists have created a Buddhist Declaration on Climate Change (see https://oneearthsangha.org/statements/the-time-to-act-is-now/) that urges our society to take action to avoid long-term problems as a result of greenhouse gas pollution and associated climate change. The first person to sign the declaration was the Dalai Lama.

Hindu Declaration on Climate Change

Hindu teachings recognize that man is not apart from nature but is linked to it in fundamental ways. This is the main message of the Hindu Declaration on Climate Change that was presented to the Parliament of the World's Religions in 2009. A key element of the declaration is the need for a greater global understanding of how to solve the problems. "We must transit to complementarity in place of competition, convergence in place of conflict, holism in place of hedonism, optimization in place of maximization. We must, in short, move rapidly toward a global consciousness that replaces the present fractured and fragmented consciousness of the human race." (http://www.hinduclimatedeclaration2015.org).

Clearly, all of the world's major religions have something to say on issues of climate change. They all recognize that it is an ethical issue that has long-term implications on the future of our species on our planet. They also understand that the ones that are most impacted are the ones who can do the least to solve the problem. As such, the issue becomes a social justice problem.

Take a look at your own religious traditions (if you have them). What are their teachings on climate change and the environment? If you attend religious services, what about your own church, synagogue, mosque, or temple? Are there religious organizations or clubs at your school? Are they teaching about climate change? Are they involved with any environmental issues? Would you consider them environmental activists?

Art, culture, and climate change

Artists have long commented on social issues. In our present era, artists regularly comment through their work on issues like feminism, war, social justice, and the environment. Several artists directly address issues of climate change in their work and provide a way for society to engage with the topic in deeper ways. At the same time, pop culture outlets like music and film have addressed the issue in sometimes startling ways, especially through science fiction genres that conjure possible futures for us if we do not change our ways. These cultural expressions are a way for us to reflect on our times. They allow us to evaluate our practices and contemplate the meaning of our actions on the environment. The following paragraphs review a few artists who have contributed to discussions on environmental change.

Swoon

Swoon is a New York area street artist who is known for pasting cutout images of people on buildings. She has worked internationally on a number of interesting street art projects. However, she has also been involved with several important exhibits and performances that speak directly to issues of climate change and sustainability. Her work often questions issues of how we are individually impacted by climate change issues in our life.

One of Swoon's most recent works, Submerged Motherlands, focused heavily on the impact of climate change on individuals. It is a multistory installation built around a paper and fabric tree. It includes paper cutout images of people as well as small homes and shanties that are in some way impacted by sea level change. The artist was quoted as being influenced by the experiences of individuals living in low-lying areas during Superstorm Sandy, which impacted low lying areas of the New York region as well as the submergence of Doggerland, an area of Great Britain that was submerged 8000 years ago. The installation is vast and includes objects that were used in past performances (most notably some boats made of found objects that were used to sail from Slovenia to Venice during the Venice Biennale). Interacting with the installation gives one a sense of loss yet intimate connection with the people that experienced flooding and their homes. It makes one think about the impacts of climate change on the individual.

Raul Cardenas Osuna and Toro Labs

The Mexican-based Toro Labs, led by artist Raul Cardenas, brings to the discussion of climate change a mix of art, planning, architecture, community intervention, and community activism. The Toro Lab group is very interested in how art can be involved for greater public and social good. They have completed a number of innovative projects that address important social issues like nutrition and housing. You can learn more about them by going to www.torolab.com

One of their projects, One Degree Celsius, focused specifically on the challenges of climate change in cities. The work was, in part, focused on Tampa, a coastal city in Florida in the United States that is likely to see great changes as a result of sea level rise and overall changes in temperature. Cardenas Osuna focused on the kinds of interventions that could be made in cities like Tampa to decrease the temperature by 1 degree Celsius. He interviewed a number of climate change experts and designed a number of plans for community interventions such as billboards made of plants and a mobile science lab. The idea of artists engaged with community interventions to make improvements to society is a fascinating addition to the discourse on how best to address climate change.

Isaac Cordal

As a sculptor, Isaac Cordal uses small sculptural installations to engage viewers on a number of issues, including the environment. One of his best-known works is a series of pieces called Waiting for Climate Change (Figure 5.10). They show individuals, often people of power, in semi-flooded situations. One of the most striking of the installations is called "Politicians Discussing Global Warming". It depicts a cluster of politicians talking while submerged up to their necks in water. The scale of the installation is small. It is placed within a puddle in a square in Berlin.

What is important about this piece is that it graphically represents the challenges of trying to move forward on issues of global climate change. One gets a sense that politicians are happy to talk about climate change and greenhouse gas pollution, but are unable to develop real change. Due to the unending discourse, the politicians themselves eventually are inundated. This piece, like many of those of Cordal, makes one think about one's role in trying to solve the climate change problem.

Figure 5.10 Artists have much to say on the issue of climate change. This image shows the piece called Politicians Discussing Climate Change by the noted artist Isaac Cordal (Image courtesy of the artist).

Have you ever seen any art that makes you think about climate change? What kinds of art are present in your community or campus? Do any of the pieces speak to you about the environment or sustainability? Do you have an art museum on campus or your community? What kinds of exhibits have they had in the last few years? Have any of them had environmental themes? Does your university or school offer any courses on art and the environment? What kind of art (painting, sculpture, performance, etc.) speaks to you the most about the environment? Do you make any art? How do you think art influences how you and others think about the environment and the world in general? Can you think of any artists that influenced you? Can you talk to any art professors or students on your campus to find out what they think about art and activism? Do they think that art has a role to play in the ongoing discussions we are having on climate change?

6

Water

Water is one of the key requirements for life. We are all made up of large quantities of water and we all require water to drink to survive. Yet, there are concerns over the future of the Earth's water resources. Do we have enough clean water for the future? This has been called into question due to our consumption trends and the way that we regularly pollute water across the planet. We dump tremendous amounts of waste of different types directly into surface and groundwater systems. We treat water as if there was a clean, unending supply.

The reality is different. We have limited water resources on the planet and there is concern over the long-term survivability of some regions of the world due to pollution or excessive water withdrawals. Plus, there are questions as to the ownership of water. Is it a public resource or a private commodity?

Yet there is hope. We are developing new and innovative ways to protect and conserve water. This chapter will look at all of the issues associated with sustainability and water. We will see that much needs to be done to ensure the long-term protection of this valuable resource.

Sources of Water

Water comes to us in a number of different forms. If you recall from the discussion of the water cycle in Chapter 2, water is present on the surface and atmosphere of our planet as a solid (ice and snow), liquid (liquid water), and gas (water vapor). It is also present in all three phases in the subsurface, although we most commonly think of subsurface water within the context of groundwater. While most of the Earth's water (97%) is found in the oceans and is saline, the focus of this chapter is the fresh water that is found on or within the continents. It is from this water that societies derive useful water resources.

Of course, there is a growing question as to who owns this water. Does it belong to the public or an individual? Can water rights be sold to corporations in exchange for the development of infrastructure or should water always be a public resource? These are very important questions that many areas of the world are facing at this very moment.

There are a number of different uses of fresh water that is extracted and some of it is used for household purposes. These may include drinking water, clothes washing, showering and bathing, kitchen uses, and lawn and garden care. Some of it is used for the hundreds of possible industrial uses of water, from cooling to its use in manufacturing. There are

Introduction to Sustainability, Second Edition. Robert Brinkmann.

Companion website: www.wiley.com/go/Brinkmann/IntroductiontoSustainability

also many agricultural uses of water, including irrigation and agricultural food processing. Water is also used for recreational purposes, such as its use in water parks, boating, and ice skating.

Water is extracted from groundwater by using wells and pumps. In many cases water pumping does not exceed natural water recharge of groundwater and there is little concern for the long-term viability of groundwater resources. However, in some cases, groundwater pumping far exceeds the natural replenishment rate. In coastal areas where this happens, saltwater intrusion becomes a serious problem. In other areas, land subsidence or sinkholes can be vexing problems associated with over-pumping of groundwater. Of course, one of the biggest problems is that communities will run out of potable water if they over-pump and over-use it. The recent droughts in California and the floods in Texas demonstrate that water is challenging our infrastructure in many ways.

Surface water is another source of fresh water. This water can come from lakes or streams (Figure 6.1). In these instances, water is collected from the water bodies directly using some sort of intake system. The water is typically filtered before it goes through processing to ensure its safety. Many lakes in the world, particularly some of the larger ones like The Great Lakes of southern Canada and the northern United States provide a steady supply of fresh water for shoreline communities and cities. Rivers too provide important sources

Figure 6.1 How we manage our water resources has a huge impact on our daily lives and on the lives of others throughout the world.

of drinking water. In the United States, the extensive Mississippi River supplies drinking water to communities from the far northern portion of that country to the most distal delta regions in the far south. It must be noted that sewage treatment facilities also dump water directly into the Mississippi River throughout its length. Some have stated that a glass of water in New Orleans has been through several kidneys before making its way to the delta.

In some rare instances, other sources of water are found. For example, most of us have heard of the towing of icebergs to desert areas to provide water. However, this is not really a viable option for the long-term sustainability of a region. Some desert cultures have also harvested dew for small-scale drinking or irrigation purposes.

In some instances, streams are dammed in order to ensure a steady supply of fresh water. The benefit of this approach is that a reservoir ensures a reliable water supply, even if the natural steam flow is seasonally low. This is particularly true of streams that receive considerable flow of water during the snowmelt from mountains in alpine regions or during monsoon season in tropical and semitropical areas. The excess water is stored in the reservoir for human use during drier seasons of the year. Some of the oldest dams in the world found in the Middle East took advantage of these seasonal fluxes of the hydrologic cycle.

Each region of the world has different water management strategies for using and protecting the water that is extracted for human consumption. The water we commonly use comes to us largely from surface water bodies and groundwater. But these resources are quickly being depleted. According to the United Nations, by 2050 nearly 40% of the world's population will live in areas with diminishing and limited water supplies. Why is this?

Of course, the reason for this problem is that water is not distributed evenly across the planet. Some areas of the world have tremendous water resources and it is unlikely that they will be stressed in the next several decades. However, many millions of people live in vast arid and semiarid regions of the world where water consumption is outpacing the natural replacement rate for water. These people are living far outside of the natural water budget and unless something changes in consumption patterns, at some point their water supply systems will cease to provide potable water. When this happens, what will these people do? How will they get enough water to survive, much less grow crops? Will they migrate? Will we build a water pipeline? What can we do now to ensure that we never reach the point that the water systems fail?

As we will see, worldwide consumption trends do not evenly match worldwide water supplies. Indeed, as already mentioned, many areas are utilizing way too much water. There are other areas of the world where there is an abundance of water resources. In some instances, we have captured excess water and piped it great distances for use in thirsty areas. However, more and more, those that control these rich water resources are loathe to develop them for distant areas. Plus, there are significant ecological circumstances to the shipment of excess water from wet areas to dry ones (see text box).

It must also be noted that the presence of water does not equate to potable water. This is especially true in vast areas of Africa and Asia where there is a steady supply of water, but limited resources for the development of safe drinking water. In these places, water quality issues, as well as lack of infrastructure, leads to a variety of public health problems.

Water management in thirsty Florida

One would think that the state of Florida in the US would have an abundant supply of water. It is not unusual for areas of the state to receive two meters of rain a year during wet years. However, Florida has significant water supply problems, particularly in coastal urban areas like Tampa and Miami.

These places are only a few meters above sea level. All of Miami relies on one aquifer that rests just a few meters under the foundations of most of the homes in the area. In the Tampa Bay area, years of groundwater withdrawal along the coastal communities caused saltwater intrusion and land subsidence.

In addition, state officials have wisely divided Florida into distinct water management districts, roughly corresponding to broad watershed delineations (Figure 6.2). Each district must utilize water resources within its region. Thus, thirsty Tampa and Miami must find ways to utilize the water resources they have or limit development.

The Tampa Bay region dealt with its problems of salt water intrusion and over-pumping of groundwater by forming a public-private water management agency called Tampa Bay Water, responsible for wisely developing the region's water resources. Today, they are responsible for supplying drinking water to most of the communities in the Tampa Bay area.

When Tampa Bay Water formed in 1998, it had a myriad of problems to contend with. Upon extensive discussion with leaders in the region, it was decided that the

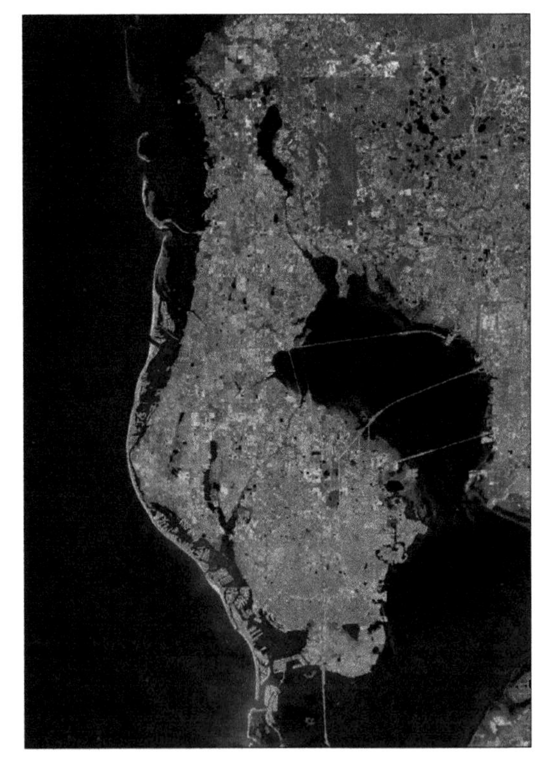

Figure 6.2 Pinellas County, Florida, home of St. Petersburg and Clearwater, as seen from a satellite shows how vulnerable the state is to issues such as groundwater intrusion.

organization would venture into bold new areas by developing a huge desalination plant and by constructing a large above-ground reservoir.

The desalination plant was designed to be the largest in North America. It draws water from Tampa Bay and uses electricity to run the water through an extensive reverse osmosis system. Initially there were problems with the plant. No other facility had utilized water like that found in Tampa Bay. It is brimming with microorganisms and the reverse osmosis membranes became clogged with organic matter. However, after numerous modifications, the desalination system is working well. Today, desalinated water accounts for roughly 8% of the region's potable water supply.

The other major project of Tampa Bay Water was the construction of a large above-ground reservoir system, called the C.W. Bill Young Reservoir. This is a rather unique reservoir in that it is not constructed within the normal confines of a river's channel. Instead, water is pumped into the reservoir from nearby surface streams and canals. The water is extracted from the streams during the summer rainy season. During this time, the reservoir is filled. Water is released from the reservoir for public consumption during the dry season. It can supply 25% of the region's water for six months.

As could be imagined, the construction of a nearly two square mile, above-ground reservoir, that holds 15.5 billion gallons was not without controversy. There was some concern over its safety, particularly after cracks were found in its earthen lining. Nevertheless, the reservoir and the desalination plant were innovative approaches to managing the region's tight water supply.

There is one area of Florida with an abundant supply of water, the Florida Panhandle. Here, the Apalachicola River flows to the Gulf of Mexico. The delta area of the river is home to a wide range of important natural habitat as well as unique economic activities such as oyster harvesting and the production of the highly prized tupelo honey. This sparsely populated area, rich in water resources, came to the attention of those interested in furthering the development of south Florida near the turn of the twentyfirst century.

In order for areas of south Florida to significantly expand, it needs access to more water. A commission was established to consider the development of a pipeline from sparsely populated north Florida to support growing areas of the lower portion of the peninsula. Once news of the commission's charge reached the public, there were immediate outcries against the proposal. The state had distinct water management governance regions and the pipeline would go against existing agreements. In addition, there were significant concerns about the ecosystems of north Florida and what would happen to the beautiful delta region and the local economy if the water system was in some way disrupted. Plus, others were worried about expanding the population of south Florida beyond its carrying capacity. Many felt that the region was already overpopulated and didn't want to see it expand any further.

At the end of the day, the pipeline was never built due to broad public concern over the project. Florida's waters are largely managed within distinct watersheds.

(Continued)

This stands in stark contrast to California. Highly populous Southern California is located in an arid and semi-arid environment. Decades ago, the population far outpaced its usable water resources and water had to be pumped over great distances. Today, Southern California receives water from hundreds of miles away. It has grown far beyond its carrying capacity. The region is highly unsustainable from a water resources perspective. This also makes the area far more vulnerable than Florida to everything from climate change to terrorism. Florida must utilize water resources within local drainage basins and communities must innovate or limit growth. In contrast, California pumps water vast distances to supply cities in environmentally dry areas. Which approach is better?

Consumption trends

According to the United Nations, 69% of the world's fresh water is used for irrigation, 19% for industry, and only 12% for domestic use. Withdrawals of water for these uses are increasing. It is expected that by 2050, there will be a 55% increase in the use of fresh water throughout the world. Overall, fresh water withdrawals have tripled over the last few years.

Unfortunately, in many areas, the water usage is already beyond sustainable levels. In Europe, for example, the World Business Council for Sustainable Development reports that 60% of large (over 100 000 in population) European cities are withdrawing groundwater at rates faster than they can be recharged.

Water use per person is a good indicator of the amount of water consumed by populations around the world. The largest consumers per capita of water are the countries in North America, Central Asia, Australia and New Zealand, and some countries in South America and Europe. Some of these countries have an abundance of freshwater (Canada for example) and in many places are not over-using resources. However, some countries, like Australia and Iran, have limited water resources and are not utilizing their resources sustainably.

As noted earlier, most of the world's fresh water is utilized for agricultural purposes. Thus, we see tremendous amounts of water used in arid and semi-arid regions where there are scant water supplies. One example of where this is happening is in the Great Plains of the United States. Here, the natural landscape consists of dry prairie lands, originally housing a number of different tribes of Plains Indians who had limited populations on this harsh landscape, which receives less than 30–60 cm (12–24 inches) of rain a year.

As the population spread across the United States in the nineteenth and twentieth centuries, the Great Plains were seen as a potential agricultural wonderland—if a water supply could be found to support crops. A vast aquifer, called the Ogallala Aquifer, was discovered under the Great Plains and it was tapped to provide irrigation to support the growing population of the plains.

Soon, however, it was discovered that the use of the aquifer wasn't sustainable. Usage far outpaced recharge in this dry environment. Today, the aquifer supplies drinking water to approximately two million people and supplies irrigation water to vast stretches of agricultural land. Yet this water will not last forever. Some of the water that is extracted fell as rainfall millions of years ago—indicating the vast amount of time it took to store this water. Today, it is believed that the Ogallala Aquifer will begin to run out within 25 years.

This will be a serious threat to the region's agriculture. Indeed, the aquifer provides water for roughly 30% of all of the irrigated land in the United States in a swath that extends from North Texas all the way through portions of New Mexico, Oklahoma, Colorado, Kansas, Nebraska, Wyoming, and South Dakota.

The problem with the Ogallala Aquifer is well known and documented, but withdrawals continue to outpace replacement. This is happening to aquifers all over the world. What is troubling is that over-consumption is continuing without really trying to adequately solve the problem. This is one of the great tragedies of our modern era—our inability to effectively manage and deal with large regional environmental over-consumption. The Great Plains, like many areas of the world, have a very short time to try to figure out how to deal with these problems before the curtain closes on their ability to maintain their standard of living.

While agriculture and industrial uses account for 90% of all fresh water used, it is important to note that household consumption has its own issues. In the United States, the average American household uses 320 gallons of water per day. About half of this is for indoor use such as showering and bathing, toilets, clothes washing, and cooking. The other half of this consumption is for outdoor use, most of which goes toward lawn watering. Indeed, nearly a third of all residential fresh water goes toward watering landscape. The US is one of the only countries in the world that spends huge amounts of money purifying water only to have a third of that water sprayed on lawns (Figure 6.3).

Figure 6.3 Arid areas often have landscaping that is in tune with the local ecology.

Contrast this consumption pattern with Yemen, an arid country in the Arabian Peninsula with 28.25 million people. Here, household water consumption is approximately 16 gallons per day. Plus, water is running out. The capital of Yemen, Sana'a, gets all of its water from groundwater sources. The aquifer is dropping by up to 8 meters a year. Yemen, one of the fastest growing countries in the world, is truly in a water crisis as its population expands—even though individuals only use 16 gallons per day.

Clearly, water use varies from place to place and there are different consumption issues depending on local water sources and consumption patterns. As time passes, water will be one of the key resources that many regions will need to protect or develop. There is concern among some that access to fresh water will lead to greater global conflict as those who don't have water try to gain access to resources outside of their borders. That is why it is crucial to develop ways to conserve and protect existing water resources in order to ensure a safe steady supply of water for future generations throughout the world.

Sources of water pollution

Water pollution is a serious issue that threatens safe water supplies as well as ecosystems. There are six major sources of water pollution. Each will be reviewed below.

Agricultural pollution

Given the importance of water to the development of agricultural products, it should not come as a shock that pollutants that derive from agricultural operations are one of the major contributors to pollution. A wide range of activities causes agricultural pollution, as can be seen in Table 6.1. They include release of sediments from plowing, addition of fertilizers, spreading manure, runoff from feedlots, runoff of salts and nutrients from irrigation, release of sediments from clear cutting trees, and release of nutrients, pesticides, and herbicides from basic agriculture, silviculture, and aquaculture.

Most water bodies around the world are in some way impacted by agricultural runoff. Sediment, nutrients, pesticides, and herbicide pollution are very common in surface streams and lakes in agricultural areas. While much progress has been made to improve this form of pollution in many parts of the world, serious water quality issues still remain.

Industrial pollution

Another major source of water pollution is industry. There is a wide range of chemicals that are directly released into water, including metals, nutrients, and organic chemicals. Given the far range of industrial activity around the world, there is quite a variety of industrial pollution present on the planet. However, some countries have very strict controls on industrial pollution. For this reason, many industrial nations have far fewer pollution problems than lesser-developed countries with limited environmental regulation.

It should be noted that heat is also a significant pollutant from industrial processes. Many activities, including power generation, require significant amounts of water for cooling processes. The heat collected in the water is often released to surface water bodies where it can change the overall biogeochemistry of aquatic and near-shore ecosystems.

Table 6.1 Agricultural impacts on water quality (Food and Agriculture Organization of the United Nations, 1990).

Agricultural activity	Impacts surface water
Tillage/ploughing	Sediment/turbidity: sediments carry phosphorus and pesticides adsorbed to sediment particles; siltation of river beds and loss of habitat, spawning ground, etc.
Fertilizing	Runoff of nutrients, especially phosphorus, leading to eutrophication causing taste and odor in public water supply, excess algae growth leading to deoxygenation of water and fish kills.
Manure spreading	Carried out as a fertilizer activity; spreading on frozen ground results in high levels of contamination of receiving waters by pathogens, metals, phosphorus and nitrogen leading to eutrophication and potential contamination.
Feedlots/animal corrals	Contamination of surface water with many pathogens (bacteria, viruses, etc.) leading to chronic public health problems. Also, contamination by metals contained in urine and feces.
Irrigation	Runoff of salts leading to salinization of surface waters; runoff of fertilizers and pesticides to surface waters with ecological damage, bioaccumulation in edible fish species, etc. High levels of trace elements such as selenium can occur with serious ecological damage and potential human health impacts.
Clear cutting	Erosion of land, leading to high levels of turbidity in rivers, siltation of bottom habitat, etc. Disruption and change of hydrologic regime, often with loss of perennial streams; causes public health problems due to loss of potable water.
Silviculture	Broad range of effects: pesticide runoff and contamination of surface water and fish; erosion and sedimentation problems.
Aquaculture	Release of pesticides (e.g. TBT (Tributyltin)) and high levels of nutrients to surface water and groundwater through feed and feces, leading to serious eutrophication.

Storm water pollution

Storm water is water that runs off of cities and streets into surface water bodies. It gathers non-point surface pollutants, such as urban dust, and carries them through storm water sewers into streams, lakes, and groundwater systems. The problem with storm water pollution is that its chemistry is highly variable. When rains fall, the first flush of storm water contains the greatest amount of pollutants, thereby pulsing surface water with a toxic soup of pollutants (Figure 6.4).

The second cause of the highly variable nature of storm water pollution is the highly variable range of activities that occur throughout the world. Storm water captures the surface chemistry of the planet. Thus, in places where there is a great deal of agricultural fertilization, the storm water will carry fertilizer wastes. In areas where there is considerable release of metals, there will be a flush of metal pollution in the storm water. For this reason, storm water is a particularly difficult pollution source to manage.

Figure 6.4 When it rains or when snow melts, all of that water is carried overland to surface water bodies, often through storm water sewers. When snow melts, high levels of pollutants, particularly salts, are carried in the meltwater.

Sewage

Another important pollution source is sewage. Human waste contains high amounts of nutrients and organic materials that can cause significant pollution problems. Approximately 25% of the world's population does not have access to sanitary sewer systems. In these places, human waste can enter surface or groundwater systems. But even in places with sewage treatment, there is a range of processing that takes place. Some areas conduct very thorough processing and some areas do not. There is always some degree of waste that enters water systems, even in the best of circumstances.

In addition, sewage lines can sometimes receive industrial pollution that is not suitable for sanitary sewage treatment. In these cases, industrial pollutants can be released into surface water bodies due to inadequate treatment.

Leaking underground tanks

Leaking underground tanks are another significant source of water pollution, especially groundwater pollution. Tanks that hold hazardous liquids are often buried underground for safety and accessibility reasons. Most gas stations around the world, for example, contain buried underground gasoline tanks from which the gas is pumped at a filling station. However, over the years we have learned that tanks can leak—especially older tanks that were metallic and poorly constructed. Some of these tanks have corroded and caused devastating groundwater pollution problems.

Many areas of the world now require that underground storage tanks meet particular environmental requirements to avoid leakage. A number of tanks have been removed and replaced throughout the world and a great deal of effort has gone into cleaning up places that were found to have leakages of hazardous chemicals.

Landfills

Societies around the world produce huge amounts of garbage every day. We have found a number of ways to deal with this waste, from burning to storing it in large landfills or dumps. When garbage is stored in landfills, care must be taken to avoid water flowing through the landfill. It can pick up a variety of chemicals that can enter the groundwater. It is like water flowing through a coffee filter. It starts clean, but ends up as a soupy toxic mix called leachate.

Eutrophication

Eutrophication is a widespread problem of oxygen depletion of surface water bodies due to the addition of excess nutrients from fertilizer runoff or the addition of sewage effluent. Each year, millions of tons of fertilizers are added to agricultural fields, lawns, and gardens. When it rains, some of the fertilizers run off into surface water bodies to add nutrients to aquatic systems. Nutrients from raw sewage or sewage treatment plants can also make their way to the same systems.

Natural aquatic systems evolved over many thousands of years with little change to the overall annual nutrient input from natural processes. When fertilizers and sewage effluent is added to these environments, they become highly unnatural and the ecosystems change. When more nutrients are added, they cause excess plant growth.

In eutrophic settings, phytoplankton growth expands and non-native species can move in that thrive in high nutrient environments. Often the water becomes more turbid and the system becomes less ecologically diverse.

When the phytoplankton and plants die, dissolved oxygen becomes tied up in the decomposition process, thereby removing it from the water. This causes a situation caused hypoxia. Widespread fish kills have occurred as a result of significant hypoxic conditions in some surface water bodies. In some places, toxic phytoplankton blooms have occurred that devastate ecosystems (Figure 6.5).

(Continued)

Figure 6.5 This eutrophic surface water system in Guyana is used as a sewage outflow.

Clearly eutrophication is a serious modern environmental problem caused by fertilizer runoff and the release of sewage effluent. However, organizations around the world are working to reduce these problems. In Long Island Sounds, for example, a number of organizations are working to reduce storm water pollution and nutrient releases from septic systems. In areas with a great deal of concern over runoff from dairy or cattle operations, extensive manure collection operations are in place.

In some situations, buffers are designed to slow down any water moving overland prior to its entry into streams or lakes. These buffers often store water in wetlands or ponds to try to limit the addition of nutrients directly into low nutrient ecosystems.

Over the years we have improved landfill technology. Today, many areas require thick impenetrable landfill linings to avoid water from entering groundwater systems. But older unlined landfills remain problematic on our landscape. Over the years, unlined landfills have caused significant groundwater and surface water pollution.

While there is a variety of sources of water pollution, it must be noted that it derives from two main types of sources: point and non-point pollution. Point pollution originates from a single identifiable source. Non-point pollution comes from a number of dispersed points that may not be easy to identify or that are too numerous to consider managing at a single spot. An example of a point source of water pollution is an outflow pipe from a sewage treatment plant. In contrast, a non-point water pollution source might be an agricultural field that has had applications of fertilizers and pesticides. Another example of a non-point pollution source is a suburban or urban neighborhood that has a myriad of land uses and a range of pollution types.

The management of point sources of pollution is easily done when compared with non-point sources. Once a source is identified, it can be controlled in some way. However, non-point sources are very difficult to manage and are far more complex overall.

In recent years, we have become concerned about broad new forms of pollution that have entered water systems: pharmaceuticals, plastics, and radioactive materials. A wide range of pharmaceutical products from hormones to antidepressants are making their way into surface streams and groundwater from sewage treatment plants and animal holding pens. While the use of these chemicals is highly regulated in most areas, we have not found effective ways to prevent them from entering ecosystems as they pass through the body of humans and other animals. We are finding them more and more in our drinking water.

Plastic is another pollutant that has been receiving significant attention recently. Plastic does not easily break down in the environment. We use it once and then throw it out where it can enter the waste stream for decades. Some of this plastic finds its way into surface water bodies. Large areas of the oceans have seen rafts of plastic waste floating at the surface. Sea creatures such as turtles and gulls have died from ingesting plastic bags and other plastic items that have been mistaken for food. Indeed, it is estimated that half of all sea turtles and all sea birds have eaten plastic. Just think of all of the plastic bags you have thrown away in your lifetime and try to imagine where they might be right now. Many areas of the world have banned plastic bags due to problems with pollution of waterways.

One particular type of plastic, little plastic pellets, are one of the most problematic forms of plastic pollution that have been found lately. It is believed the small plastic pellets are finding their way into surface waters from wastewater treatment plants throughout the world. The pellets are used in a variety of products from facial scrubs and soaps to toothpastes. The pellets have been discovered in a number of water bodies, including the North American Great Lakes.

One last form of emerging pollution is radiation. In the previous chapter, I wrote about the problems associated with nuclear energy and some of the accidents associated with nuclear power plants. The Fukushima Daiichi disaster of 2011 caused widespread release of nuclear materials to the nearby ocean. Some of this radiation is still detectable and there is concern over the long-term impact of nuclear materials entering aquatic ecosystems in this area.

Water management and conservation

Water management is an activity that focuses on the organized use of water resources. It can include development and planning of water resources. Often water managers work for local, regional, or national governments in order to protect and equitably distribute water.

Given all of the problems associated with water usage, it is worthwhile examining how different parts of the world manage and conserve water. As we will see, there are a number of different options utilized to deal with local and regional water issues.

National and regional water conservation and management

For many years, water has been looked at as an unending resource that could be pulled from the ground or out of water bodies forever. There was little concern over individual consumption. More could be produced in some way. If we couldn't get it locally, engineers

Figure 6.6 It wasn't that long ago that wells were commonly used as the main source of drinking water throughout the world. Most of our drinking water infrastructure is relatively new. This is a well near my uncle's trailer on his hunting property in 1976.

could find ways to move it around from wet areas to thirsty areas. There didn't seem a need to conserve or limit use (Figure 6.6).

Now, however, things are very different. Many have predicted that the next international conflict will not be about resources like oil or minerals, but about water. We can no longer think of water as a limitless resource. We have to be cleverer in the ways we manage it.

Water usage is a very local activity, but some parts of the world have developed national or regional water management and conservation activities that support one or more of the three "E"s of sustainability (environment, equity, and economic development). Before examining some conservation efforts, let us first examine some regional water development projects.

Water as a tool for regional development

Take for example the Tennessee Valley Authority (TVA). It was established in 1933 specifically to promote economic development in a region of the United States greatly impacted by the Great Depression. The TVA sought to promote development within the multi-state region that makes up the Tennessee River watershed.

The TVA was designed as a federally owned corporation to develop a number of water resources in order to provide hydroelectric power to a region of the United States that was largely undeveloped. While the effort wasn't directly a water conservation effort, it did lead to the development of water resources and access to electricity for millions of people. The authority also provided a great deal of expertise to local governments and families seeking to improve living conditions.

In our modern era, the TVA is seen less as a development organization and more as a power utility. They have built a number of fossil fuel and nuclear power plants.

It must be noted that the TVA has been criticized for having a heavy hand in its projects and that it was not particularly sensitive to environmental or cultural issues as projects moved forward. The Appalachian region has changed greatly since it built its 29 hydroelectric dam projects.

Another important water development project of regional significance is China's Three Gorges Dam project. This project began in 2004 and was more or less completed in 2008 with continuing improvements taking place today. The dam is located on the Yangtze River in Hubei province in central China.

Like the TVA projects, the dam was built as a large development project. It was designed to increase energy production, expand shipping lanes, and control flooding. Today it is the largest dam in the world by energy production.

Also like the TVA projects, the dam has seen its share of controversy. Over a million people were displaced and many important historic sites were flooded. There are also concerns over a number of environmental issues. Nevertheless, the dam is a cornerstone of China's aggressive development process.

These two projects illustrate how nations can use water as an organizational tool for development. While the projects are not necessarily for the development of water resources for consumptive purposes, they do illustrate the organizational power of water in regional development. Many areas of the world have utilized water in similar ways to promote regional development projects.

Water supply management

As noted in the text box about Florida, some areas of the world are going through extreme measures to ensure that they are able to provide a steady supply of water to consumers. This approach provides high-quality water to consumers no matter what the eventual use of the water might be. In cities across the world, more water treated at great expense to ensure that it is safe to drink is used to flush toilets, wash cars, and water lawns than is used for human consumption—often in places where there are extreme shortages of water!

We have found two main ways to manage these water supply problems. They are called hard path water management and soft path water management.

Hard path water management

The hard path water management approach focuses on managing demand. What this means is that water managers increase the price for certain types of water users in order to limit the demand. There are a number of ways in which this approach can be implemented. For example, costs can be distributed evenly across users by increasing costs per volume of water consumed.

However, in reality, most users use modest amounts of water. Particular homes or businesses use far more water per person than others. There are some people who just love green lush lawns and continuously water their yards to ensure that they look the best they can. There are also some businesses, such as car washes or agricultural processing industries that use very large volumes of water. As a way to reduce consumption of these heavy users, managers have opted to increase the costs of water for them in an attempt to manage demand. This demand management approach limits the cost of water to the average consumer and creates an environment whereby heavy consumers are driven to reduce consumption in order to reduce their own costs.

This hard path of water management can be controversial. Heavily consuming but important industries that employ many citizens often complain that they are unable to bear the increased costs of water. Plus, many wealthy influential people with large lawns can be vocal critics of this approach. In addition, there is often the assumption that the increased revenue will be used to increase water resources in some way, thereby not really solving the problem. While water managers using this approach do also include consumer education about water conservation within this system, the broader focus of the hard path of water management is to reduce water consumption by heavy users while also advancing technology to improve efficiency and find new resources for consumption.

It is also important to stress that hard path water management really does not look holistically at water use to solve over-consumption.

Soft path water management

Hard path water management strategies are fine approaches in some situations. However, some have argued that soft path water management strategies are more appropriate. According to noted water resource experts Oliver Brandes, David Brooks, and Stephen Gurman, the soft path of water management is different from other water management strategies because it: (1) treats water as a service, (2) makes ecological sustainability the most important criterion for assessment, (3) matches the quality of water delivered for use, and (4) plans from the future to the present. Let's take a look at each of these issues.

Water as a service. We look at water as a general need. We have it for any use we choose to make of it. However, if you think about it, water is used inefficiently for many things. In many parts of the world, water managers allow water to be used for anything. But, consider water for irrigation. The water is being used to grow some type of crop. However, if we think about the water for growing food, there is an opportunity to engage with the land owner to limit the amount of water used for irrigation by better crop management. If open ended or seemingly limitless amounts of water are used for activities like irrigation, there is no incentive to make more intelligent decisions about water use for particular purposes.

Water for ecological sustainability. We often ignore the ecological impact of our water management decisions. For example, the Aral Sea largely dried up when agricultural planners from the Soviet Union diverted huge amounts of water from tributaries of the sea into irrigation canals to water cotton plantations in very dry land. This clearly was not the best situation for the overall ecology of the region. Traditional fishing has declined significantly, along with the general ecology. In the soft water management approach, it is important for water managers to recognize the critical importance of ecosystems prior to making any decision that would harm the region.

Water quality for appropriate use. In some parts of the world, clean drinking water is delivered by truck to homes. It is stored in rooftop tanks and used only for human consumption. Dirtier water is used for flushing toilets or irrigation. Friends of mine from these parts of the world are flabbergasted when they visit me in the United States and learn that we spray treated drinking water onto lawns and agricultural fields. They see this as a waste, and it is—a very expensive waste. Soft water supply management encourages the development of a more diversified water supply that provides safe and healthy drinking water where appropriate and less clean water for other uses. Indeed, water can be reused in such a system, whereby wastewater from one use becomes an input for another use.

Planning from future to present. In most traditional water management strategies, water is produced for current consumptive trends. Water managers try to meet demands regardless of impact. However, in the soft path water management system, managers seek to work with communities to envision the future they want. If Aral Sea fishermen were part of the water management strategy decision-making for the development of the cotton plantations, would the Aral Sea now be in such bad shape? In other words, how can water managers work with the public to soundly ensure a safe and steady supply of water and a healthy ecosystem, while advancing the future desires of the population?

Water management and innovation

There is no doubt that we are quite good at finding ways to get access to water, even as we find ourselves vulnerable as we become more and more reliant on complicated water management systems. Just think about the complex web of canals, pipes, and aqueducts that supply southern California with its water. Water managers created a modern engineering wonder of the world by moving water from northern California and the Sierra Nevada mountains to the dry arid landscapes of Los Angeles.

Thus, we are good at building water plumbing systems like dams, canals, and pipes. However, this comes at huge energy costs. Roughly 80% of the day-to-day costs of our complex water management system comes from the cost in energy of the massive pumps that are needed to move water from place to place. Thus, our worldwide water supply engineering marvel is quite vulnerable to energy costs and energy shortages.

One of the more innovative new approaches to water supply management is desalination. This was discussed in some detail in the Florida example. However, it is worth stressing that the production of water comes at a very high energy cost. Thus, in places where desalination is in place, residents are exchanging water for carbon pollution. Plus, there are often waste problems associated with the salty brines produced in the desalination process.

As of late there has been a great deal of interest in using wastewater from sewage treatment plants to alleviate water shortages. These "toilet to tap" initiatives mix highly treated sewage effluent with existing ground or surface water prior to further treatment for tap water. These initiatives have met with limited public approval due to the obvious "ick" factor. Yet in many areas of the lower Mississippi River Valley and other rivers of the world, this is already occurring as a matter of practice. Many sewage treatment plants release their effluent directly into streams that are just upstream from water collection sites for drinking water of downstream communities.

Thus, the "ick" factor in some communities about some of the toilet to tap projects is a bit misplaced. Some communities have opted to pump treated sewage directly into deep underground aquifer systems (Figure 6.7). Others spray it directly onto the ground in an irrigation scheme. Yet others use it as a special watering system for suburban and urban

Figure 6.7 Using treated sewage for drinking water supplies has an "ick" factor. However, treated sewage is used to supply astronauts with a steady and sustainable water supply in space.

Figure 6.8 Water is often managed at the local or regional scale. This is a photo of a local water management office. How is water managed in your community?

lawns. Regardless of approach, such innovate uses of wastewater has improved the water budget in some communities.

Architects and designers have also been active in finding new ways to save water in homes and buildings. Low-flow faucets, showers, and toilets are commonly in place in new and renovated buildings. Some buildings have built-in water collection systems that collect precious rainwater for use in toilets. In some areas, cisterns are making a comeback for local storage of rainwater for lawn or garden irrigation or for use in clothes washing or toilets.

Water quality

Water quality is often highly regulated and there are rules that protect surface and groundwater resources (Figure 6.8). In addition, drinking water and wastewater are tightly controlled in order to protect the public health of citizens. For example, the US EPA and most other water agencies in developed countries regulate a wide array of drinking water contaminants such as microorganisms, metals, and organic compounds.

There are a number of waterborne diseases that can cause death and a variety of illnesses. These include protozoan infections like cryptosporidiosis, parasitic illnesses like schistosomiasis, bacterial infections like cholera or dysentery, and viral infections like SARS (severe acute respiratory syndrome). Each of these can derive from coming into contact with or ingesting contaminated water. There are many examples of outbreaks of waterborne diseases that have caused great distress in developed and developing nations. Indeed, while these illnesses are most common in places without high quality water treatment, there have been highly publicized outbreaks in some industrialized nations. That is why there is such a high priority on drinking water quality standards throughout the world. A list of contaminants regulated by the US EPA is shown in Table 6.2.

Water managers also regulate a wide array of inorganic chemicals. They include things like arsenic, lead, cadmium, and mercury. These elements can derive from natural geologic conditions. However, in most contaminated water supplies, they have an anthropogenic origin. Arsenic is an element that has received considerable attention in recent years because it has been found to cause a variety of health problems at low exposure levels. While it can come from a variety of geologic sources, it can also derive from some fertilizers, wood preservatives, paints, metal operations, and the burning of coal in power plants.

John Snow and the development of epidemiology

In the early 1800s London was struggling with serious water pollution problems. The city had grown tremendously and city managers were not all that effective at developing sound water and sewage systems. Water intake systems in the Thames River for drinking water were not all that far from drainage ways from streets that brought a range of pollution, including sewage, to the streets.

In densely populated regions of the city, outhouses used by the burgeoning flats and apartments were adjacent to wells. In the midst of this setting, the city experienced a series of outbreaks of major diseases, including cholera.

For many decades prior to the outbreaks of the 1880s, many believed that cholera and other diseases were caused by miasmas, or bad air. Noxious materials were believed to be present in the air that caused illness and harmed the body. Most at the time believed that breathing polluted air was responsible for many illnesses.

(Continued)

Table 6.2 List of water contaminants regulated by the EPA.

Microorganisms	Disinfectants	Disinfection by products	Inorganic chemicals	Organic chemicals	Radionuclides
Cryptosporidium	Chloramines	Bromate	Antimony	Acrylamide	Alpha particles beta
Giardia lamblia	Chlorine	Chlorite	Arsenic	Alachlor	particles and
Heterotrophic plate count (HPC)	Chlorine dioxide	Haloacetic acids	Asbestos	Atrazine	photon emitters
Legionella		Total Trihalomethanes (TTHMs)	Barium	Benzene	Radium 226 and
Total Coliforms			Beryllium	Benzopyrene (PAHs)	228
Turbidity			Cadmium	Carbofuran	Uranium
Viruses			Chromium	Carbon tetrachloride	
			Copper	Chlordane	
			Cyanide	Chlorobenzene	
			Fluoride	2,4-D	
			Lead	Dalapon	
			Mercury	1,2-Dibromo-3-chloropropane (DBCP)	
			Nitrate	o-Dichlorobenzene	
			Nitrite	p-Dichlorobenzene	
			Selenium	1,2-Dichloroethane	
			Thallium	1,1-Dichloroethylene	
				cis-1,2-Dichlorethylene	
				trans-1,2-Dichloroethylene	
				Dichloromethane	
				1,2-Dichloropropane	

Table 6.2 (Continued)

Microorganisms	Disinfectants	Disinfection by products	Inorganic chemicals	Organic chemicals	Radionuclides
				Di(2-ethylhexyl) adipate	
				Di(2-ethylhexyl) Phtalate	
				Dinoseb	
				Dioxin	
				Diquat	
				Endothall	
				Endrin	
				Epichlorohydrin	
				Ethylbenzene	
				Ethylene dibromide	
				Glyphosate	
				Heptachlor	
				Heptachlor epoxide	
				Hexachlorobenzene	
				Hexachlorocyclopentadiene	
				Lindane	
				Methooxychlor	
				Oxamyl	
				Polychlorinated biphenyls (PCBS)	
				Pentachlorophenol	
				Picloram	
				Simazine	

Table 6.2 (Continued)

Microorganisms	Disinfectants	Disinfection by products	Inorganic chemicals	Organic chemicals	Radionuclides
				Styrene	
				Tetrachloroethylene	
				Toluene	
				Toxaphene	
				2,4,5-TP (Silvex)	
				1,2,4-Trichlorobenzene	
				1,1,1-Trichloroethane	
				1,1,2-Trichloroethane	
				Trichloroethylene	
				Vinyl Chloride	
				Xylenes	

It is not really a surprise that some would feel this way. Nineteenth century London was not a pleasant place. The industrial revolution was in full swing with all of the impacts of environmental pollution. Chimneys in coal burning factories and homes spewed dark clouds of smoke that permeated the air and impacted the breathing of many. Plus the smells of the human and animal wastes were notorious.

Yet there were some who believed that some of the diseases of the time were carried in local water supplies. The most well-known of these individuals was John Snow. Snow was a physician and surgeon from York, England. He became interested in outbreaks of diseases early in his career and was part of the first society of physicians interested in studying epidemics. This group, the Epidemiological Society of London, became the founders of the field of epidemiology.

What is interesting about this group is that they were interested in understanding the causes of diseases and how they are spread, and also with the communication of the results of the research with the government. In many ways, the advent of this group brought together the ideas we now know of as public health. Most major cities of the world today have workers involved with public health issues such as HIV, diabetes, and mental health.

As noted above, John Snow did not believe that cholera originated in the air. He felt there was some other reason for the spread of the disease. In 1854, during a severe outbreak, he mapped over 600 deaths and illnesses from the disease. What he learned by doing so was that the outbreak was centered around one well in the Soho district. Everyone who died from cholera had drunk water from the well.

Upon further investigation, it was determined that an outhouse was adjacent to the well that likely contaminated the water supply. After Snow's report, London became much more cautious about the use of well water and the mixing of drinking water with human waste.

In the United States, relatively new regulations have required drinking water suppliers to reduce arsenic levels to 10 parts per billion. The old rules were set at 50 parts per billion. This significant shift caused a great deal of consternation among water suppliers, particularly small suppliers that didn't have the infrastructure in place to treat water so that it met these new rigorous standards. As a result, the rule was phased in and funding was given to some utilities in order to meet the requirement.

This example illustrates that local public water suppliers have direct public health impacts and that these impacts can be mitigated by technological solutions. Yet some contaminants are not as easy to manage as others. There are a range of organic chemicals that are especially troublesome for water managers.

Our society is creating new organic chemicals all the time that are not tested for their overall impacts on human health. In addition, we are releasing a variety of prescription drugs, hormones, and other chemicals through human and animal wastes. We have started to regulate some of these chemicals in our water supply, but there are far more that remain unregulated.

Some of the regulated organic chemicals include well-known compounds like benzene, chlordane, styrene, toluene, trichloroethylene, and vinyl chloride. Some of these compounds are banned, but linger in the environment today.

Organic chemicals are released as waste products in manufacturing or are leaked from underground tanks. Some are also released in our everyday garbage or from day to day activities. Regardless of how they find their way into our groundwater, they cause significant health problems when ingested at relatively low levels. Some are cancer causing and others attack particular parts of the body. As a result, these and many other chemicals that cause health problems are regularly screened in many public drinking water supplies around the world.

Water management in karst landscapes

There are some geologic areas of the world, called karst landscapes, that are especially difficult for water management (Figure 6.9). Karst areas are underlain by soluble rocks like limestone, gypsum, or salt. As water percolates through, it can dissolve the rock to produce large openings, or pores. Some of these openings can be quite large and become caves or caverns. In most instances, the large pores are interconnected and the rock is almost like Swiss cheese in the subsurface. Water often fills these pores to make highly productive aquifers. Karst aquifers store far more water than most other aquifer types. In many places, sinkholes or sinkhole lakes connect the surface with the subsurface aquifer systems. As you can imagine, there are a number of environmental issues associated with these unique landscapes.

Figure 6.9 Karst areas, such as this one in Croatia, have challenges with water management because of the complex underground geometry of karst aquifers.

In karst areas, there are few streams because most rainfall is diverted directly into the subsurface during precipitation events. This is different from most other areas of the world where water flows overland into creeks, rivers, or other types of streams. In karst areas, water drains through sinkholes or cracks in the rock near the surface directly into the subsurface karst aquifer.

Because there are so few surface streams and because water drains so quickly into the subsurface, karst landscapes tend to be susceptible to drought. When karst areas go through long periods without rainfall, the surface soils can dry. Florida in the US and the Yucatan Peninsula in Mexico are both karst areas that are susceptible to extreme droughts due to the lack of surface water. Indeed, given the high rainfall in both of these areas, one would expect the land to be covered with lush, tropical vegetation. Instead, the surface has extensive areas of scrub vegetation that can survive long periods without rainfall. Both areas also contain ecosystems that evolved with fire. Lightning strikes regularly start fires in both of these settings. Given the dry conditions, unique ecosystems have emerged in these places that are associated with drought and regular burning. Of course in the modern era, fire suppression has significantly altered these areas.

Karst areas are also susceptible to regional flooding. In most areas of the world, streams carry excess water from one place to another so that flooding tends to be localized along the floodplain of a river. In karst areas, flooding is much more widespread. When extreme rainfall events happen, the karst aquifer can fill and there is no place for the water to go. It overflows the aquifer and standing water will fill low areas until the water evaporates or the aquifer drains. Because most karst areas do not have a network of surface streams, the water tends to stay in place or have limited flow. While flooding in karst areas is not a common occurrence, it is regionally devastating when it happens.

Another problem in karst areas is groundwater contamination. Because the network of pores in the subsurface is interconnected, contaminants that originate in one location can spread very rapidly across the entire aquifer in a very short time period. Bowling Green Kentucky, for example, has one of the most problematic interconnected karst aquifer systems in the world. It is a cavernous system that has both air- and water-filled passages underneath the city. Over the years, there have been times when explosions have occurred due to the build-up of gasoline fumes. The gasoline leaked from subsurface tanks into the aquifer. Some of the gasoline evaporates to cause highly explosive conditions in the caverns. Of course, the water under the city has had serious contamination problems and the city and surrounding region have had to work hard to prevent surface contamination in order to protect the aquifer.

Karst landscapes cover about 20% of the Earth's surface. However, because of their productivity, about 40% of groundwater withdrawals are from karst areas. They are located in many areas of the world and there are extensive karst areas in Europe, China, Russia, North America, the Caribbean, and western Asia. Major cities are built on karst landscapes including Orlando, Florida and Paris, France.

However, many more organic chemicals are emerging with today's fast-paced technology. As a result, they are not screened in our water supply. For example, pharmaceutical products and hormones are finding their way into groundwater and surface water systems. We do not fully understand the long-term implications of these materials entering the public water supply. There have been indications that there are problems with some organisms exposed to water contaminated with some of these materials. But we do not fully know how they are impacting our public drinking water supplies and overall human health.

Water supplies are also regularly checked for radiation. There is a range of radioactive materials that can find their way into water supplies due to natural processes. Natural radiation is all around us and some areas have higher radiation levels, particularly radon, than others. In places with high natural radiation, exposure rates in drinking water can exceed recommended levels. In rare instances, radioactive waste has been seen to cause high levels of radiation in water supplies. There is especially concern in locations where there have been nuclear accidents such as the Chernobyl and Fukushima incidents. But, by far, the greatest concern is from natural deposits.

There are also a range of fine-grained plastics that are finding their way into surface and groundwater systems. While most water supply systems will remove these materials through filtration, there is concern over their entry into freshwater systems. Recently, water samples from the Great Lakes revealed that fine-grained plastics are becoming quite a common occurrence in water. The Great Lakes supply the water for most of the cities along the shore including Chicago, Detroit, and Toronto.

Understanding drainage basins

One of the best ways to understand water systems and water management is to look at water within the context of a drainage basin. A drainage basin, sometimes called a watershed, is an area of land in which all the water that falls as rain ends up within a particular stream. All streams, no matter their size have a distinct drainage basin. For example, the Amazon River in South America has one of the largest drainage basins in the world. Even small creeks and streams have drainage basins, albeit small ones.

Drainage basins are separated from other drainage basins by drainage divides. These areas between drainage basins are the high points of land, sometimes mountains or hills, which separate distinct hydrologic environments. For example, the Mississippi River drainage basin is distinctly different from the neighboring Colorado River drainage basin. Drainage basins are often places that separate distinct cultural groups from each other. If you have ever travelled to the Appalachian or Carpathian Mountains you will find distinct differences from one valley community to another.

Drainage basins are very useful organizational tools for water management. Societies have evolved to work within the confines of the water budget drainage basin. For example, in the Mississippi River drainage basin, there is an abundance of water in most areas. Water is not a limiting factor for irrigation, drinking water, or industrial use. However, in the nearby Colorado drainage basin, the arid landscapes provide distinct limitations for most water uses.

Drainage basins out of synch

In our modern era, we have gotten out of synch with the limitations of water use within drainage basins. Many dry areas have over-developed and are over-utilizing the water within their region. They are pumping out water in aquifers that took thousands of years to develop and they are pulling water from distant watersheds. Such situations are not sustainable in the long or short term. There are serious concerns about over-development in many arid and semi-arid regions of the world. Places like Phoenix, Arizona, Los Angeles, California, Zaragoza, Spain, and Amman, Jordan all have serious water supply problems due to over-use of existing aquifers.

Drainage basin pollution

Management of pollution within drainage basins is a standard environmental methodology. Both non-point and point pollution is best managed within drainage basins in order to ensure that pollution does not reach outside of the impacted system. For example, if sediment from a mine remains within one drainage basin, it can be managed to limit the impact to one system and management strategies can be developed to mitigate the impact. However, if the tailings are transported to another system, the impact is double.

Drainage basins have distinct streams and streams can have tributary streams. The term stream is a generic term that refers to any body of flowing water within a channel. There are a number of different terms that are used for streams depending on their size and location. Some terms include river, creek, rill, wadi, arroyo, wash, and rio. Each part of the world has their own nomenclature and the term stream is used in most technical writing about these systems. Sometimes, individual countries have regional terms for streams that vary from place to place. What names do you give for streams in your region?

Stream profile and base level

Another important aspect of streams is the concept of stream profile and stream base level. A stream profile is a graph of the elevation of the stream for its whole length. At a stream's headwaters, the elevation is highest at the drainage divide between drainage basins. However, the stream's elevation drops considerably over short distances. Here, streams often have waterfalls and rapids within linear V-shaped valleys. Eventually, the stream profile flattens and gradually declines in elevation over great distances. Here, streams cut meandering paths within very wide valleys.

Eventually, streams enter a relatively still body of water such as an ocean or a lake. The energy of the stream at this point is dissipated. The level at which the stream enters these bodies is called the base level. For a drainage basin like the Mississippi River or the Thames, base level is sea level.

Lakes

There are many regions of the world where the landscape is dotted by hundreds of lakes. There are many ways that lakes form, but most of them are created from geologic processes of glaciation and karstification. When glaciers melt, sediment is transported by melting water across the vast melting landscape. The sediment fills in low areas around ice blocks and can even cover the ice. When the ice melts, holes, or kettles, are left behind. If the depressions are near the water table, they become kettle lakes. Large areas of northern Europe and Asia have lakes that formed in this way. Minnesota, known as the "Land of 10 000 Lakes," and the other lake districts in these northern reaches, have so many lakes because of the glacial melting that occurred about 10 000 years ago. Some very large glacial lakes, like the Great Lakes of central North America on the US and Canada border, occur due to extensive gouging and erosion by glacial ice.

Lakes that form as a result of karstification are present in sinkholes. Karst landscapes occur in areas of soluble bedrock. When rock dissolves in contact with groundwater, cavernous voids form underground. On occasion they collapse to form depressions. When the depressions fill with water, they can become lakes.

Urban sinkholes: evidence of failing infrastructure

Geologists use the term sinkhole to describe a depression that forms from ground collapses due to the solution of soluble bedrock. However, the public does not follow this definition. Many, particularly those in the media, use the term to mean any type of ground collapse or instability that leads to the sudden formation of a depression.

There are many examples of these types of "sinkholes" that form in cities as a result of the failure of underground sewage and water pipes due to aging or poorly managed urban infrastructure. The pipes, of course, are usually under or near roadways. Their collapse can lead to snarled traffic or damage to vehicles.

In Winnipeg, Canada, the urban sinkhole problem has become so severe that the "Winnipeg Sinkhole" has its own Twitter account. Many, not just in Winnipeg, utilize the formation of these sinkholes to highlight problems with aging infrastructure in cities. The engineering of water supply and waste water systems is very advanced in many areas of the world. However, once it is installed, it is often difficult to maintain these complex systems. Many parts of the world are finding that maintaining their engineered water and wastewater systems expensive undertakings.

Karst sinkhole lakes can be found all over the world where one finds soluble rocks. Some of the best known sinkholes in the world are in Mexico. Here they are called *cenotes*. The ancient Maya felt that the sinkholes were a pathway into the underworld. They would drop precious goods into the lakes as gifts to their gods. Other important sinkhole regions are found in Florida, Texas, Kentucky, England, Cuba, Jamaica, China, and Spain.

Sinkhole lakes can form suddenly from ground collapse. Thus, the presence of sinkhole lakes is indicative of a potential hazard for a community. Each year, millions of dollars of property damage occur as a result of ground instability in karst areas. On occasion, deaths do occur when sinkholes form suddenly.

There are many lakes that form tectonically, or as a result of shifting of the Earth's plates. Perhaps the best known tectonic lakes are those in the Rift Valley of Africa. Here, the African plate is splitting apart causing a linear drop of land that forms the long valley. The lakes fill in especially low portions of the valley. Some of the more notable lakes are Lake Victoria, Lake Tanganyika, and Lake Malawi. The lakes are among the deepest in the world. The rift valley is growing by about 7 cm per year. If it keeps expanding at that rate, the lakes will eventually grow into broad oceans.

Seas

The term sea is used for a number of different types of water bodies and thus is a relatively imprecise term. For example, some large freshwater bodies of water are called seas in some parts of the world while in others they are termed lakes. The planet also has relatively large salt water bodies, such as the Mediterranean Sea and Red Sea that in many ways are extensions of oceans. Thus, the term sea is a local term that does not have universal definition.

Oceans

Oceans cover approximately 72% of the Earth's surface. They are found associated with the lowest portions of the Earth overlying a narrow layer of oceanic crust (continental crust is several times thicker than oceanic crust).

The oceans remain the most mysterious part of our planet. While explorers have probed the depths of them, we have much to learn. We do know, however, that they are fragile ecosystems that can be damaged by our actions. Over-fishing, pollution, oil spills, ocean acidification, and a number of other problems have been well documented.

Education on Ocean Ecosystems

Jacques Cousteau is largely responsible for the broad interest in oceans around the world. In the middle of the twentieth century, Cousteau began a well-publicized campaign to document information about the oceans. His ship, Calypso, was used to travel to many remote locations where he made documentary films about ocean ecosystems and the unusual plants and animals that can be found in them.

His documentaries were commercially successful and he earned a great deal of fame for himself for his efforts. His efforts earned him a chair of the French Academy in 1988. Yet Cousteau's legacy is that he brought information about remote areas of the ocean to the television screens of people all over the world. He raised awareness on a number of issues such as ocean pollution, over-fishing, and biodiversity.

Who else can you think of that is well known for their work in educating people about ocean ecosystems?

7

Food and Agriculture

We all eat food, some of us more than others (Figure 7.1). But in recent centuries, our world has moved from a largely agrarian one that involved the majority of people with food production in some way to a world that is largely urban and industrialized. Fewer people are producing food for more people than ever before. And this change is accelerating.

When I was a boy growing up in rural Wisconsin in the 1970s, in the northern Midwest of the United States, many of my friends were from farming families. I spent time on their farms and sometimes helped with mucking out barns or bailing hay. Yet by the time I went to college in the 1980s, many of these family farms were gone. They were bought out by large farming operations that had distinct ties with large corporate agricultural businesses. The story is similar in other parts of the world. Large farming operations are extremely efficient at producing volumes of crops and utilizing some of the most interesting technology that has been developed in recent years. They have transformed agriculture from a family operation into one that is high tech and carefully planned. Agricultural output of key crops such as corn, rice, and wheat have increased, and concerns over widespread starvation has declined significantly since the well-known famines of the twentieth century in parts of Africa and Asia.

At the same time, there has been a significant negative reaction to this new technology. Many who are concerned about the state of our modern agriculture have developed innovative approaches such as organic farming and community sponsored agriculture. This chapter will review the current state of worldwide agriculture, summarize several modern agricultural innovations, and review reactions to the high-tech agricultural movement.

Development of modern agriculture

For millennia, agriculture was rooted in small rural farms that produced food for families, with excess sold to markets (Figures 7.2 and 7.3). While certainly more productive than hunting and gathering, agriculture remained a relatively small-scale activity until the development of agricultural machines and improved crops in the industrial revolution. The technology for plows and harvesters improved rapidly as a result of the development of engines and mass production, as did the selection of seeds for crops with the expansion of the understanding of evolution and genetics. In addition, the development of modern fertilizers in the late nineteenth and early twentieth centuries greatly enhanced productivity.

Introduction to Sustainability, Second Edition. Robert Brinkmann.

Companion website: www.wiley.com/go/Brinkmann/IntroductiontoSustainability

Figure 7.1 What you eat really does matter to the planet. What did you eat today? How did your food get to your plate? Who grew it and prepared it? What impacts did growing, transporting, and preparing your food have on the environment?

Figure 7.2 A small rural farm in Wisconsin.

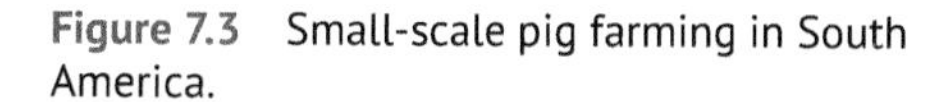

Figure 7.3 Small-scale pig farming in South America.

Overall, the industrial revolution saw significant expansion of farms and farm output. This output, along with improved health conditions, helped expand world populations throughout this era.

The rapid expansion of agriculture, along with the development of refrigeration, freezing, and industrial food processing, as well as the advent of the steam locomotive, trucks, and rapid shipping, led to the ability to transport food vast distances without spoiling. Areas of the world that did not have access to fresh fruits and vegetables or meat could have it easily shipped into markets. We were no longer dependent on the local farmer, but instead our food became part of a globalized network of food suppliers tied into large farms. Many of these farms were managed by college educated farmers and farm managers who had the ability to draw considerable output and profit from the farms through mechanization, mass production, and high-tech innovations. These advances largely occurred in the industrialized nations of the west throughout the nineteenth century and the first half of the twentieth century.

While these advances were important, they were not enough to stop famine throughout much of the twentieth century. The causes of famine are complex, but it is generally the result of food shortages that lead to widespread hunger and high numbers of deaths. In some cases, famine is caused by crop failures, droughts, or natural disasters. In other cases, famines are caused by war, a disruption in the food supply system, or other human-caused events. Several well-documented cases of famine occurred throughout Asia and Africa in the twentieth century that received a great deal of attention in western countries.

Thus, agricultural scientists worked hard in the middle of the twentieth century to improve agricultural technology to enhance output and efforts were made to export western food production technology to developing countries. This era of the worldwide expansion of agricultural technology is often called the green revolution. It helped to greatly expand food production in developing nations and it helped to reduce mortality in these areas. Since the green revolution, the population in many of the developing countries has expanded significantly.

In recent decades, a number of new advances in agriculture have significantly impacted the agricultural world. These include the advance of genetically modified crops and the advent of factory farms for meat production.

Genetically modified crops (GMO foods) are grown from seed that is genetically altered to improve its overall production. The DNA of the crop is engineered for a number of reasons. One of the most common reasons that a crop is modified is to make it resistant to particular diseases. Thus, genetically modified crops tend to be much more productive than traditional crops.

Most scientists agree that there is limited known health risk from the introduction of genetically modified crops in the food chain. But there are a number of unknown factors about their long-term impact on human health and the environment. We have been consuming genetically modified foods for a very short time period and we do not fully know if there are long-term impacts. Plus, we are essentially releasing new organisms out into nature and we do not know what will happen to ecosystems as a result of their release. Many of the new crops have built-in DNA materials that make them resistant to particular diseases. These diseases are themselves life forms. How will this strange mix of DNA transform with time and impact natural systems? While we are seeing little impact now, we just don't know how the new melange of genetically modified organisms will impact us in the future. Anti-GMO activists have coined the term, Frankenfood, to draw attention to their concern over GMO crops and products.

But it is important to note that there is also broad criticism over some of the cultural issues associated with GMO foods. Because the seeds are genetically engineered in high-tech laboratories, the seeds are patented and owned by large corporate entities. In many parts of the world the seeds cannot be harvested at the end of the season for replanting. Many farmers have been sued for trying to save seeds from genetically modified crops. They have to be purchased each year.

This reduces the ability of farmers to be self-reliant—a historic trait common in the culture of agriculture. The farmers become reliant on a corporation for seeds and other supplies instead of local vendors or distributors. In other words, agriculture is turning more and more into a big business operation that is seeking to draw considerable profit from the agricultural sector.

As a result of this, some countries have banned genetically altered crops from being grown in their country and some have banned crops or products containing GMO materials from entering their country. Some countries require products that contain GMO materials to be labeled. Still others ban particular GMO crops and not others. The United States is the largest producer of GMO crops, although the Brazil, Argentina, and Canada are also large producers. Dozens of countries have some type of ban in place. The European Union does allow GMO crops, but individual nations can ban them. Several, such as Poland, Austria, Hungary, and Greece have done so.

There is no doubt that GMO crops are part of the modern agricultural system, but they are highly controversial. Many areas of the world are trying to move to ban them or at least require labeling of products that contain genetically modified materials. Are you comfortable eating GMO crops? Would you grow them in your garden?

Meat production

Meat production has changed drastically over the last several decades. In the past, animal husbandry and butchering was a highly distributed activity that was not heavily regulated (Figure 7.4). However, this has changed. In some countries, meat production has become very centralized and regulated. Plus, many high-tech approaches to reproduction and feeding animals have expanded production. Overall, we've seen an approximately 20% increase in worldwide meat production over the last decade (Table 7.1). Most of this production is in the form of beef, chicken, and pork. Let's take a look at some of the sustainability issues associated with meat production as well as some of the critiques of the worldwide meat industry.

Activities associated with meat production cover approximately 30% of the land's surface. This includes livestock grazing, as well as activities to grow food to feed animals. With growing populations and greater demand for meat, we are running out of room to produce meat for the general public. It takes far less land to feed people on a meat-free or limited meat

Figure 7.4 How we raise meat has a profound impact on the lives of animals and the workers associated with raising and butchering animals. Have you ever butchered an animal for food? Would you?

Table 7.1 Top meat producers in the world (2017)

Rank	Chicken (billion)	Beef production (% of world production)	Pork production (% of world production)
1	China 4.8	US 20	China 48
2	Indonesia 2.1	Brazil 15	EU 21
3	US 2.0	EU 13	US 10
4	Brazil 1.4	China 11	Brazil 3
5	Iran 1.0	India 7	Russia 3
6	India 0.8	Argentina 4	Vietnam 2
7	Mexico 0.6	Australia 3	Canada 2
8	Russia 0.5	Mexico 4	Philippines 1
9	Pakistan 0.5	Pakistan 3	Japan 1
10	Turkey 0.3	Turkey 3	South Korea 1

diet. Thus, many have advocated trying to encourage people to reduce the consumption of meat to limit the overall impact of meat production on the planet.

Plus, meat production accounts for approximately 15% of greenhouse gas emissions, including approximately 40% of the world's methane production. The gases come from animal flatulence, as well as the emissions from fertilizers and waste byproducts. While many have advocated changing energy usage to combat global climate change, others have encouraged reducing meat production to reduce greenhouse gas emissions around the world.

Modern meat production is now highly centralized and high tech. In the past, meat was raised in relatively small-scale operations that allowed animals to have some degree of comfort and freedom. Now, a considerable amount of meat is produced in large factory farms that provide little comfort to the animals. Let's take a look at three ways that animals are managed using factory techniques: piggeries, feed lots, and chicken houses.

Piggeries

Piggeries are indoor facilities for housing large numbers of pigs through their entire life cycle. The indoor facilities allow farmers to regulate temperature and moisture conditions, therefore negating the need for outdoor mud wallows required to cool the body temperature of pigs. The piggeries confine the pigs to very small spaces, but generally keep pigs healthy until they are butchered. Waste is typically removed from the piggeries regularly through automated systems for storage outside of the building in manure lagoons. The lagoons themselves are problematic in that they can contaminate surface or groundwater and smell quite bad, thereby leading to regional air quality concerns.

But the piggeries are very efficient at producing pork. They are so efficient that most conventional pig farmers in Iowa have gone out of business. They cannot compete with the improved economy of scale of factory farming of pigs. In the United States, approximately 50 producers control 70% of the nation's pork production.

Some have criticized pork production from an animal rights perspective. The pigs are living in highly unusual circumstances. They are indoor, are confined, and have limited access to anything natural like mud or unprocessed food. Concern over the treatment of pigs has caused some areas to ban certain practices associated with piggery activities. For example, voters in the state of Florida in the US banned the crating of pregnant sows. Some piggeries used confinement crates for pregnant females that do not allow them to even turn around.

Feed lots

Feed lots, or concentrated animal feeding operations (CAFOs) are used in the beef industry to fatten up beef prior to butchering. Once grazing cattle reach a certain weight, they are transported to CAFOs from great distances. The CAFOs are located near the source of food that includes a variety of grains and beans. Most CAFOs in the United States, one the largest meat producers and consumers in the world, are located in the Midwest and Great Plains.

Once at the CAFOs, the animals spend up to four months eating grains specially designed to fatten the animals. Thus, modern beef production no longer relies on a life cycle of animals grazing on grasses. Instead, the animals eat mainly grains for the last few months of their lives.

Like piggeries, there are tremendous waste problems. The manure and urine is highly concentrated in these areas and there have been several instances of water pollution and concerns about air quality. There are also concerns over the release of pesticides, hormones, and antibiotics that are used at these facilities. Because the animals are confined in close spaces, they require antibiotics to keep them healthy.

Factory butchering operations are typically located near CAFOs to minimize loss of weight from the transportation of animals sent to slaughter. Modern slaughterhouses can process thousands of animals a day. Workers typically have one specific job to do in such settings and there have been concerns over worker safety and occupational health in the large butchering operations.

The large operations have called into question the safety and security of the modern food system. There have been several cases when tons of meat have been recalled from a single slaughterhouse due to concerns over meat contaminated with *E. coli* or other materials. Plus, the slaughterhouses produce tremendous amounts of waste that burdens a region already burdened with waste from the CAFOs.

Chicken houses

Chickens are also raised in intensive ways. For example, a modern indoor chicken house can raise up to 20 000 birds in an area roughly 130 meters long and 7 meters wide within a month and a half. This highly productive form of chicken production has similar environmental costs as piggeries and feed lots: there is a tremendous amount of waste produced and concerns about air quality.

Overall, these industrial agricultural systems are highly productive. While there are distinct environmental concerns, there are also concerns over the ethics of raising animals in

Figure 7.5 My father butchering hogs in the 1970s. Today, most of our families do not butcher our own meat.

such confined spaces. Many organizations have advocated for the rights of these animals and their protection.

Some organizations have filmed the insides of the piggeries, chicken houses, or the butchering floors of the factory meat production systems and released them to the public on YouTube or Facebook sites. People for the Ethical Treatment of Animals has been particularly effective at communicating the conditions of factory farms to the public. This has raised the ire of the meat industry, which has been very successful in promoting local and state laws to protect their industry. Several new laws have been implemented to prevent undercover filming at farms or meat processing facilities (Figure 7.5).

Even with the advent of this new technology, there are questions as to whether our current food production system is sustainable. We have a world population of over 7 billion people and we are expected to reach 8 billion by 2025. While we have done a relatively good job of feeding 7 billion people, we cannot continue to do so in the way we are doing it today. We are running out of a number of key resources that are important to maintain agriculture for future generations. So, while we have greatly enhanced our ability to produce food, we are doing so at a cost to the environment and at costs to future generations.

One of the key questions for future generations will be how to feed the world's growing population while protecting the world's environments. We have made great strides with technological approaches to food production, but there are questions as to the ethical treatment of animals and the use of genetically modified organisms in our current era. How would you modify the current food system to make it more sustainable long into the future? Is it even possible to meet the world's food demands and protect the environment?

World agricultural statistics

The world produces huge amounts of food every year. While the previous section highlighted some of the issues with modern food production, particularly meat, it is useful to examine the overall amount of food produced in the world. The top agricultural products by crop type are listed in Table 7.2. The same table lists the top crops produced by crop.

As can be seen, cereals and grains are the most common crop type. Maize, rice, and wheat are the major grains produced. Interestingly, sugar cane is the top worldwide product.

The major food producing countries are also some of the most populous. A glimpse at Table 7.3 shows the major food producing countries of the world. As can be seen, China,

Table 7.2 Major crop types and crops ranked (World Food and Agriculture Organization).

Rank	Top crop type	Million tonnes	Crop	Million tonnes
1	Cereals	2.263	Sugar cane	1794
2	Vegetables and melons	866	Maize	883
3	Roots and tubers	715	Rice	722
4	Milk	619	Wheat	704
5	Fruit	503	Potatoes	374
6	Meat	259	Sugar beet	271
7	Oilcrops	133	Soybeans	260
8	Fish	130	Cassava	252
9	Eggs	63	Tomatoes	159
10	Legumes	60	Barley	134

Table 7.3 Major food producers in the world.

Rank	Country	In Tonnes
1	China	616 251 000
2	US	475 959 000
3	India	297 850 000
4	Russia	117 744 000
5	Indonesia	102 933 000
6	Brazil	84 129 000
7	Ukraine	65 218 000
8	Argentina	61 148 000
9	Canada	58 793 000
10	France	54 652 000

the United States, and India make up the top three regions in terms of food production. But it is important to note that China produces almost a third more of the crops of the US and double the crops of India. Indeed, that country alone produces a substantial proportion of the world's food output.

Food deserts and obesity

One of the concerns many have about our modern food system is access to fresh and healthy food. This is especially an issue in some urban areas where there is a preponderance of fast food restaurants and very few grocery stores. Such areas are called food deserts.

Many have mapped these places and found that they tend to be in low-income and rural areas. In recent years, there has been an expansion of obesity and diabetes problems in many of these neighborhoods—largely thought to be caused by poor diets. Residents in these areas do not have access to healthy food and thus shop in convenience stores and eat at fast food restaurants. While many of us eat at fast food restaurants occasionally, it would be difficult for us to maintain a healthy lifestyle if we ate fast food every day. Most nutritionists would find a diet that consists of fast food and food from convenience stores hazardous to our long-term health.

In 2004, the documentary filmmaker Morgan Spurlock documented a fast food diet in his film *Supersize Me*. For 30 days, he ate three meals a day at McDonalds. He consumed over 5000 calories a day and gained 24.5 pounds. His cholesterol increased and he had a number of other health problems. For a month he ate what some people eat all year.

His efforts demonstrated that the food that some fast food companies produce is overall unhealthy to the public. Lawsuits against some fast food companies proceeded that attempted to blame health problems on the food sold in the restaurants. The growth of health problems in the US and other countries was blamed largely on fast food restaurants. Lawyers for the cases used similar tactics to the US tobacco lawsuits that highlighted that the product produced by fast food restaurants was unhealthy and that the companies knowingly sold unhealthy products to their customers. The lawsuits were not successful and since then several state legislatures have made it illegal to sue fast food restaurants for product liability.

In a highly controversial move, New York's Mayor Bloomberg banned the sale of sugary drinks over 16 ounces in volume in restaurants in New York City. The ban did not apply to grocery stores or fast food markets. The ban was an attempt to try to improve the health of New Yorkers by limiting their consumption of sugar in fast food meals. The city pays a tremendous amount of money for healthcare for the poor and the effort was an attempt to improve the overall public health.

While his actions were praised in the public health community, the mayor's efforts were met with much derision. His actions were seen as heavy handed and many criticized New York as a "nanny state" that oversaw too many personal decisions of its residents. In 2013, the ban was struck down by the courts. It is unlikely if those concerned about the quality of food in cities will give up any time soon in trying to eliminate unhealthy food from communities.

The efforts of Bloomberg and Spurlock are but two seeking to change the way we think about food. There are many areas of the country with limited access to healthy food in their neighborhoods. Think about the location of grocery stores or other places to buy fresh fruit and vegetables near your home and school. How easy would it be for you to purchase healthy food? Is it easier for you to walk to purchase healthy food or purchase food from a convenience store or fast food restaurant?

While Mayor Bloomberg has worked hard to try to remove unhealthy food from New York, others around the world are seeking to improve the quality of food in schools for children. In the United States, public schools often receive excess food supplies from the government, much of it processed and not particularly healthy. At the same time, dwindling school budgets have forced schools to limit labor costs by purchasing prepared or easy to prepare foods. In addition, schools have also added vending machines that sell sugary sodas and candies, and fat-laced snacks to pupils. What this means is that school children have been getting less fresh food and more processed food than ever before.

But this is changing. A number of activists have urged changes to school food systems to include more fresh fruit and vegetables. At the same time, others are encouraging students to be more active during recess and after school. While some children are not pleased with the changes, it is starting to make a difference. Schools have noticed distinct weight reductions since implementing better food options for children.

Another leading food advocate is Mrs. Obama, the former First Lady of the United States. When her husband, Barrack Obama was elected President in 2008, Mrs. Obama decided to build a garden on the White House grounds. The property has a long history of gardens. Indeed, its first occupants, John and Abigail Adams planted the first garden to supply food for themselves and their guests. Many other early presidents followed suit. The gardens came into cultural prominence during World War I and World War II when the presidents highlighted the importance of self-sufficiency during times of food shortages and national crisis.

The Bronx Green Machine

The Bronx, a borough in New York City, is home to one of the poorest and most densely populated school districts in the United States. It is an unlikely place to find a strong focus on food sustainability. But thanks to high school teacher, Stephen Ritz, the Bronx is one of the leading centers of the urban food movement (Figure 7.6).

Ritz became concerned about the future of the children in his classes. They live in large apartment buildings, have little connection to nature, and very little access to fresh and healthy food. Instead of grocery stores, local bodegas, convenience stores, and fast food restaurants dominate the neighborhood from which his students come to school. Many are children of immigrants and many are from single parent families; many are homeless. It is an area where one would think there would be little hope for the future.

However, there is tremendous hope due to the activity of Mr. Ritz and his students in a project called The Bronx Green Machine. They have built extensive expertise in the growing and management of small indoor hydroponic growing systems. He and his students grow tremendous amounts of food for children in his school and even provide

(Continued)

Figure 7.6 Stephen Ritz of the Bronx Green Machine speaking at Hofstra University.

services to other schools and organizations seeking to learn how to build indoor and outdoor systems, including green walls.

Ritz uses agriculture to teach a number of academic skills. Math, reading, and a variety of other activities are built, working with the hydroponic systems. Students have improved their academic performance through the activities built into the horticultural work. In addition, Ritz is providing tremendous life skills for the students. Because they essentially run a business out of the classroom through their consulting work and food distribution, students learn everything from food production to sales and marketing.

Ritz and his students are helping to transform a neighborhood that seems hopeless into an area of great hope. His efforts have drawn some high-profile attention, including segments on the Food Network, the Today Show, and Ted Talks.

However, the Obamas are the first to develop a garden and promote gardening in the modern era. Mrs. Obama started the garden to provide healthy food for the Obama's two daughters, Malia and Sasha. She used the garden to educate many local school children about the importance of eating fruit and vegetables, and an overall healthy lifestyle. The garden has been featured in books, television specials, and magazine articles. As a result, Mrs. Obama has become a well-known advocate for healthy eating and local gardening.

It is worth noting that the garden did not only produce fruit and vegetables. The garden also supported the White House beehive, which produced honey for the First Family. This has helped to promote local beekeeping. Mrs. Pence, the wife of Vice President Pence, installed a beehive on the ground of the residence of the Vice President.

Sustainable alternatives to the industrial food movement

Vegetarianism and veganism

Many cultures around the world have been vegetarian for centuries. But in the modern era, there is a resurgence of interest in vegetarianism and veganism for health and ethical reasons. Many people believe that these diets promote greater health and others choose to be vegetarian or vegan because they do not wish their food choices to impact the lives of animals.

Vegetarians choose not to eat meat. Some do eat other animal products like milk, cheese, and honey. Some people eat fish as well, but they are not really vegetarians, but pescatarians. Some health professionals have advocated vegetarian diets for the health benefits, although many take up vegetarianism for ethical issues. They are uncomfortable with eating the flesh of animals. Others become vegetarians because they do not trust the modern food system.

Vegans do not eat any meat products. Not only do they not eat meat, they also do not eat any animal products like dairy, honey, or animal oils. The vegan diet has been advocated by a number of health professionals to reduce heart disease, diabetes, and cancer. A number of well-known personalities are vegans. Former President Bill Clinton, who was famous for his voracious fast food appetite while campaigning for office, became a vegan after a series of heart problems. Some have argued that vegan diets lead to a lack of vigor and are unhealthy. However, many famous athletes extoll the virtues of a vegan diet for better health and performance. They include ultramarathoner Scott Jurek, professional baseball player, Path Neshek, professional hockey player, Michael Zigomanis, former professional boxer Mike Tyson, and professional basketball player, Jahil Okafor.

The group People for the Ethical Treatment of Animals is one of the organizations that is leading the worldwide effort to encourage vegan and vegetarian diets. They have sponsored a number of famous racy advertising campaigns extolling the sexual benefits of a vegan diet. One of the main spokespersons for PETA is the actress, Pamela Anderson.

One approach that many are advocating for those not ready for a full vegan or vegetarian diet is the practice of meatless Mondays. This essentially restricts meat eating to six days a week, thereby reducing the impact of a meat diet by one-seventh. While it might seem like a small amount, it adds up. The Environmental Working Group notes that if all Americans skipped meat one day a week, the equivalent carbon savings to the planet would be the same as taking off 7.6 million cars off the road. This approach allows one to explore vegan or vegetarian eating without making significant changes to one's life.

Interestingly, the new approach to vegetarianism and veganism has a long history. Many religions have restrictions on meat consumption or the type of meat consumed. Even the ways animals are butchered and prepared have ritual in the Muslim and Jewish faiths. Halal and Kosher meat products are prepared in particular ways with the approval of a religious leader. There is growing interest in the role of food in our lives and it is connecting us with the past in interesting ways.

If you are living on or off campus, a vegan or vegetarian diet may seem difficult. However, there are ways to make it work. Talk to your campus food service provider and find out about their vegetarian or vegan options. Most campuses offer several every day for breakfast, lunch, or dinner. If you live off campus, find the local health food store or search for options in your local grocery store. Plus, most campuses have vegetarian or vegan clubs where you can learn more about how to live a vegan or vegetarian lifestyle. Before diving in all the way, make sure that you study the types of foods you will need to eat to maintain a healthy and balanced diet.

Organic farming

Organic farming is a form of agriculture that relies on natural forms of agricultural practices, thereby removing the use of non-natural pesticides, herbicides, and fertilizers. The organic movement evolved in the twentieth century in a reaction to the development of synthetic fertilizers and other chemical additives to farm fields and crops. Many of the founders of the movement believed that traditional forms of agriculture were in greater harmony with nature and that the synthetic chemicals were hazardous to not only natural ecosystems, but also to consumers of the crops.

Since its advent, the organic movement has sought to develop very clear guidelines for what is and is not an organic crop or farm. As concern over food crops grew in reaction to the environmental movement of the 1960s and 1970s, consumers demanded greater labeling of products in order to understand what was in them. Many products were called "natural" or "organic" without have any clear guidelines over the meaning of those terms. In reaction, greater emphasis was placed on food labeling and in the precise definition of the meaning of the term "organic".

The first standards for organic products were established for several countries in the European Union in the early 1990s. Great Britain, Japan, and the United States followed. Now there are over 140 countries involved with organic production using clearly defined guidelines developed by the International Federation of Organic Agricultural Movements.

As noted, organic agriculture operations follow very strict rules about everything from prior land use to weed management. Some in the sustainability movement, most notably Joel Salatin, have criticized the organic movement as too rule based and government regulated.

Small farm movement

As noted above, large agricultural operations have prospered in recent decades. They are taking advantage of high-tech approaches to agricultural operations and produce huge amounts of food. However, this approach has been criticized as somewhat dehumanizing. Small farms are disappearing as is the diversity of crops that can be produced on smaller, carefully tended farms. In many parts of the world, the traditional agricultural lifestyle is succumbing to large mechanized operations where farmers are workers in a corporatized environment.

There has been a significant reaction to this in the development of the small farm movement (Figure 7.7). Many are advocating for the development of small productive diversified farms as key ways to preserve not only the agricultural lifestyle, but the agricultural practices of the past. Many are concerned as to the staying power of the large operations and fear losing traditional practices as rural residents migrate to cities to find jobs.

Figure 7.7 A small rooftop farm in Brooklyn. Are there small farms in your neighborhood?

In recent years, many agriculturalists have bucked the big ag trend and gone to the country to start farms. Many of these operations are quite small and are meant to support a family with excess sold at farmers markets or farm stands. Still others seek to develop small farms into highly profitable operations focused on niche and local agricultural products. Joel Salatin, mentioned above, is perhaps one of the most famous small farmers in the country. He was featured in a number of documentaries and is known all over the world for his promotion of the small family farm.

Salatin raises chicken and grass-fed beef on his property using methods long in practice in the past, but that seem innovative in today's high-tech world. He essentially moves his animals around to different pastures so they do not do damage to the grasses on which they feed. By doing this, he is able to produce high-quality grass-feed beef and chickens that sell at prices higher than market value. Thus, he is able to earn an income by producing a high-quality product, while doing minimal damage to the land.

The ability to earn a profit while following environmentally sound agricultural practices is what the small farm movement is all about.

Many universities try to grow their own food on campus or purchase their food from small local farms. At my university, Hofstra University on Long Island, New York, our student garden has supplied some small amounts of food to our food services. However, we are a relatively small urban/suburban campus and we do not have extensive agricultural fields like some other universities. But our food services operations do try to purchase food from local sources first. They work with local distributers to ensure that they are getting food from local Long Island farms. Each summer, the food service operators provide tours for anyone interested in visiting the Long Island farms from which we derive a large portion of our food.

Does your family purchase local food? Do you know the farm from which your food comes? What about your campus? Does your campus food service derive food from gardens or fields on campus? Does your campus food service purchase food from local farms or have a policy to purchase locally when possible?

Locavores

The locavore movement is another component of the small agricultural movement. Locavores are different from vegetarians or vegans in that the type of food that they consume is not important to them. What matters is the location where the food is grown. They seek to consume food that is grown or raised near where they live.

Locavores have opted to focus on local food for a number of reasons. One of the most important reasons that is often given is that the food is fresh. We all know that fresh food tastes better. Who doesn't love a freshly picked apple or tomato? There is something special about how they taste. Another reason that one may choose to be a locavore is that you are opting out of the industrial food system. Local food is often grown by local farmers and consumers can get to know who is producing their food. There is a sense of trust that develops between consumers and farmers in the production process. Locavores also note that they are reducing the impact of greenhouse gases on the planet by reducing the travel time of food to the grocery stores. As we know, grocery stores contain food from all over the world that in some cases is shipped vast distances at considerable cost. Locavores break down this transportation chain and instead focus on what is local. Locavores also note that they are supporting the local economy by choosing to buy and eat local food. Their money is not shipped vast distances and instead is shared locally to promote a local economic system. They are trying to break down the globalized aspect of the modern food system.

There are a number of ways that locavores can obtain food. More and more grocery stores are opting to label locally produced foods and highlighting it within their shelves (Figure 7.8). In addition, the increase in farmers markets and community-sponsored agriculture (which I'll discuss below in more detail) has significantly expanded the opportunities for locavores to purchase locally grown food and other products.

Locavores have also helped to revive the public's interest in food preservation. Canning, pickling, and freezing of farm and garden produce was very popular in the middle of the twentieth century. In recent years, many locavores have looked to these methods to preserve local food for consumption during the cold winter months. A number of community centers offer classes on food preservation and there are a number of new books on the topic that have been published in recent years. Many have become interested in canned and pickled farm and garden produce as a form of small economic development options marketed to locavores in particular communities.

There are some clear challenges to solely eating local food. Most areas do not produce the range of fruits and vegetables that one would expect to find in the modern diet. For example, it would be difficult to include bananas in a locavore diet in Moscow or to include Pacific salmon in a locavore's diet in Atlanta, Georgia. What most locavores try to do is to purchase

Figure 7.8 Everyone likes local food! Here I am in a selfie with canned quince jam made with quince I picked from a tree on my campus. How can you get involved with local food in your community?

as much of their food as possible locally. Many recognize that they cannot get everything they want from local sources.

Locavores can also be challenged by the realities of the season or environment. The Inuit of the Arctic regions of the world developed a distinct diet of local food around fish and marine mammals in a very challenging setting. Today, while these local foods are still part of their native diet, many Inuit have opted to purchase food on the globalized food market. However, in many more temperate areas of the world, there are more varieties of meats, fruits, and vegetables that can be found locally. But there are challenges. While finding a variety of food is not as daunting as the situation of the Inuit, some find it difficult to buy all their food from local sources.

There is some disagreement whether locally produced food counts as local food or if it has to be locally grown. For example, a local, privately owned bakery may produce wonderful local baked goods using materials purchased in the global food market. The flour may come from Canada, the eggs from Florida, the milk from California, the yeast from Chicago and the fruit from Chile. Is this food local? Some would argue yes, because it is locally made with the bulk of the profits from its manufacture going to local individuals. Others would argue that it is not a local food because it is made from materials not locally grown. What do you think?

Farm to table

Another interesting advance in the local food movement is the growth of farm to table offerings in restaurants. Due to the public's interest in sustainable food and locally derived food sources, restaurants have focused on developing farm to table initiatives. In some cases, restaurants grow their own food on their own farms, while in other cases, they develop a relationship with local farmers.

In some cases, chefs and farmers work together to develop local food that is interesting to their customers. The farmers in some way become artisans by developing unique crops that chefs can highlight in menus. In addition, the farmers work to derive a variety of crops that would be needed by a single restaurant.

This farm to table movement has achieved some degree of economic success. All over the world, the farm to table restaurant movement has been met with acclaim by food critics and customers. However, a new farm to table movement, called the midnight supper club, is on the rise.

A midnight supper club is an event sponsored by a chef or home cook in their home or other area for a small number of guests. While in many instances the chef will feature local food, in other instances, they will feature unique or special menus. In other words, a midnight supper club does not have to always feature local food, but they tend to.

Midnight supper clubs often don't have the required licenses to be an official restaurant and many of them operate secretly. In many instances, they are by invite only. Typically, a chef will put word out that he or she is cooking a multi-course meal and send out word as to the cost of the meal. There will be limited seating (typically ten or less). Midnight supper clubs are rather popular in large cities like New York, Paris, and Tokyo. However, they are expanding all over the world as the interest in food has increased.

Have you ever been to a midnight supper club? Would you ever host a midnight supper club event at your home focused on local food?

Farm to table in rural France

French cuisine is one of the most highly regarded in the world. It is known for its delicious sauces and scrumptious methods of preparing meat, vegetables, and desserts. Years ago I encountered a small *salon de the* in Aubazine (near Brive la Gallard in south central France near the Dordogne valley) called *La Maison de Leontine* that specializes in local French cuisine. I've never forgotten the meals that I ate there as they are so distinctly of that particular region of France.

A *salon de the* is a small restaurant open in the late afternoon. Run by Pierre Guyomard and Simon Pittaway, *La Maison de Leontine* is also open for small groups of diners in the tradition of a *table d'hote*. For a fee, Pierre, the chef of the operation, prepares a multi-course meal featuring local food cooked in the classic French tradition. Local wines and aperitifs are served with the meal.

The region is known for a variety of agricultural products, including walnuts, apples, duck, pork, and a range of fresh vegetables. Meals at *La Maison de Leontine* usually include some variety of these locally derived products (including sips of locally derived liquors made of herbs and walnuts). Pierre's meals usually come with a story about where the food comes from and how it is made. He knows the butcher, the baker, and the farmers where he buys his food. He also preserves jam and preserves fresh food for use in the winter months.

While *La Maison de Leontine* is not as large as a typical restaurant featuring local food or as funky as midnight supper club, it does preserve local food traditions in a unique setting.

Community sponsored agriculture

Community sponsored agriculture (CSA) is another trend in the sustainable food movement that has gained significant attention in recent years. It started in the second half of the twentieth century in Japan and expanded to North America and all over the world in recent years. A CSA farm is a subscription farm whereby consumers purchase farm shares. All of the produce grown on the farm is shared with the shareholders. On average, a typical CSA farm has about 100 shareholders.

CSA farms are often run very simply. They typically have a paid farmer or farm manager and a number of volunteer or part-time staff. Some of these farms are organic, but most are conventional farms. There is a high cost to getting officially certified as an organic farm. Many CSA farms, to limit their costs, practice organic farming, but do not pay for the certification. Thus, even though they may be organic in practice, they are still considered conventional farms because they did not pursue a certification.

What is interesting about community sponsored agriculture is that they provide far more than just food to subscribers. Most offer a variety of other activities for their members. For example, one can take classes on a wide variety of subjects from preserving or cooking food to building a backyard garden. Many farms also offer their space for rent for special events and provide educational tours of their facilities. Some also host special concerts, lectures, flea markets, or fund-raising events to supplement the income from the membership fees.

In other words, CSA farms often are informal community centers for people interested in local food and alternative agriculture. They are not just farms, but sources of information and community connection. They also help to diffuse the CSA model to other farms. Research has shown that CSAs are often located in or near cities with universities or colleges. They tend to be popular with highly educated people. Indeed, college towns often support more than one CSA.

CSA membership usually costs several hundred dollars a year for a full year-long membership. This is a one-time fee that is paid at the start of the year. Produce is usually available to pick up once or twice a week from the farms or from pick-up locations more convenient to members. Some CSAs offer a delivery service. The produce available changes throughout the year. In the early part of the growing season, spring crops like lettuce and peas are common within the weekly shares. Depending on climate, as the season progresses into summer, crops like tomatoes, beans, cucumbers, and squash arrive. In the fall, autumn crops like potatoes, winter squash, and fall greens become available. Regardless, there is a progression of available crops throughout the year and each week is a surprise for members. That is part of the reason that many like CSA memberships. They are confronted with new types of fresh vegetables each week and they get to try new recipes on their friends and families.

Some CSAs provide much more than vegetables. Some have orchards or vineyards and offer a variety of fruits and nuts throughout the year. Others offer bee products like honey and beeswax candles. Some even offer meat and fish products that they raise on the site.

Will Allen's good food revolution

When Will Allen moved to Milwaukee in the 1970s, he never anticipated becoming one of the most famous urban farmers in the world (Figure 7.9). Allen started his professional life as a basketball player in Europe after playing college ball at the University of Miami. As a young African American man growing up in rural America, he had strong roots with the land. He was not part of the twentieth century urban diaspora of African Americans from the rural south to urban areas of the north. While some of his relatives became caught up in this broad cultural shift, his family moved to the north to work a small farm in Maryland.

Figure 7.9 Will Allen dedicating a student garden on the Hofstra University campus.

However, once he was in college, he lost contact with the land and became much more interested in pursuing a professional career. After his time in Europe, where he saw how strongly agriculture was integrated into the urban landscapes of many cities, he moved to Wisconsin to work as an executive manager for a number of different companies, including Kentucky Fried Chicken. However, his agricultural roots called to him and he started to farm on the outskirts of Milwaukee, Wisconsin.

He grew crops that would be of interest to those purchasing produce at farmers markets and roadside stands. He also grew crops that he thought would be of interest to a variety of different ethnic groups. With time, he became concerned about the lack of fresh food in many urban areas of Milwaukee and about the lack of connection between people and the land. He believed that re-awakening agriculture interest could help to build community connections and return people to a concern about the environment.

He developed a number of innovative agricultural operations on an old greenhouse property and parking lot in Milwaukee. He turned brewery waste into soil; he used agricultural wastes to feed fish in aquaculture tanks, and he developed innovative compost systems to help heal the land. Through it all, he focused on providing fresh healthy food to consumers that didn't have access to food. He also helped the communities by providing jobs, training, and internship opportunities.

His efforts have been transformative. He demonstrated that abandoned urban property can be transformed into agricultural operations that can help to feed a community. Will has retired from managing the farm he started. However, the dream of sustainable agriculture continues.

Membership costs may be out of reach for some. Yet many CSAs are aware of the need for fresh food for people with low incomes. To address this issue, CSAs offer reduced or free memberships in exchange for volunteering on the farm.

Community gardens

In many cities, urban dwellers, particularly those living in apartments or condominiums, do not have access to property to grow garden crops. As a result, many cities have set aside land for community gardens. This is an older concept that has been around for many decades if not centuries.

In our modern era, community gardens are usually managed by a civic organization or local government. Each year, for a small fee, local residents may rent a small plot of land on which they can grow crops of their choice. Garden rules vary, but typically there are rules about the types of fertilizers and pesticides that can be used. Some community gardens try to use only organic materials. In addition, there are usually rules that require that the garden must be maintained regularly or else the plot is taken away. In some cases, garden tools and watering hoses are shared.

Garden size varies considerably from just a small footprint to rather large allotments that are more akin to a small hobby farm. These large gardens are rather rare. Most community garden allotments are just a few meters by a few meters in size.

Most point to the 1970s as the advent and expansion of the modern community garden movement. For example, the oldest community gardens in most cities of the United States date to the early 1970s.

Community gardens start for a number of reasons. Sometimes, it is a group of serious gardeners seeking to find land on which to grow crops. In other cases, the initiative is government driven in that the leaders of a local community seek to provide a service to residents.

Farmers' markets

Farmers' markets are another traditional approach to agriculture that has seen significant growth in recent years (Figure 7.10). While many of us like to buy local or grow our own

Figure 7.10 Farmers' markets, like this one in France, often showcase local food and are a great way to support local food producers.

food, this can be difficult if we do not have access to gardens or local food sources. This is where the farmers market can really help out.

Farmers' markets are regularly scheduled time periods where food producers can bring their produce to a designated spot in a community for sale. There are many well-known farmers' markets all over the world that draw crowds to purchase high-quality and unusual farm produce, meats, and food products such as jams, honey, and breads.

Farmers' markets are usually run by local governments, non-profit organizations, or volunteers. They charge sellers a small fee for the right to put up a stand. This fee funds the management of the farmers' market operation.

Many farmers' market managers have very clear rules as to what can and cannot be sold in the market. Some require that the food be locally grown (no selling of food purchased in the global food network) and have a particular quality. Some markets allow vendors to sell agricultural products like cheese and honey. Some allow limited vending of craft or home-made items. Farmers' markets commonly occur once a week, often on weekends, with limited hours.

In my local farmers' market in Long Island, I can find a number of seasonal vegetables, fish straight off the boat, locally made cheese, breads, and flowers. It is only open on Saturday morning. When I wake up, I run down to the market to buy all kinds of wonderful things for the week. I usually run into friends and acquaintances who are also seeking to find some seasonal local food treasures.

Everywhere I travel, I find that each place has their unique range of goods available. In Santa Rosa, California, I was blown away by the variety of peaches and plums as well as the array of flowers for sale. In Helsinki, Finland, I encountered a range of pickled fish and caviars prepared in a variety of treatments. What can you find in your local market? How do farmers markets reflect traditional and modern food production in your community?

Beekeeping

Beekeeping has greatly expanded in recent years in urban and suburban areas. For generations, beekeeping was a relatively rural activity. There are two reasons that bees are important for agricultural purposes: they are important pollinators and they produce honey. Apiaries are kept near areas of important crops that need bee pollination to thrive. They are kept near orchards and other fruiting crops where they not only provide the pollination service, but also create a sweet product for sale on the market.

Of course some honey production is located away from farm fields. In such places, the bees are used strictly for honey production. In the American South, one particular honey, Tupelo honey, is produced in the swamps near the Gulf of Mexico. It commands one of the highest prices for honey on the global market. A small network of beekeepers competes for the best location for placing their hives to gain access to the tupelo forests.

In recent years, more and more people have become interested in beekeeping in their communities. At my own university, one medical school professor built a beehive on our campus. He gathered the honey and sold it to support the expenses associated with beekeeping on campus. He taught a number of classes to the local community about beekeeping and he started new hives in a local park. He is but one of many individuals expanding the range of urban and suburban beekeeping throughout the world.

Some have had concerns over the impact of honey bees on the local community. Historically, beekeeping has been banned in cities. Many people are allergic to bee stings and can suffer great distress or even death if they are stung. However, beehives are usually very safe and the bees will usually only sting if they feel threatened. Because of the resurgence of interest in beekeeping, many communities are re-thinking their zoning regulations and allowing beekeeping back into their communities with distinct restrictions. Hives typically have to be a certain distance from neighbors and they also need to be licensed.

The modern beekeeping movement is important because bee populations are declining significantly due to something called colony collapse disorder or CCD. CCD occurs suddenly when a bee colony falls ill and dies. The disorder is most pronounced in North America, but it is known in other parts of the world as well.

There isn't one definitive reason that has been identified for CCD. Some have suggested that it is caused by moving hives around to farm fields for use in pollination. A very large number of bees are moved around the country by beekeepers to pollinate fields and orchards as they come into flower. There has been concern over the stress placed onto the colonies due to the transportation and constantly changing geography.

Some studies have suggested that common mites and parasites that plague beehives are causing the problem. But others suggest that environmental stresses, such as changing climate or changing land use are to blame. Some think that a range of these problems is leading to overall colony stresses and thus their collapse.

However, there is emerging research that suggests that the problem is caused by the use of neonicotinoid pesticides. The European Union recently noted that this class of pesticides poses a risk to bee populations and banned the use of some neonicotinoids. It has been shown that bees exposed to relatively small amounts of the neonicotinoids have difficulty in returning to a hive.

Regardless of the cause, there is no doubt that bee colonies in North America and many other parts of the world are undergoing stress and there has been a striking reduction in their population. The problem is so severe that there is concern over the long-term viability of some crops. There is a shortage of bees available for pollination of fields. This is one of the reasons that there is so much interest in urban and suburban beekeeping.

Many are also interested in beekeeping because they believe that raw honey from local sources reduces the impact of allergies. This has not been scientifically proven, but many believe that it works wonders at reducing a variety of allergy symptoms.

Regardless of the benefits, local beekeeping has grown in recent years and is likely to expand throughout urban and suburban areas around the world.

The urban chicken movement

One of the newest movements toward local food production is the urban chicken movement. Many people living in urban or suburban areas are raising chickens for egg and meat production. This was unheard of in much of the western world until recently. Chickens are seen as loud and smelly creatures that are not desirable within close living situations. However, that viewpoint is changing quickly.

Many small urban farmers have worked with local governments to change zoning and land use rules to allow chickens back into the city. The new rules require a limited number of chickens and often ban the loud roosters who make so much noise at sunrise.

St. Louis, Missouri, for example, allows homeowners to have four backyard chickens. Tampa, Florida allows one chicken per 1000 square feet of land. The coop also has to be set away from a neighboring dwelling. There are hundreds of communities across the United States that have adopted chicken keeping laws to advance the ability of homeowners to gain access to fresh eggs and to provide a source of meat. While there have been concerns over smell or reduction of property values, these problems have not seemed to be serious issues.

Guerilla gardening, freegans, and other radical approaches to food

There are a number of other more radical approaches to worldwide food problems. Many of us see food as a commodity that can be bought and sold. Many earn their living in growing, transporting, selling, and preparing food for consumption. However, many people also see food as a basic human right—especially when there are huge amounts of food waste and excess production in many parts of the world.

As a child, I grew up hunting and fishing for our family's proteins and gathering nuts and berries in the woods for pies, jams, and wine. We also collected mushrooms and any other thing we could eat or utilize. We carried on a long tradition of hunting and gathering.

Our diets, however, were supplemented by regular trips to the grocery store and by a very extensive garden.

Today, most people are far removed from these activities and rely heavily on purchasing the vast majority of food. Yet, some are returning to hunting and gathering food and are re-learning some of these lost arts.

Plus, a number of food movements have emerged in recent years in reaction to the view of food as a basic human right. These approaches include guerilla gardening and freeganism.

Many are involved with guerilla gardening, which is a movement focused on planting gardens on unsightly unused public or private property. Many guerilla gardens focus on planting edible plants or try to improve the look of the land by planting attractive gardens.

Around the world, there are guerilla gardening clubs and individuals who seek to improve the visual impact of their communities by taking abandoned or unattractive property and turning them into areas that are full of beautiful plants. Some of the garden plots are small – the size of a median or a small unplanted area near a sidewalk. However, some guerilla gardeners have taken on large strips of land where they plant extensive gardens.

The guerilla garden movement has met with mixed success. Some property owners and local governments welcome the improvements, while others have taken out gardens and cited the gardeners for trespassing. In addition, guerilla gardens are often placed in inhospitable areas with poor soils and limited water supplies. Thus, in some cases, guerilla gardens need special long-term maintenance.

Food and sustainability in Serifos

Food and sustainability in the Greek Islands is a challenge in the same way that it is for many of the world's island tourist destinations. Serifos, an Island in the western Aegean Cyclades group, is no exception.

As you can imagine, life on a Greek island is lovely for tourists, but a sustainability challenge for locals (the first photo in this chapter showed food in Serifos). It is difficult to sustain a modern life when there are limited resources. One has to make do with the materials, goods, and services available on the island or one has to import them at great cost.

Food, of course, is a huge issue for island communities.

Serifos has a dry Mediterranean climate. This means that it has warm and dry summers and cool and wet winters. American readers and travelers would find the climate north of Los Angeles a fair comparison to that of many of the Greek Islands.

If you are a lover of olives, lemons, capers, and wine, you will be able to grow all that you need in abundance. Sheep also do well in this climate. Your diet will be limited to the types of things you see on a traditional Greek menu. Items may include a Greek salad with olives, lemon chicken, and roasted lamb. These are delicious dishes when you have options, but they would get boring if they were all you had to eat. If you

(Continued)

wanted apples, bananas, peas, or broccoli, you would need to import them along with your beef.

During the tourist season in the summer, imports come in regularly via ferry from Athens. Transport trucks load up with produce in Athens' main port of Piraeus and are driven onto ferries to supply the many restaurants that feed the tourists and the stores that cater to the local community.

In the winter, when the population drops to 1600 and when the Aegean Sea is rough, deliveries are much less common. Many rely on their gardens, small olive and fruit orchards, sheep, and chickens for their food. They also preserve produce and meats. Salted capers, canned jams, cured olives, and salted fish are regular parts of the traditional winter diet.

As we saw in the previous chapter, water resources are also crucial for human sustainability, but on islands, water supply is truly a matter of life and death. Water management issues are especially problematic for sunny vacation islands like Serifos because flocks of tourists come during the dry sunny times and utilize water resources just when they are most stretched within the natural hydrologic system.

In 2004, the Greek government built a unique double dam that captured water flowing from a single stream that split into two valleys. It collected water that flows in the winter in one of Serifos' few major intermittent streams. Water is stored for use in the summer when water consumption peaks with the tourist season.

In the past, homes in Serifos were built with cisterns that collected rainwater from buildings. There were also wells that collected groundwater. The new dam now makes the use of wells and cisterns unnecessary in many areas of the island.

Like many tourist islands, Serifos is a vulnerable place. It has to import all of its energy and manufactured products. If tourists do not come, residents are not able to earn money to bring in supplies and services to the island that will allow them to thrive during the down season. Tourism is thus a double-edged sword from a sustainability and economic development standpoint for some regions of the world like Serifos. On one hand, tourism provides economic growth and expanded wealth. On the other hand, tourism strains natural resources and makes the island more vulnerable to environmental and social problems.

Think about the tourist areas you visited. How has tourism improved the lives of people? How has tourism strained their resources? Do you think that tourism makes local people more vulnerable to environmental, economic, and social problems?

Freegans are people who gather edible food that is thrown away. There is a tremendous amount of waste food present in the modern food system. For example, many restaurants or grocery stores throw out food that is not used within relatively narrow time windows. Freegans capitalize on this waste by collecting it for their own consumption or by sharing it with others. While it may sound gross to eat food that is discarded, freegans are able to utilize excess food so that it does not enter landfills.

Freeganism became a "thing" in the 1990s, in part as a critique of the modern corporatized food system. Many started regular collection of waste food for distribution in soup kitchens and homeless shelters. For example, an organization called Food Not Bombs focuses on the collection of excess food from a variety of sources for the distribution to the hungry. They serve vegan or vegetarian meals in public places in order to highlight problems associated with food production and the overall tragedy of homelessness.

While Food Not Bombs is a strong organization focused on ensuring that food that would normally go to waste is used, many other organizations and individuals are active in the freegan movement. Freegan organizations collect huge amounts of food around the world for distribution to the hungry. Could you find food in your location by searching the waste stream of restaurants, grocery stores, and bakeries?

8

Green Building

Buildings are part of our cultural landscape. They are an expression of who we are as a people and are reflective of our interests, our technology, and our values. They are also functional dwellings, commercial spaces, factories, and schools. While they tend to be relatively temporary features of the landscape, they have a major impact on the environment. Where and how they are built varies considerably from place to place and some buildings are more sustainable than others.

Take, for instance, the small cabin my father built in northern Wisconsin many years ago (Figure 8.1). It was built of wood he bought from a local mill and he utilized a number of recycled wood and metal building materials in its construction. While not an officially designated green building, he utilized a number of the tenets we now recognize as green in its construction. If we compare this cabin with many modern buildings, we would find that the impact per square foot or the impact per person is much higher in most of our modern buildings. However, this is changing and there is greater interest in green design, architecture, and construction.

The broad field of green building and community design has received a great deal of attention in the last decade among those interested in sustainability. This chapter, and the next one on transportation and community design, highlight some of the key elements associated with this phenomenon. This chapter focuses on green building rating systems, green site selection, green building design and construction, and the technology that goes into building. As we will see, buildings have a great deal of interaction with the environment (Table 8.1).

LEED rating systems

Many have heard of the term "LEED" as it is applied to green buildings. The acronym stands for Leadership in Energy and Environmental Design. It was developed by the United States Green Building Council in the 1990s and has gone through a number of modifications over the years.

LEED rated buildings are built for three main reasons (Figure 8.2). First, many building owners wish to build green because they believe in trying to live as sustainably as possible. Through the rating system, green buildings provide peace of mind for the owners and tenants that they are doing what is environmentally appropriate. Second, many choose to

Introduction to Sustainability, Second Edition. Robert Brinkmann.

Companion website: www.wiley.com/go/Brinkmann/IntroductiontoSustainability

Figure 8.1 My father and brother putting up a wall on our cabin in the 1970s. Many of the materials that were used to build the cabin were recycled.

Table 8.1 Issues associated with buildings and the environment.

Planning Stakeholder assessments (for large projects) Green design Green certification Site selection Passive design Green construction Materials Foundations Renewable energy (solar, wind, hydropower, biofuels)	Energy conservation (appliances, heating, cooling) Landscaping Green roofs Organic building materials (straw bales, wool insulation) Green refurbishment Sustainable building services Heating Cooling Insulation	Venting Electricity Water Water treatment Water use reduction (toilets urinals, sinks, showers, landscaping) Hot water Building drainage Sewage treatment Wastewater management Building management systems and other smart technology	Carbon reduction Sustainable project management Project benchmarking Building lifecycle management Green demolition Historic preservation Facilities management (cleaning, food, travel, recycling, space use, information technology)

build green buildings because they are required to by local laws or organizational policy. For example, many universities have policies in place that require them to build LEED certified buildings on campus. Finally, many opt to build green buildings because it provides a competitive edge in business or for attracting tenants. Organizations utilize the LEED name to draw attention to their organization. LEED buildings are still relatively uncommon and locating your business or home in one provides a degree of specialness.

Figure 8.2 This is Hofstra university's first LEED Certified building. It is part of its medical school complex. How many LEED buildings are on your campus?

One of the most important advances of the LEED system is its embracing of different types of construction. Instead of just rating a single type of building, LEED now is able to rate and evaluate new construction, homes, schools, retail space, healthcare buildings, and core and shell construction. LEED also has evaluation tools for interior design, building operation and maintenance, and neighborhood development.

LEED evaluates all projects on a point-based system. Indeed, it is one of the best-known quantitative sustainability measurement tools in the world. While it was started in the United States, the LEED system is used across the world.

Building projects can earn points in a number of different categories. In most instances, 100 points are available over a number of different categories that include site selection, water use, energy and atmospheric health, materials and resources, indoor environmental quality, innovation, and regional priorities. Once points are established for a project, the building is evaluated by individuals approved to certify within the LEED rating system.

The highest rating a building can earn is LEED Platinum (80 points and over). Gold LEED buildings earn 60–79 points, Silver LEED buildings earn 40–59 points, and certified buildings earn 40–49 points. There are thousands of certified buildings all over the world. You may be sitting in one reading this right now! Try to find one in your area and discover how it got its rating.

There is no doubt that in many cases, the costs of building a LEED building are higher than building a conventional building. Specialized architects and designers familiar with the requirements for green buildings are needed in all phases of project development. Plus, some of the materials are more expensive or difficult to find in some areas. Some have questioned whether the costs of building a LEED building are worth the expense.

However, many believe that the long-term savings in energy and water use is worth the up-front investment. Plus, many believe it is just the right thing to do to try to improve the environment and to support innovation in green building.

Site selection

One of the most significant ways that we can impact the environment is by choosing where to build structures. Take, for instance, the case of the mass migration of people from the northern part of the United States to the Sunbelt states that extend from California to Florida. The state of Florida doubled in population roughly every 20 years in the last century. This migration caused wide expansive development across the state. Florida could have developed high-density communities, but instead opted to construct low-density suburbs that extend throughout much of peninsular Florida.

The US Environmental Protection Agency (EPA) suggests that anyone building a new structure utilize elements of smart growth. We will spend more time discussing smart growth in an upcoming chapter, but for now it's worth noting that the EPA encourages building buyers or builders to consider the following questions when selecting a property (http://www.epa.gov/greenhomes/HomeLocation.htm):

- Is the community designed with people in mind?
- Does the community provide adequate transportation choices?
- Are services and amenities within reach?
- Can you work, live, and play in the neighborhood?
- Has the land been previously developed?
- Does the neighborhood design preserve open space?

If you think about community design throughout much of the world, you will recognize that there is tremendous variety from place to place. What about in your own community? How would you answer the above questions? See the text box to learn a bit about my answers to the questions.

Brownfield development

Brownfields are pieces of land that may or may not be contaminated due to past land uses. There is an overall perception that the land is somehow problematic due to the actions of previous owners. They are often vacant properties in commercial or industrial areas.

Does your community design meet the needs of residents?

I used to live in Manorhaven, New York. This is a village in Long Island with a population of about 8000 people. It is the most densely populated village in all of the state of New York. Half of the population lives in rental properties. The homes in the village were built on an old sand mining site just a meter or two above sea level in the early- to mid-twentieth century. The flat land allowed the development of homes within a grid-like structure. In recent years, many of the original homes have been destroyed to make room for larger duplexes to house commuters to New York seeking to find a respite from city life.

1 *Is the community designed with people in mind?* Of course all communities are built with people in mind to house them, but the development of Manorhaven took place nearly simultaneously with the development of Fordist (mass production) approaches to the construction of housing. The most iconic Fordist housing development is nearby Levittown, New York. Therefore, the housing in my community was built to provide inexpensive homes to the burgeoning middle class in the twentieth century.
2 *Does the community provide adequate transportation choices?* The community is serviced by a bus line that provides direct access to a major Long Island transit hub. However, like most American suburban communities, most people in my community drive to their destination. What about in your community? How do people where you live get to work or school?

In some cases, brownfields may be found in residential areas due to concern over illegal dumping or past activities on the site.

Some of the past land uses that may be associated with a brownfield site include things like underground tanks from gas stations, buried industrial chemicals, and broad contamination from industrial processes. The activities associated with the possible contamination occurred in the past. However, the property is difficult to re-develop due to the costs associated with clean-up. All of the previous property owners since contamination occurred may be liable for these costs, which can be substantial. It is the cost of clean-up that limits the desire of property owners to initiate a re-development process.

As a result, many brownfield properties end up vacant (Figure 8.3). Property owners do not start a clean-up process because they are afraid of the costs; they just don't know what will be found once they start. Plus, some clean-ups can take years, thereby limiting economic value of the property. Some sites eventually become owned by the public if a property owner stops tax payment on the property. Either way, the property is a drain on public resources since it is not generating property or sales tax.

Brownfield properties have a negative impact on communities. No property owner wants to be next to a vacant property that may be contaminated. Brownfield sites are usually poorly maintained and lead to declining property values in a community. A poorly maintained property can lead to overall neighborhood decline and loss in community economic activity.

In recent years, a number of innovative public and private agencies have worked to develop policies and programs to encourage the clean-up and re-development of brownfield properties. For example, the US Government, along with state governments, has provided special grants and expertise to encourage public–private partnerships to re-develop brownfield sites. These public–private partnerships have grown in popularity in the last two decades because private companies often do not have the resources needed to deal with the complex issues associated with brownfield clean up.

Figure 8.3 Brownfields are sites that are difficult to develop because they may or may not be contaminated.

A great example of a brownfield site that is transforming into a mixed-use center is the Brooklyn Navy Yards in the New York City Borough of Brooklyn. The Navy Yard has served as a major US Navy facility for over 200 years. It was established in 1801 due to its strategic location on the East River across from Manhattan. It was used to build and repair Navy ships until its closure in 1966. From 1969 through to the present, it has been used for a variety of industrial and commercial activities.

The expansive site, covering approximately 200 acres, has been home to a huge variety of industrial activities that were responsible for a range of contamination problems on the site. The state of New York as well as New York City worked with several tenants to clean up much of the site to allow for a new phase of re-development. While there is still some contamination present, the site now hosts a range of green businesses and manufacturers, and a large studio.

While the Brooklyn Navy Yards are perhaps one of the most complex brownfield sites in the world, there are many smaller sites that impact, say, a commercial street corner. These sites may not be as large as the example here, but they do have deleterious effects on communities.

In New York City's notable sustainability plan, PlaNYC, one of the city's goals was to clean up all contaminated sites in the city. This is a major economic undertaking, but with huge economic benefits. Development, and associated tax benefits, will occur faster if property owners can develop their property without fear of finding contamination on the site.

Other aspects of sustainable building siting

Of course, choosing a place to build is much more than ensuring that the building does not impact green, undeveloped properties. The design and selection of the building site impacts the behavior of building users and visitors.

For example, points are earned in the LEED system for building where there is access to mass transit. Ensuring that building residents have access to buses or trains eliminates traffic and congestion as well as pollution from cars. Similarly, building bicycle storage, and changing and showering rooms for bicycle commuters also eliminates the need for cars and parking spaces. Plus, adding electric car charging stations and eliminating traditional parking spaces solidifies the communication of the building as a green, transit-friendly space.

In preparing the site for construction, it is also useful to consider storm water removal and remediation to ensure that the site is not contributing to storm water pollution or excess runoff that can lead to flooding. Along with this, sites can be prepared to try to protect or restore natural habitat and preserve open land.

One of the concerns with urbanization lately is the vast amount of light pollution that is emitted into the sky. This excess light impacts the lives of nocturnal animals and has been proven to cause migratory and sensory difficulties. For example, sea turtles are often distracted by the lights of buildings and make inappropriate spatial decisions because they cannot distinguish the artificial light from moonlight. For this reason, LEED encourages the limiting of light pollution.

Water use

Another important way that LEED buildings earn points is by limiting the water use in the building. Most of us have experienced showers, baths, or toilets that utilize way more water than needed. Many plumbing fixtures, particularly older ones, were designed for comfort, not conservation. However, now, with innovative technological advances, there are a number of important ways that buildings can be designed to reduce water use:

1 *Reduction of water flow for faucets.* In the past, faucets had basic on/off mechanisms that allowed users to turn on water for washing hands or dishes. It is easy to walk away from these types of faucets and leave water running for long periods of time. Now, however, we have a variety of technologies to reduce water use through motion sensing or push button timed water release.
2 *Reduction of water flow for showerheads.* Old showerheads utilize as much 19 liters of water per minute. However, with new technology, low-flow showerheads use less than 9 liters per minute.
3 *Reduction of water for urinals and toilets.* Old toilets utilize over 16 liters of water per flush. New toilet technology significantly reduces toilet water use to 4 liters of water. Low flow or waterless urinal technology is also a key component of large volume buildings such as schools or office buildings. Some buildings have water collection devices to store rainwater for use in toilets or urinals.

4 *Reduction of water for landscaping and other outdoor uses.* Water use for landscaping can account for up to 30% of home energy use. LEED building landscape design seeks to eliminate the use of water for irrigation. In places where that is not possible, irrigation should be kept to a minimum by using native vegetation and drought tolerant landscaping. Buildings can also be designed to collect rainwater for irrigation uses.

There are, of course, specialized building uses that require more water than others. In some cases, interesting innovative technologies have been installed to advance water sustainability in these unique settings. For example, car washes can re-use water multiple times in order to reduce water use.

Energy and atmospheric health

LEED provides credits for reducing air pollution and reducing overall energy use in the building. This is done by ensuring that energy efficient equipment is used in the building, by utilizing green energy, and by limiting the use of refrigerants or ensuring that they are managed appropriately.

In 1992, the US Environmental Protection Administration started a program called Energy Star to try to reduce energy use in the US. They developed a product grading program to provide guidance for consumers seeking to utilize energy efficient products. Energy Star grades a wide variety of products from televisions to heating units. They also rate these products for a wide range of uses including residential, commercial, and industrial settings.

Energy Star seeks to reduce energy use, while also ensuring prime use of the product. In other words, the program seeks to ensure that the product is still able to meet its design needs while also limiting energy use.

According to the EPA, products are selected if they meet the following specific guidelines (http://www.energystar.gov/index.cfm?c=about.ab_index):

- Product categories must contribute significant energy savings nationwide.
- Qualified products must deliver the features and performance demanded by consumers, in addition to increased energy efficiency.
- If the qualified product costs more than a conventional, less-efficient counterpart, purchasers will recover their investment in increased efficiency through utility bill savings, within a reasonable period of time.
- Energy efficiency can be achieved through broadly available, non-proprietary technologies offered by more than one manufacturer.
- Product energy consumption and performance can be measured and verified with testing.
- Labeling would effectively differentiate products and be visible for purchasers.

Since it started, the Energy Star program has saved huge amounts of energy. For example, clothes washers now consume roughly 70% less energy than their older counterparts. Energy Star products are available widely and are typically the first choice of consumers.

The ASHRAE (American Society of Heating, Refrigerating, and Air Conditioning Engineers) also provides clear guidance for green infrastructure in new and existing buildings. They have published a variety of standards that are encouraged in green buildings, particularly related to heating and cooling. They have also provided innovative approaches for managing refrigerants to protect the atmosphere.

Figure 8.4 The energy management system at One Bryant Park, the Bank of America Building. This is the first skyscraper to achieve LEED Platinum status.

One can also earn points in a LEED system by using a building energy management system (Figure 8.4). Such a system provides heating and cooling when the building is utilized. For example, turning down a thermostat in a school at night and over the weekends makes a great deal of sense. However, if one relied on individual teachers or users to remember to turn the heat down at night, the program would be marginally successful due to the busy lives of teachers. A building energy management system allows one to program the exact times to heat and cool space. It can also set times for individual rooms in case there are some rooms that are used at night or on weekends.

Most home thermostats now utilize this energy management capacity by setting times when to heat or cool space. Most of us live very busy lives and spend only a few of our waking hours at home during the week. These systems encourage us to save energy throughout the day while also saving considerable amounts of money.

Buildings can also earn points in the LEED system by using green energy systems. There are a number of ways to do this through the use of passive solar design to promote natural light and heating and cooling, to the use of active solar or wind energy systems. For example, some buildings utilize small solar panel systems to heat the hot water used in buildings, while others have much more aggressive photoelectric and wind energy systems that supply substantial amounts of electrical energy. Geothermal energy systems are also becoming common supplementary heating and cooling elements that are built into the design of new green buildings.

Of course, it is easy to say one is utilizing these green approaches to buildings, but there needs to be some type of verification. For this reason, LEED encourages the measurement and verification of energy consumption so that the building users understand the impact of green technology on the building.

Materials and resources

The types of materials that are used within a building can greatly impact the planet. If you've seen pictures of Versailles or have been lucky enough to visit the French home of their last royal family, you've seen some of the richest design elements and interiors ever utilized during that era (Figure 8.5). In today's globalized world, we can all build our own homes with materials that are lovely, but that have detrimental impacts on societies. We do not recognize that many of us are building our own versions of Versailles in our own neighborhoods. The LEED system strives to measure the use of materials to ensure that they are having limited impact on the environment. One of the downfalls of the French monarchy was its

Figure 8.5 How we design a home has a huge impact on the environment and society. This interior of the Versailles palace outside of Paris shows some of the design excesses that led, in part, to the downfall of the French monarchy.

excesses. In today's society, we cannot build our own Versailles without having undesirable social impacts.

LEED focuses heavily on several aspects: materials reuse, recycled content of construction materials, locally derived materials, rapidly renewable materials and certified sustainable wood, and waste management. Let us take a look at each of these.

Material re-use

When a building is demolished, the easiest thing to do with the waste is to landfill it. It is cheap to haul away materials and forget about them. However, it is not the wisest choice from a sustainability perspective. A considerable amount of demolition debris can be re-used. Bricks, concrete, wood, and plumbing or wiring fixtures can all be re-used or re-purposed.

Many local governments have added strict rules regarding the use of demolition debris in the construction of new buildings. For example, the City of Los Angeles developed a green building code in 2005 that requires building demolition projects to recycle at least 50% of the materials generated (see http://dpw.lacounty.gov/epd/cd/).

Rules like those in Los Angeles have led people to reconsider the demolition process. Instead of thinking of tearing down a building and dealing with the waste as a pile of mixed debris, a better way of thinking about re-using materials is to deconstruct an edifice. This allows one to carefully collect similar materials for appropriate uses. In some cases, the foundation or walls of the building can be re-used in new building construction.

Recycled content of construction material

Another aspect of green building is the use of recycled components in the construction material. For example, counter tops and tiles can be made with a variety of waste materials including glass, mirrors, and ash. Agrifiber, a material that can replace wood in most building settings, is material that is made with waste agricultural products and residues like hulls, husks, and stalks. These materials can replace plywood or cabinet wood. In some cases, ash from burning fossil fuels is used for concrete or cement. Plastics and rubbers can be transformed into a variety of building materials.

The availability and variety of recycled content construction materials has increased in recent years to the point that a home or building can be built largely with re-used and recycled materials. It just takes a bit more time and money to make this happen. However, because of the interest in these materials, the costs are going down or are competitive with traditional materials. In some cases, particularly in cabinetry, cement, and tiles, recycled content is already becoming the norm.

Locally derived materials

Of course, one of the challenges for building buildings is finding suitable supplies. Many of the modern "big box" hardware stores are a study in globalization. The materials are from all over the world. It can be a challenge, therefore, to find locally derived materials at a competitive price.

LEED challenges builders to find materials that are extracted, processed, and manufactured near the building site. Generally, this is considered to be within 500 miles of the construction site. However, some transportation methods are lighter on the environment than others. Therefore, special consideration is given to the type of shipping that is used to deliver materials. Shipping by boat allows one to divide the distance travelled by 15, shipping by rail allows one to divide the distance by three, and shipping by inland waterway allows one to divide by 2. Thus, if you are shipping steel 3000 miles by boat and by truck 200 miles, your actual distance for LEED local certification would be $(3000/15) + 200 = 400$.

Using this standard, the steel purchased from 3000 miles away can be considered a regional material because the mileage impact is 400 miles. This is an interesting perspective in that the locally derived materials focus clearly on the impact of transportation on the construction of the building and not on the local economy or the labor or social issues of the procurement of the materials.

Regardless, in some areas, it is relatively easy to purchase building materials within 500 miles of a construction site. However, in some more remote areas, this could be a challenge.

Renewable materials and certified sustainable wood

Another way to earn points in the LEED building system is to use renewable materials or certified sustainable wood. Renewable materials are generally considered plant-based materials that are grown sustainably or that are made from agricultural or forestry by products. A wide variety of construction materials are made from renewable materials, including insulation, wood-like products, flooring, and fibers.

Certified sustainable wood products are those that are certified by a third-party certifying agency and that meet a number of key criteria. The forests are generally audited to ensure that there is protection of biodiversity, wildlife habitat, and water quality. They are also reviewed to ensure that the harvesting is sustainable and that there is rapid re-forestation. The certification process provides a sense of trust that the wood is derived using sound forest management practices.

The certification process was put in place to try to avoid bad forest management practices that hurt the world's ecosystems. In today's globalized marketplace, there are numerous examples of inappropriate clear cutting of forests and subsequent soil erosion and habitat loss. A large housing project, for example, could devastate a small forest and all of its inhabitants.

There are numerous certifying agencies that not only review forest practices, but also ensure a sound chain of custody to provide assurances that the materials available to builders are derived from well-managed forests. These organizations charge for their services, but provide a way for builders to advance the green building agenda.

Waste management

Reducing construction waste is another important aspect of green building. LEED encourages removing more than 50% of construction waste from the waste stream by recycling it or finding other uses for it.

Summary

Overall, the use of building materials is one of the most important aspects of green building and we have made very significant strides in greening the construction industry. Just in the last several years, a wide range of new materials have entered the marketplace. Most new or renovated buildings contain a variety of recycled or re-used materials. Next time you are in a hardware or appliance store, take a look at the variety of green sustainable materials for sale. Also, next time you are near a construction site, take a look at how they are managing their waste. You'll probably find that their practices are far greener than they were in the past.

Indoor environmental quality

In recent years, there has been considerable interest in the health impacts of working or living in buildings. Some people believe spending time in some buildings has made them sick. This phenomenon, called sick building syndrome, is based on real or perceived issues with air quality. Some air quality problems include well-known pollutants like carbon dioxide or mold. However, a wide range of other pollutants can cause problems. Regardless of the source, the LEED indoor environmental quality credits are highly focused on ensuring that the air and living space are clean and comfortable.

The reason that there is so much concern over indoor air quality is that there are a number of health problems associated with breathing contaminated air. Some may be minor, such as slight allergies or coughing. However, some may be quite problematic and lead to severe cardiovascular problems such as cancer and emphysema. Many of these issues are preventable with correct building design and construction, which is why LEED emphasizes indoor air quality.

Indoor air quality is managed with several categories of the LEED system that can largely be grouped as ventilation and air delivery monitoring, construction indoor air quality management, the use of low-emitting materials, indoor chemical and pollution source control, controllability and design of lighting and temperature systems, and access to daylight. Each of these will be briefly summarized.

Ventilation and air delivery monitoring

One of the key elements of any building design is to ensure that there is solid ventilation and clean and healthy air. Points are earned in the LEED system for sound ventilation and air quality monitoring. While it might seem smart to constantly ventilate a building, the reality is that ventilation is largely needed only when a building is occupied. There are sensors available that monitor CO_2 and air outflow to ensure that ventilation occurs when needed. In addition, building occupants can also turn on ventilation systems when they wish. The use of constant ventilation systems leads to greater air conditioning and heating energy costs. That is why the sensor-based systems of ventilation and air delivery are the standard.

Construction indoor air quality management

Construction can be a messy business. Large amounts of dust are produced and hazardous chemical vapors are often released by glues and paints. That is why it is important to ensure that the construction process does not cause any health problems for workers or users of a building. LEED rating systems require that a construction indoor air quality management plan be in place.

The plan ensures that any potentially hazardous processes are conducted at times when other workers are not present or that they occur in separate, well-ventilated areas of the construction site. The plan also must take into consideration the type of safety equipment that might be needed for workers, such as dust masks or filters.

Use of low-emitting materials

Of particular interest to those interested in indoor air quality is the use of low-emitting materials in the construction or styling of the building. Many materials can emit dangerous gases to the atmosphere that can impact workers or future building occupants. Building materials and supplies such as paint, caulk, sealants, composite wood, and glues all can emit volatile organic compounds (VOCs). Many VOCs have been linked with a number of long-term health problems.

Furnishings, such as carpets, drapes, and furniture can also emit VOCs (Figure 8.6). In recent years there has been a concerted effort to try to reduce the use of VOCs in a variety of construction materials, supplies, and furnishings. It is now rather easy to find them and they are quickly becoming the industry standard.

Figure 8.6 Some carpets are more sustainable than others.

Indoor chemical and pollution source control

This LEED credit focuses on trying to keep pollution from entering the building and trying to keep pollution sources in buildings under control. To keep outdoor pollution out of the building, credits are given for long matting systems to eliminate dust and sediment on shoes and clothes. Janitorial storage areas and other chemical storage areas must be vented and have doors that close automatically to limit contamination.

Controllability and design of lighting and temperature systems

One thing that all building users have in common is that we never agree on lighting and temperature. One person may be cold and one warm. One may like abundant artificial light and another not. This LEED point system allows for variable needs, while encouraging conservation of energy. For example, the optimal condition in a lighting system is to turn all lights off at night automatically, while allowing individuals who are working late to turn on selected lights for safety and individual work.

Obviously in large rooms with shared space by many people, there is a limit to the individualization of lighting and temperatures. However, LEED encourages room thermostats and lighting controls that can supersede energy management systems.

Access to daylight

If you work in a large building, one of the perks as one rises in rank is access to better views and greater natural light. LEED seeks to democratize access to views and light by encouraging greater access to both of these things. Points are given if a building is designed to ensure that 75% of the regularly occupied spaces have access to sunlight and if 90% of the regularly occupied spaces have access to views. Thus, the design of the building with abundant window space along with building siting are significant issues that must be addressed in order to earn these credits.

Summary

Overall, these areas focus attention on ensuring that there is a healthy indoor environment for users of the building. Years ago I had an office in a poorly ventilated building without windows. LEED indoor environment credits ensure that buildings like that are not constructed in the future.

Innovation

The innovation credit for green buildings is fully up to the developer to assess what is innovative for the particular place and time. Each green building is unique and the innovation credit could focus on a myriad of issues from construction materials to heating or cooling. Often, the innovation credit is given for advancing new ideas within green building. For example, there has recently been great attention on innovative ways to build solar and green roofs and walls into buildings. Credits can be given for new ways of approaching these themes.

Innovation credits can also be given for unique ways that users interact with the building. For example, a building that utilizes a number of heat generating computer servers could collect the heat for re-use somewhere else or a community center could focus on interacting with and enhancing local transit options. The most important aspect of this credit is that there are unique opportunities to address the specific issues associated with the building, occupants, activities, or situation that can facilitate sustainable development within the realm of the space.

Regional priorities

LEED regional priorities vary from place to place, depending on the particular needs of the region. In Long Island, where I live, the main ways to earn regional credits are by optimizing energy performance of buildings, by creating on-site renewable energy, and by creating public transportation access. In suburban Florida, similar goals are in place along with building connectivity and density in the low-density suburban landscape. In contrast, the main regional priorities in India are creating on-site renewable energy and increasing the recycled content of buildings.

Expansion of green building technology

The focus on green building has significantly enhanced the overall technology available to builders and building renovators. It was not that long ago that low-flow shower faucets, energy saving windows, or motion-activated faucets were invented. We now expect green technology to be in place throughout the built environment. It seems odd to us when we encounter some of the old full-flow showers or sink faucets that are artifacts of a more wasteful time.

One of the benefits of the LEED system is that it normalizes these innovative technologies and promotes innovation. Things that seemed odd a few years ago, such as toilets that use grey water or xeriscaped (the use of locally adapted plants—particularly drought tolerant plants) landscapes are now common. Green high-tech innovations that were rare or unknown a decade ago are now used in most appliances, and heating and cooling systems.

Some of the more challenging innovations, such as green roofs and walls, are a bit more difficult to advance into mainstream buildings due to some of the maintenance issues associated with them. They often require special and expensive plumbing and costly maintenance that some building owners do not want to take on as part of the building's overall budget (Figure 8.7). Nevertheless, these innovative and striking projects do catch the eye and the imagination and provide inspiration to those who interact with the building space.

Other green building rating systems

There are a number of different types of building rating systems in use throughout the world similar to the LEED rating system.

Figure 8.7 This green wall in Paris is beautiful, inspirational, and environmentally sound.

BREEAM

In the United Kingdom over 250 000 buildings have been certified using BREEAM (Building Research Establishment Environment Assessment Method) methodology since 1990. BREEAM is similar to the LEED System. Indeed, the LEED system evolved from BREEAM. Today, it is the main assessment tool in several European countries, including the UK, Norway, and the Netherlands. Individualized sustainable building rating systems have been developed for the United Kingdom, Germany, the Netherlands, Norway, Spain, Sweden, and Austria. There is also a broad international rating scheme.

According to the BREEAM Website (http://www.breeam.org/about.jsp?id=66), the benefits of building a BREEAM Building are (direct quote):

- Market recognition for low environmental impact buildings
- Confidence that tried and tested environmental practice is incorporated in the building
- Inspiration to find innovative solutions that minimize the environmental impact
- A benchmark that is higher than regulation
- A system to help reduce running costs, improve working and living environments
- A standard that demonstrates progress toward corporate and organizational environmental objectives.

Like LEED, BREEAM provides a number of different rating systems for distinct situations: new construction, refurbishment, communities, in-use, and sustainable homes.

BREEAM new construction

BREEAM for new construction evaluates building development within nine broad themes: management, health and wellbeing, energy, transportation, water, materials, waste, land use and ecology, and pollution. Each of these themes in some way addresses issues of environmental, social, or economic sustainability. Each theme is subdivided into subcategories, such as responsible sourcing of materials, to earn points.

BREAAM refurbishment

A BREEAM measurement tool for refurbishment (or renovation) exists for both domestic and non-domestic buildings. The renovation of buildings provides opportunities to make significant improvements in building sustainability. The building can become much more energy efficient, thereby decreasing fuel costs for owners and tenants. Renovation can also make the building safer from fire, flood, and unwanted egress. Furthermore, renovations can improve the indoor air quality of the building, thereby improving the health of residents and workers. Points are earned for refurbishment in areas of energy, water, materials, pollution, health and wellbeing, waste, management, and innovation. Each of these categories is broken down into more detailed ways to earn points. However, there are minimum standards in safety, energy efficiency, water use, responsible sourcing of materials, and flood reduction that must be met.

BREAAM communities

BREAAM for communities is used for new developments. It focuses on infusing sustainability policies and practices within the initial planning process to ensure that appropriate policies and procedures are in place to promote social, environmental, and economic sustainability of the community. The scheme provides a framework for working out a holistic planning process from inception to construction that brings together a range of stakeholders to ensure that all involved with decision-making in the community are following appropriate and collaborative processes.

The focus of BREEAM communities is to infuse economic, social, and environmental sustainability at all levels of a community development. Forty distinct issues are addressed in the certification process, which include things like transportation access, housing, and pollution. The City Council of Bristol in the UK requires a BREEAM communities assessment on all new significant developments. The scheme is not only used in the UK but also in Sweden, Norway, Belgium, and Turkey.

BREEAM in-use

BREEAM for in-use buildings is an assessment tool for buildings that are already occupied and in use. It provides a way to make improvements in the sustainability of a building without a full renovation by assessing day-to-day sustainability issues that are in play throughout the life of the building. The assessment process has three main parts. The first part focuses on assessing and improving the performance of the building, given the existing conditions of the space. The second part looks at building management. Here, issues associated with the operation of the building are addressed, including energy and water use, waste generation, and assessment of other specific consumable materials. The third part of the assessment centers on the way the occupier manages the building policies to engage staff and get results on established goals.

BREEAM sustainable homes

BREEAM for sustainable homes is part of a broader UK initiative to enhance the development of green buildings within the country. It is similar to the BREEAM for new buildings criteria. Points are earned in the areas of energy, transportation, pollution, materials, water, land use and ecology, health and wellbeing, and management. Within each category, points can be earned in these areas:

- *Energy*: dwelling emission rate, building fabric, drying space, ecolabelled goods, internal lighting, and external lighting
- *Transportation*: public transportation, bicycle storage, local amenities, and home office
- *Pollution:* NO_x emissions, reduction of surface runoff, renewable and low emission energy sources, and flood risk
- *Materials:* environmental impact of materials, responsible sourcing of basic building materials, responsible sourcing of finishing elements materials, and recycling facilities.
- *Water:* internal potable water use and external potable water use
- *Land use and ecology:* ecological value of site, ecological enhancement, protection of ecological features, change of ecological value of site, and building footprint
- *Health:* daylighting, sound insulation, and private space
- *Management*: home users guide, considerate constructors, construction site impacts, and security.

For all of the BREEAM assessment schemes (new construction, refurbishment, communities, in-use, and sustainable homes), evaluations are made by third party experts in the system. Once completed and approved, the building attains BREEAM status and can use the listing in promotional materials. This is one of the more interesting aspects about green buildings. They are very popular with the public. People like the idea of working and living in green buildings. In both the LEED and BREEAM literature, the positive marketing aspects of their labels are highlighted. The certification helps to set the building apart from others and can be used as a promotional and marketing tool. The green appellation granted by these certifying agencies not only helps the environment, but helps to brand particular developments, buildings, governments, businesses, and non-profits by the choices they make about the use of their space.

PassivHaus

Germany is one of the world's leaders in alternative energy production and it is no surprise that many in this nation focus on finding ways to significantly reduce energy consumption in buildings. A rating scheme, called PassivHaus, was introduced in the 1990s. This system focuses exclusively on the energy use of the buildings, not the other broad categories as defined by LEED and BREEAM.

PassivHaus makes use of thoughtful building design to create a space that is extremely energy efficient. This is done by ensuring high standards of building insulation and creating a very air-tight envelope around the building to prevent leaking air. It also seeks to ensure that there is good indoor air quality through appropriate ventilation while recovering heat. One of the key elements of PassivHaus success is passive design, which uses building design to take advantage of the site to reduce energy consumption. Windows are placed strategically to draw in or reduce sunlight, depending on the need for cooling or heating, as well as lighting and shade.

In many parts of the world, summertime solar heating through windows is a significant problem that leads to high energy consumption through air conditioning use. Buildings can be designed that enhance natural ventilation through windows or vents. However, buildings can also be sited to reduce the impacts of afternoon lighting. External window shades or buffers can also limit the impact of the sun's rays, as can appropriate landscaping.

Another aspect of passive design is the smart use of expanded wall, floor, or roof space to create a thermal mass. Thick walls made of concrete, stone, or brick will absorb heat and store it for long periods of time without it entering or leaving the building. What that means is that the building materials provide a way of buffering the impact of the temperature differentials from the building to the outside. When it is very cold externally, the heat from the internal areas of the building becomes stored in the walls and limits the expansion of cold air into the building space. Likewise, when it is very warm outside, the heat from external areas is stored in the walls of the building.

This approach of storing and trapping heat within walls is used to great effect in warm arid areas. Here, buildings are built of thick walls to trap and store heat from the burning desert sun. At night, when temperatures drop and it becomes very cold in the dry desert air, the heat is released from the walls into the building. This effectively reduces the need for internal heating and cooling in homes. This traditional form of thick-walled building construction is a common indigenous practice in arid areas throughout the world. It is coming into greater favor in the green building community to try to reduce energy use for heating and cooling.

Green building policy

Many parts of the world are requiring communities to adhere to green building policy. Some are very specific about their approaches and require a particular standard, such LEED or BREEAM. As noted earlier, the town of Bristol in the United Kingdom requires BREEAM for community development for major developments. Others may pick and choose some aspect of green building, such as the use of energy efficient appliances or insulation, to insert within their building code.

Many universities are making it a standard policy to use some type of green standard in all buildings. At my campus, Hofstra University, there is a policy to ensure that all new buildings will be built to at least LEED Silver standards. The policies by campuses and communities help to advance technology and infuse sustainability principles within the broader culture of their region.

There are also many professional associations and green construction and design businesses that promote sustainability and green building policies. They seek to advance the field of green building in order to advance technology and often advocate that governments develop strong green building codes. Many of these organizations and businesses also promote broad education and training on sustainability in the building and real estate business. Do an online search for "green building meetings" in your community and see what opportunities there are for you to interact with the green building community. Often, these groups support networking and mixer events specifically for students in order to assist them in achieving educational goals and finding employment opportunities in the green building and real estate business.

Critiques of green building

It is important to note that there are many critiques of the green building movement.

Many argue that the classification schemes do not take into account the overall use of the building (Figure 8.8). For example, the headquarters of a major polluting industry could build a green building where they manage their polluting operations within a very green space. One could also build a green jail, slaughterhouse, or chemical weapons facility. Those in support of the green building movement argue that the building would have built anyway, so why not build it in the greenest way possible. They believe that it is not the architect or the designer's responsibility to evaluate the use of the building.

Plus, unsavory individuals like despots or robber barons as well as unsavory organizations could build certified green buildings. Because most major certification agencies focus on the benefits of marketing green buildings to stakeholders and clients, some believe that the green building community has an obligation to evaluate the ethical dimensions of the owner or the occupants. Again, as noted above, most in the green building community would argue that their role is to build and design green buildings and not to evaluate their use.

Many schemes do not take into consideration the overall per occupant footprint. The richest person in the world could build the biggest building in the world. He or she could

Figure 8.8 The headquarters of the National Cave and Karst Research Institute was built with the greenest technology available and includes the world's only intentional bat roost.

live in the house alone. And it could be a certified green building with a huge per person footprint. Is it truly a green building when occupants do not use the building efficiently?

Others argue that the cost of certifying a green building is prohibitive. Many builders state that they are already building to green standards and that it is not worth driving up the costs to get the buildings they construct certified. Clients, particularly clients concerned with construction costs in a time of economic decline, do not want to pay for the expense of certification. Some builders argue that the process of certification slows down construction and is an extra expensive step. Others argue that standard building practices already pass many of the criteria established by certifying agencies and that the improvements that must be made to pass certification are not going to have that large of an impact.

The greenest building and historic preservation

There is a statement that many sustainability experts use regarding green building: the greenest building is the one you do not build. Thus, it is important to stress that historic preservation is an important consideration when thinking about sustainability (Figure 8.9). Existing buildings required considerable energy and resources to construct as well as the time and talent of many individuals. By tearing down a building, one is destroying all of the effort. If one can preserve a building, one is saving resources for the future. In many ways, historic preservation is like energy efficiency when considering green sources of energy. While many of us like green buildings and renewable energy sources, some of the best strategies for sustainability are actually in energy efficiency and historic preservation.

Figure 8.9 Preservation of historic buildings in Greece, along with standardized building form and design, help to retain the character of a community.

Many nations, regions, and cities have policies regarding historic preservation to try to prevent widespread destruction of the past that was initiated to promote "modernization" by building things like Interstates or expansive sports or entertainment complexes. The modernization efforts often wiped out whole communities and destroyed many important historic buildings.

In many places, rules are in place to protect buildings that are of a particular age or style. Some communities use a 50-year scale so that buildings that are built today will be protected 50 years into the future if they have architectural or community value. Regardless of the rule, historic preservation is also a key component of economic development strategies. Think about how people are drawn to historic neighborhoods or buildings in today's world. They have much more character than strip malls or modern shopping centers.

One program that has worked hard to promote historic preservation, particularly in smaller rural towns, is the National Main Street program run by the National Main Street Center with is a subsidiary of the US National Trust for Historic Places. The Main Street Program focuses on improving US communities at their core—the main street. The program started over 30 years ago at a point in American history when many of the commercial centers of small towns were decaying due to out migration to larger cities and competition from larger commercial developments in the form of big box stores, malls, and strip malls on the outskirts of town. During this era, many main street downtowns declined, which was evidenced by boarded up stores, abandoned commercial districts, and depopulation of entire communities.

To try to reverse this trend, the Main Street program initiated a series of programs that focus on the centers of these small communities, largely on building connections among various stakeholders around issues of historic preservation and economic development to promote excitement and visibility of small downtowns.

Innovation in the headquarters of the US National Cave and Karst Research Institute

The US National Cave and Karst Research Institute (NCKRI) is a non-profit research organization that is funded by the US Government, the State of New Mexico, and the City of Carlsbad, New Mexico to fulfill the mission of advancing cave and karst research initiatives in the United States. The headquarters are near the banks of the Pecos River in southern New Mexico in the City of Carlsbad. When the building was designed, it was important to the stakeholders of the organization that the building expressed a scientific and environmental vision that would be appropriate for a major national research initiative.

Thus, the building was designed to LEED standards by using a variety of innovative technologies associated with energy and water. It was also built on a brownfield site associated with past rail transport. The building collects rainwater from the roof and stores it in cisterns within the walls of the building for use in irrigation. The storage of water makes sense since Carlsbad is a very arid community with annual rainfall of about 13 inches per year.

The building also uses a unique geothermal heating and cooling system that utilizes the temperature of the Earth to heat or cool air that is pumped through pipes under a

(Continued)

nearby parking lot. The geothermal system saves a tremendous amount of energy that would normally be needed for air conditioning.

There are some very innovative design elements to the building. For example, spelunkers, or cave explorers, need to know how to rappel or climb up and down ropes. Rappelling stations were built into the building to train cavers on how to climb ropes within a safe above-ground environment.

Perhaps the most innovative element of the building is the bat roost that was built into the building. Nearby Carlsbad Caverns, a US National Park, is home to 17 different species of migratory bats with total populations that range from 400 000 to over 700 000 annually. During certain times of the year, they fly in and out of the cave to seek insects for food. Their diurnal flights are very popular tourist attractions.

Carlsbad Caverns is not the only habitat for bats in New Mexico. Bats will find any cool, dry, safe environment to roost for sleep and raising their young. They can find homes in chimneys, trees, sheds, rock overhangs, and a number of other areas.

The designers of the NCKRI headquarters decided to build a bat roost into the headquarters of the building. A narrow passageway was created that allowed access into an interior space of the building that was cool, dry, and safe. Cameras are installed in the space to allow researchers to film the bats in order to learn more about their day-to-day lives. This is the only bat roost ever intentionally built into a building in the world. Many homes can be plagued with bats as a nuisance. However, in this case, the bat roost provides an innovative approach that uniquely expresses the mission of NCKRI within the overall building design.

Bank of America Building and the Empire State Building: two LEED certified buildings in the heart of New York

New York City is home to two very interesting LEED certified buildings. One of them, the Bank of America Building, built in 2010, is quite new and the other, The Empire State Building, dates to 1934.

The Bank of America Building was the first skyscraper in the United States to earn a LEED Platinum rating. The building is home to a number of well-known tenants, including former US Vice President Al Gore's firm, Generation Investment Management.

The building is known for its many unique design features. In terms of building materials, it uses very energy efficient floor to ceiling windows that allow the maximum amount of daylight into the building to limit the need for lighting. The concrete used in the building is made of approximately 45% slag, a blast furnace by product.

The building also uses a number of hi-tech features that ensure that the building is comfortable for users. Lights dim and increase automatically as needed to ensure that there is adequate lighting. Carbon dioxide sensors ensure that air is appropriately ventilated.

One of the more innovative features of the building is its cooling system. During the evening hours when energy costs are low, large tanks of ice are produced in the basement of the building. During the day, when workers want the space cooled, the ice slowly melts. A heat exchange system is used to cool air that is vented throughout the building. Individuals are able to modify their workspace heating and cooling to insure individual comfort.

The building also produces its own electricity in a cogeneration power plant that produces useful heat as well as electricity for the building. Waterless urinals are installed throughout the entire building and water collected in the roof is used to flush toilets. The tower even has its own beehives!

There have been critics of the Bank of America Building. Sam Roudman, writing in the *New Republic* (http://www.newrepublic.com/article/113942/bank-america-tower-and-leed-ratings-racket) was quite critical of the certification because the building actually uses more energy than other comparable office buildings in New York City. Yet the criticism is not very fair. The Bank of America hosts its very energy-intensive trading operations in the building. The trading activities require extensive IT support and energy gobbling computer servers. So, while the building is an energy hog due to the tenants, the building overall is a marvel of technology and a great example of the types of innovation that can be implemented in skyscrapers around the world.

When the 103-story Empire State Building opened in 1931, few worried about energy efficiency. Today, however, many are looking at ways to transform older buildings into more efficient spaces to save money and to reduce carbon outputs into the environment. In 2011, the Empire State Building was awarded LEED Gold status for its extensive green renovation. The cornerstone of the renovation was the addition of a number of energy-saving initiatives, including the replacement of every window in the building with extremely energy efficient glass. The energy improvements save approximately $4.4 million dollars a year and cut carbon by approximately 7000 metric tons a year.

A number of other initiatives were initiated during the renovation for the building's long-term maintenance. The owners have committed to using green cleaning and pest control methods and to use recycled paper products. The building management company also implements a unique green lease requirement that binds lease holders to maintaining certain green standards.

These two buildings are examples of how some of the most complex, yet iconic, buildings in the world can become more sustainable. The Bank of America used state of the art design and innovation in the construction of a new building, whereas the Empire State Building used new technology in an already existing space.

Since it started, nearly 250 000 historic buildings have been renovated in this program, helping to create over 115 000 new jobs in these historic places.

Think about your own community. Does it contain any historic buildings? Are they preserved? What interesting historic buildings have you visited? Where would you rather shop or spend time, in a historic district or a strip mall? Why? Are there any "hot" downtown main street districts in your community that contain historic buildings? Have you been to failing small town districts? Why do you think they are failing?

Small house movement

In many parts of the world, home sizes have been increasing dramatically. In the United States, for example, the average home size is over 2500 square feet, up almost 1000 square feet since 1980 (Figure 8.10). This amount of space is unprecedented in human history, even as family size is decreasing. We all want, in the words of Virginia Wolfe, a room of our own.

Yet the cost of that space comes at a tremendous personal, environmental, and social cost. We have to pay to build, maintain, heat, cool, and furnish the space in order for it to retain value. The expanded house also takes up far more space and requires more resources to build than homes in the past, so the environmental costs of their construction are very large. Plus, the larger homes create a sense of personal and community isolation. We set ourselves alone from our families within our homes, plus we separate ourselves much more clearly by financial wealth by creating neighborhoods with McMansions separated by walls and gates from people with the means to afford smaller homes. In other words, the large homes are symbolic of modern consumptive trends that are harmful for broader societal integration. As a result, many have advocated for returning to small homes to lessen the impact of our residential lifestyle on the environment.

The small house movement originated in the 1970s at the apex of the mid-twentieth century environmental movement. At the time, many advocated going "back to the land" to live a more simple life to get away from the trappings of cities and the growing consumerist lifestyle that was emerging during that era. However, this early small house movement

Figure 8.10 The re-development of a community that displaces small homes for larger homes is problematic in many areas of the world. A small home was demolished to make room for the large home on the left.

Table 8.2 Elements of design that will make small spaces seem larger.

Glass on the perimeter	Multiple-purpose furnishings	High ceilings
Skylights	Horizontal wall graphics	Next to ceiling shelving
Mirrored walls	Soft surfaces	Lighter color furnishings
Glass-block walls	Glass tabletops	Rounded corners
Curved walls	One-story floor plan with variety	Colors in corners
Angled wall planes	Appropriate human scale	Spotlights on selected objects
White or neutral walls	Smaller-scale furniture and artworks	Pocket doors
Wall photo murals (in perspective)	Built-in furniture and storage walls	Transparent/translucent screens
Single-pattern floor coverings	Small-scale floor/ceiling tiles	Core functional areas
Wall-to-wall floor coverings	Linear floor lines	Simple and precise
Large furnishings against walls		arrangements
Multiple activity rooms		
Open traffic patterns		

did not become a strong cultural phenomenon and limited itself to what some would term eccentric individuals who moved away from mainstream society.

However, the small house movement had a renaissance in 1997 when Sarah Susanka published the book, *The Not So Big House: A Blueprint for the Way We Really Live*. In it, Susanka suggests that large homes do not really provide greater contentment. Instead, she suggests that homes should be built better and not bigger. She suggests a range of design elements that can give small spaces a more spacious feel.

According to Tremblay and von Bamford (1997) a number of elements can give small spaces the feel of greater space (Table 8.2). The list includes a number of suggestions that range from creating soft surfaces, to the use of glass bricks in walls. These elements give a sense that one is in a more extensive space and provide a greater sense of scale. The use of vertical space is particularly important in small homes where there is very limited space for storage. Plus, there are many innovative space saving devices that can be used in small homes such as a pull down bed that doubles as a dining room table and an ironing board.

While many small homeowners utilize these design elements, most of them move to the smaller homes for more altruistic or economic reasons. Some have opted to move into extremely small homes, some of which are less than 120 square feet. These smaller homes can be placed on wheels so that they can be transported from place to place like a mobile home.

In many ways, the small home movement is an outgrowth of a broader environmental aesthetic consciousness that emerged with the design and writings of Frank Lloyd Wright. He believed that the design of buildings should emerge out of its space or setting. When placed in nature, the building should be in harmony with it. He designed several homes that would be modest by today's American standards, but that are in line with the square footage of the modern small home movement.

Prince Charles House: A green home for the 50+ crowd

In St. Austell in the United Kingdom, a 31-unit green, affordable housing complex was constructed in 2012 for people over 50 and those who need particular support needs, called Prince Charles House. The project was considered a green demonstration project to try to advance green technology in the area. It was built to UK BREEAM standard and achieved a rating of outstanding.

The project is unique in many ways. Few green certified buildings are constructed for this particular population. Older people are very concerned about temperature comfort and thus require particular individualized heating or cooling within units that some might find difficult to achieve in a green building. However, individualized unit heating control was a key element of the plan. In addition, the designers utilized passive solar energy and window space to give residents greater access to sunlight and warmth. The designers also utilized a number of interesting green building initiatives, including addition of photovoltaic electricity, a green roof, ventilation with heat recovery of individual apartments, a building management system, and a zone activated sprinkler system for gardens and lawns.

Waste from the construction process of the building was monitored and 89% of the construction waste was diverted from landfill. Indeed, a building needed to be demolished on the site prior to construction and 2000 tonnes of demolition material was used to create the foundation for Prince Charles House. The green building provides opportunities for engaging residents on sustainability and the environment.

In thinking about this, how would you promote sustainability among this older or disabled population? I recently gave a talk on some of the sustainability problems facing the world to residents of a retirement home and was asked what they could do to help save the world. They wanted to make a contribution, but were limited by the fact that they lived within a building managed by the company that owned the retirement community. They had little control over decisions on energy or water conservation, landscaping, food, or conservation. If they were in a building that was a certified green building, there is an expectation for greater engagement and participation by residents on sustainability issues.

While certainly design is a large issue in small homes, they benefit the environment in many ways. They use considerably less energy than a conventional home. They do not have as much space to heat or cool and they often have far fewer appliances and gadgets. For some of the tiny homes, there really is not much space for things like washing machines and waffle makers. Some of the small homes are fully off the grid and generate their power with wind turbines or solar panels. The small homes tend to have less lighting needs. They are designed to promote daylight lighting of living space and only the living areas need to have lighting options. Small homeowners also produce less waste since they do not have room to purchase "stuff" in the consumerist world in which we live.

Many of the small homes are about the size of a standard university dorm room. If you live in a dorm room, think about how you might live and raise your family in that space. How would you make it work if you had to live in such a space? What renovations or innovations would you need to make to your dorm room to make you feel comfortable about living in that size of space your whole life? How do you think your life would have been different if you grew up in a small home about the size of a dorm?

While these small homes may seem uncomfortable to many people in the United States and other places where large homes are common, it is important to note that many people around the world live in very small spaces. The small and tiny home movement is not a new thing to them. It is a reality of life.

Further reading

Cotgrave, A. and Riley, M. (eds) 2013. *Total Sustainability and the Built Environment*. Palgrave Macmillan, 308 p.

Tremblay Jr., K.R. and von Bamford, L. 1997. *Small House Designs*. Storey Publishing, Vermont, 202 p.

9

Transportation

Transportation is one of the main ways that humans expend energy. We all need to go places to work, play, or visit. The way we choose to travel has a large impact on the planet. In this chapter, we will learn about the state of transportation around the world, the impacts of different transportation options, and innovative approaches to reducing the environmental and social impact of transportation on society in general.

Everyone has a distinct environmental footprint associated with their transportation choices. As I am writing this, I am on the Amtrak Acela high speed train from New York to Washington DC. I have access to food, the Internet, and a pleasant work station. The ride is approximately three hours long. I could have taken my electric hybrid car, but it would have required far more time and I wouldn't have been able to write during the trip.

This is not my normal commute. Normally, I get in my car and drive the 20 minutes from my home to my office. There isn't convenient mass transit from where I live to where I work. I could take it in an emergency, but it would take me about two hours. When compared with my short commute, mass transit is not a viable option for me.

When I lived in New York, I also took the Long Island Railroad from where I lived on Long Island into New York City about once a month for business or recreational opportunities (Figure 9.1). This is a commuter rail line that links easily to the New York Subway system. I could also transfer to Amtrak lines at Penn Station or walk a block to the main New York City bus terminal. Penn Station is a great example of a multi-modal transit hub.

I also fly quite a bit. This year, I took two international and eight domestic flights. I also took several ferry trips. My carbon footprint is quite large due to all of this travel and my daily commuting. As we will see, there are ways to measure your transportation footprint and mitigate your transportation choices. What about your transportation choices? What is your commute like? How many trips do you take? Are there mass transit options for you?

In our modern busy world, many of us rely on travel for business or recreation. The question is, "How can we make our travel greener?"

In this chapter, we'll look at different types of transportation options and their overall impact on the environment. We'll also look at ways that communities are trying to reduce the overall impact of transportation on the environment. Finally, we'll explore ways that transportation and planning specialists are re-designing cities to make transportation more sustainable.

Introduction to Sustainability, Second Edition. Robert Brinkmann.

Companion website: www.wiley.com/go/Brinkmann/IntroductiontoSustainability

Figure 9.1 This is the train station that gets me into Manhattan. The Long Island Railroad is one of the most successful commuter rail lines in the United States.

Transportation options

Our modern transportation system is not that old. It wasn't that long ago that our ancestors relied on wind powered boats or animals as the main forms of transportation that didn't involve walking. Of course, the building of boats or the maintenance of large animal stables had distinct measurable impacts on the environment. However, no-one could have imagined a few hundred years ago the vast array of transportation options available to most people on the planet now.

Today, each of us can book a flight or a bus or train trip with a few finger strokes on our keypads. We can also buy a car and make online purchases that will be shipped by truck or air. The modern transportation system has many options, each with its own impact. In this section, we will look at cars and roads, trains and train tracks, ships and ports, and airplanes and airports.

Vehicles

Right now, it is estimated that there are over 1 billion vehicles on the road around the world (Figure 9.2). This amounts to one vehicle for every seven people. Considering the impact of manufacturing the vehicles, the need for roads, and the need for infrastructure like garages and parking lots, that is a whole lot of cars and trucks! The growth has been very rapid. Just a decade ago, there were approximately 25% fewer cars in the world. While car ownership has decreased in much of the developed world (most notably in the United States) due to greater

Figure 9.2 The billions of cars on the road need billions of dollars of infrastructure.

access to mass transit, car ownership is increasing rapidly in places like China, India, and Brazil. Everyone loves the convenience of getting into a car and going exactly where they want to go.

This ease of transit has created a car culture that is heavily dependent on very expensive public roads and parking areas. With the advent of the car, shopping districts have dispersed from central cities to the suburbs where there is ample space for parking. Plus, many of us have relocated from the dense urban centers of the world to suburbs where there is room for cars, parking lots, and garages. The car has transformed our lives, but also the geography of our planet.

Most of the world's cars are produced in China, Japan, Germany, India, South Korea, the US, Spain, and Brazil and are shipped all over the world. Factories in these places are highly dependent on a complex network of industries and high-tech shipping for parts.

In the past, car manufacturing was centered around some key manufacturing regions of the world. However, now car manufacturing is highly dispersed and relies less on big manufacturing networks. Now smaller networks of suppliers use "just in time" manufacturing processes (see text box). The impact of this is that car manufacturing is a global process that, for example, uses parts made in Mexico for a car assembled in Kentucky. Thus, it is difficult to trace the impact of the manufacture of vehicles on one community. The impacts are felt globally. In the past, global car manufacturing centers like Detroit had a huge footprint in the city. Now, car manufacturing is more dispersed and associated industries do not need to be in close proximity.

There is a range of motorized vehicles that are manufactured throughout the world, including cars, light trucks (pick-up trucks), and semi-tractor trailers.

Cars

Cars are very convenient for personal use. Many people living outside of dense cities rely on them for day-to-day travel to work, school, or shopping. Since they were first mass produced in the United States in the early twentieth century, they have been part of the world's consumer culture. They range in size from small economical vehicles to large luxury cars and SUVs. The great range in style and color has turned them from functional tools in modern society to consumer possessions that denote a particular social status. They are not just utilitarian, but are expressive of who we are as individuals.

While the developing world is very quickly increasing the number of cars on the planet, there is no doubt that the developed world has the greatest number of cars per person in the world. Table 9.1 shows the top ten countries for the number of vehicles per person on the road. Tiny San Marino leads the list with 1.3 per person. The United States is fourth on the list with 0.8 vehicles per person. Australia and Switzerland are also in the top ten with approximately 0.7 vehicles per person. Six of the ten countries with the highest rate of car ownership are in Europe. The United States, Australia, New Zealand, and Brunei round out this list. The highest rate of car ownership in Asia is in Brunei with roughly 0.7 cars per person. Barbados has the highest car ownership rate in Latin America and the Caribbean with 0.4 cars per person (Argentina has the highest rate in South America at 0.3 cars per person). Libya has the highest car ownership rate in Africa with 0.5 cars per person.

Trucks

Light trucks, or pick-up trucks, are used for personal or professional purposes. Many are used in construction, delivery, and maintenance. Some people prefer to drive these trucks rather than cars and use them as their personal vehicles. They are highly prized in rural areas where they can be used as both a utility vehicle on the farm or job, and also as a personal vehicle for day-to-day transportation. Trucks have worse fuel efficiency than cars and in recent years many who owned them for personal use have opted to buy cars that get better gas mileage.

Table 9.1 Top ten countries for vehicles per person (http://en.wikipedia.org/wiki/List_of_countries_by_vehicles_per_capita).

Rank	Country	Cars per 1000 people
1	San Marino	1263
2	Monaco	899
3	New Zealand	860
4	United States	838
5	Iceland	824
6	Liechtenstein	773
7	Finland	752
8	Australia	730
9	Brunei	721
10	Switzerland	716

Heavier trucks, sometimes called semi-tractor trailers, are utilized to haul goods long distances. The larger versions of these vehicles drive across continental distances for delivery. While trains are good at taking goods from one port or transit hub to another, semi-trucks are good at taking goods to places not served by trains or boats. In the US there are roughly 6 million semi-tractor trailer trucks in use.

Trucks are used to move goods from one place to another. They often pick up materials at ports or train depots for dispersal to distant locations via roadways. They are an important piece of the global transportation network that allows goods to be moved from one location to another very quickly. In the last 20 years, the number of freight miles carried has increased by roughly 50%, as has the amount of energy needed for the heavy truck transit.

Vehicles and fuels

Most vehicles run on fossil fuels. Even if a car is electric, it is likely using electricity produced from some type of fossil fuel. Around the world, some countries impose standards for fuel efficiency. In the United States, these are called Corporate Average Fuel Economy (or CAFE) standards. The CAFE standards date back to the mid 1970s when congress established them during an energy crisis in the United States. Since then, the CAFE standards have been used to reduce greenhouse gas and cut the national reliance on fossil fuels.

CAFE standards have been set for cars and light trucks. The change in efficiency has been significant. In 1978, the passenger car CAFE standard was 18 miles per gallon; today, it is 38. That is quite an improvement in technology and manufacturing in just a few decades. The US plans to increase efficiency to 61 miles per gallon by 2025. However, this is in question at the moment since the Trump administration wants to pullback on this regulation.

The European Union also has fuel efficiency standards for vehicles. In addition, they provide a rating system for greenhouse gas emissions. Japan, Australia, and New Zealand are among other countries with fuel efficiency requirements for vehicles sold within their borders.

Cars and trucks are responsible for a number of emissions that cause significant environmental problems. Most notably, they produce roughly one-third of all greenhouse gas emissions on the planet. There are three major ways that we can reduce these emissions: improve the efficiency of vehicles, use cleaner fuels, and drive less. See text box for a discussion of other pollutants produced by cars.

A number of new technological innovations have sought to reduce the impact of transportation on the environment. They focus on improving new fuels and new battery technology for electric cars.

New fuels

There are a number of new fuels that are now available for vehicles. They include ethanol, natural gas, cooking oil and grease, and hydrogen (Figure 9.3).

The most common of these new fuels is ethanol. Ethanol is usually derived from agricultural products like corn and sugar. Brazil uses the most ethanol of any country in the world and they get most of it from sugar cane. In the US, up to 10% of the gasoline contains ethanol, mainly derived from corn. The use of ethanol is controversial. Land that could be used for growing crops is being used to produce fuel. In addition, there are some who argue that the energy cost of producing the ethanol is greater than the energy produced. Plus, ethanol

Figure 9.3 Many buses today are fuelled by natural gas.

produces more ozone pollution than conventional fuels and there are concerns over the impact of burning ethanol in older cars due to damage to engines. Nevertheless, ethanol production and use is likely to increase as an alternative to fossil fuels.

Another important alternative fuel for vehicles is natural gas, one of the most abundant fossil fuels in the world. Many have advocated the use of natural gas for vehicles because it is far less polluting than traditional gasoline products. However, as we all know, natural gas can be a very dangerous fuel to transport. It is highly explosive and there are concerns over the safety of natural gas storage in moving vehicles. However, the use of natural gas for transportation is increasing. Many public buses run on natural gas and many fleet vehicles utilize natural gas. These fleets of buses and cars utilize natural gas filling stations that are often located near the fleet storage areas. To date, natural gas filling stations have not expanded greatly and the technology has not diffused widely.

Some have converted their cars to burn cooking oil and grease. These materials are suitable fuels if they are filtered and if engines are slightly modified to burn the material. It is easy to find information on how to convert your car to burning oil and grease. Years ago, one of my students converted his car engine so it could burn waste grease from a fast food restaurant. He had a relationship with the restaurant that he would stop by once a week to pick up waste oil for use in his car. He would strain it and then store it in his backyard in a tank. He would pump the grease from the tank into a storage container in his trunk that ran to the engine. Although he didn't pay for the grease, he worked very hard to get it and process it. His clothes also took a beating. It was well worth it to him. The exhaust smelled like chicken or French fries.

Hydrogen is the last of the alternative fuels that have made it to the fuel market. Hydrogen is a very abundant element. It is found in water and in many other compounds. However, hydrogen rarely exists in the elemental form near the Earth's surface. Thus, it must be separated from other elements to be used as a fuel. This separation involves a considerable amount of energy—indeed as much energy is required to create hydrogen fuel as one gets from hydrogen. Thus, it is important to think of hydrogen not as a reservoir of energy, but a carrier of energy.

The reason that hydrogen is of such interest to the energy community is that it produces water as its main byproduct of combustion. Thus, it is considered a zero emission fuel upon burning. However, because it takes energy to make hydrogen, one must look deeper into the manufacturing process to assess if it is truly a zero emission fuel. Many have advocated linking hydrogen production with wind farms. In such circumstances, hydrogen can be transported via fuel cells for use in cars or other energy using machines. However, if the hydrogen is produced via fossil fuels, it is really no different from any other conventional energy source.

Electric cars

Electric cars and hybrid cars utilize battery technology to reduce the reliance on fossil fuels. Hybrid cars utilize both gas and electric power for energy, whereas electric cars run fully on electric power. Hybrid vehicles utilize the breaking power to produce energy. Thus, hybrids have better fuel efficiency if they are driven in urban settings. They have worse fuel efficiency on the open freeways.

Some of the first cars ever built were electric cars. However, as time progressed, it was cheaper to develop cars around gasoline fuel. Now since gasoline is more expensive and as we are faced with a variety of problems associated with pollution from fossil fuel production and use, there is growing interest in hybrid, plug-in hybrid, and electric vehicles (Figure 9.4).

There are a number of policies in place around the world to promote the use of electric cars. Many governments give tax breaks and special rebates for the purchase of hybrids and

Figure 9.4 Many universities have plug in stations where drivers can charge electric cars. Does your university have electric car charging stations?

electric vehicles. In addition, some organizations provide priority parking and some roads provide special lanes for electric or hybrid cars.

One of the challenges in the expansion of electric cars is the infrastructure for recharging them. Most of us could charge the vehicle at home so that we could drive where we need to go during the day. However, over long distances, one cannot simply pull up for a refill. The cars need a few hours to recharge. Communities have worked to develop electric car charging stations to allow cars to recharge while they are parked at work, school, or stores.

US Senator Schumer from New York proposed in 2019 a plan for rapidly increasing the use of electric cars in the United States. The proposal is divided into three components. The first focuses on discounts for consumers who trade in gas burning cars for electric cars. Schumer predicts that this would remove 63 million gas burning cars from the road by 2030. The second part of the proposal seeks to make electric charging stations accessible everywhere with a particular focus on rural, low-income, and underserved communities. The third part asserts that the US should be the world leader in electric car and battery production and notes that a significant amount of grant dollars will be available to retool existing manufacturing sites and to advance research in improving electric car and battery technology.

While the electric cars themselves are zero emission, it must be noted that electricity must still be produced to run the vehicle. Thus, unless the electricity is produced solely by green energy sources, the cars are not truly zero emission. However, the energy produced in a power plant will have less emissions than an equivalent amount of energy produced by a gas-burning vehicle. Thus, in total, regardless of the form of electricity generation, the cars have lower emissions overall.

Automated Vehicles

In recent years, there have been major advances in self-driving, or automated vehicles. Our GPS technology coupled with driving automation opens up tremendous possibilities for the use of automated vehicles in the coming decade. Already there have been numerous tests of the vehicles. Some well-publicized accidents have provided evidence that care needs to be taken to ensure that the automation is safe. However, there are great possibilities to improve safety and environmental performance of vehicles with automation.

Rail

Our modern rail system has its roots in seventeenth and eighteenth century lines that started with the steam locomotive in Britain and quickly migrated to the United States. Today, rail lines are highly efficient forms for moving goods and people from place to place (rail transit is discussed in an upcoming section on mass transit).

Rail is one of the most energy efficient ways that we can move materials from place to place. It is many times more efficient than cars and trucks and is on a par with barge freight energy usage.

Most countries of the world have some form of freight line. The top nations by use of rail line are listed in Table 9.2. As one can see, some of the largest countries of the world, most notably Russia, China, the United States, India, Canada, and Brazil are in the top 10.

Table 9.2 Freight by country (http://en.wikipedia.org/wiki/Rail_usage_statistics_by_country).

Rank	Country	Billions of tonne-kilometers
1	Russia	3176
2	China	2696
3	United States	2326
4	India	666
5	Canada	352
6	Brazil	267
7	European Union	261
8	Ukraine	237
9	Kazakhstan	236
10	Australia	198

Rail requires a different type of infrastructure from roads. Rail lines are built on very strong foundations. The rails have a particular gauge with a set axel width and wheel style. For generations, rail gauge varied considerably around the world, making movement of train cars from one system to another difficult—even within a single country. While there are certainly differences in some places, there is much more consistency than there has been in the past.

Some parts of the world, particularly in the European Union, have worked to expand freight transport in order to save energy. They are seeking to reduce the freight miles of the less efficient trucks in order to take advantage of the more efficient rail system.

Ship transport

There are five main types of vehicle used for ship transport: bulk carriers, container ships, tankers, refrigerated ships, and roll-on/roll-off ships (Figure 9.5).

Bulk carriers

Bulk carriers are large ocean-going ships that carry huge quantities of a single product such as grain or mineral products. They are used largely to deliver materials from a point of primary production, such as a mine or agricultural region, to a point of use, such as a city or an ore processing facility.

Container ships

Container ships are key to the world's intermodal transportation system. These ships carry containers that can go seamlessly from sea to rail to truck very quickly. The largest ships can carry over 18 000 containers. Container ships are responsible for carrying most of the word's non-bulk materials carried by sea.

Figure 9.5 Even small boats can have impacts on the environment.

Tankers

Tankers are responsible for carrying liquids from place to place. The Exxon Valdez was carrying crude oil when it went aground off the coast of Alaska in 1989. However, oil is not the only thing transported by tanker ships. Cooking oil and wine are also commonly shipped using these giant vehicles.

Refrigerated ships

Perishable goods are transported by special refrigerated cargo ships. They tend to be smaller than other cargo ships because of the energy costs associated with refrigerating such large areas. These ships are used for transporting things like fish, meat, vegetables, and fruit.

Roll-on/roll-off ships

These ships, similar to car ferries, transport wheeled vehicles that can be used to quickly move goods off of a ship.

Environmental issues associated with ship transport

There are many environmental issues associated with ship transport that relate to energy, waste and pollution, and creation of infrastructure. Each will be discussed.

Energy and ships. Although the amount of energy utilized to transport goods via water is far less than it is by rail, truck, or air, ships do need to utilize and carry a great deal of energy on board. This makes them vulnerable to leaks and pollution of waterways. They also emit a great deal of pollution since they tend to burn some of the dirtier fuels (like diesel).

Figure 9.6 Channel dredging is a constant problem in many ports—it is also an environmental challenge.

Waste and pollution. Many of the ships dump their garbage and sewage directly overboard. Plus, it is not uncommon to have containers or other goods fall off ships during rough seas. There have been several well-documented container ship accidents that have caused pollution of the ocean. The Exxon Valdez is the most well-known accident; however, there have been hundreds of tanker leaks since that event.

Infrastructure. One of the most challenging aspects associated with shipping is the maintenance of port infrastructure. Shipping channels need regular dredging (Figure 9.6). Plus, newer, larger ships have a deeper draft which means that shipping channels must be cut deeper. Port dredging often disrupts coastal ecosystems and can lead to sediment or chemical pollution of near shore environments.

Perhaps the most famous example of land disruption due to ship transport is the Panama Canal. Originally completed in 1914, it has been expanded in recent years to accommodate larger ships that have had to bypass the canal due to its antiquated size restrictions. The new expanded canal can accommodate the largest cargo ships in the world.

There are significant concerns over the maintenance of water quality for some of the larger fresh water reserves in Panama as a result of the expansion. Many are also concerned over the disruption of natural ecosystems.

Air transport

One of the greatest pleasures of modern living is the ability to travel great distances quickly and safely via air. It wasn't that long ago that intercontinental voyages were dangerous undertakings that took weeks or months to accomplish. Today, we can travel from continent to continent in just a handful of hours. However, air travel has significant environmental impacts that include noise, particulate pollution, and air pollution that leads to global climate change.

By far, the country with the most passengers is the United States with around 890 million passengers in 2018 (World Bank – http://data.worldbank.org/indicator/IS.AIR.PSGR). China comes in second with about 611 million. The US also leads the world in airfreight with roughly 43 000 million ton-kilometers of freight carried in 2018.

Air travel is increasing in most countries of the world. In the United Kingdom, for example, air travel has increased 250% between 1990 and the present. In the European Union, the growth in aviation led to an 897% increase in greenhouse gases from aviation over the same time period.

The on-the-ground impact of aviation is much less than that of cars. Roads have a much larger footprint than airports. However, most commercial airports are huge operations that require long runways and extensive support buildings like airplane hangers and terminals. Runways of commercial airports can be over 5500 meters. Considering that large airports have at least two runways, this is a huge footprint. Considering road networks and parking associated with airports, they have a very large environmental impact. Because of the large amount of impermeable surfaces at airports, runoff of storm water is a significant problem. Spills of fuels have occurred at some airports as well.

Noise pollution from airplanes is a serious health risk. It can cause hearing problems, heart disease, sleep disorders, and stress. A study of the impact of aircraft noise on the health of residents living near an airport in Germany showed that the noise was linked with serious heart disease. Communities often have noise requirements for airports to try to alleviate the problem. New engine technologies have helped alleviate some of the noise problems. Some airports limit night flying in order to preserve night-time quiet.

Aircraft produce small particulate matter and water vapor that can form contrails, or man-made clouds, in the upper atmosphere; we have all seen them and there is a great deal of controversy over them. Some believe that contrails have an important impact on global climate change. They trap outgoing long-wave radiation and thus have a net warming impact.

However, research is somewhat inconclusive on the role of contrails in our global climate. Yet, there is no doubt that they are a relatively new phenomenon on our planet and we just do not know what their long-term impact might be.

Perhaps the most significant environmental impact from air travel is the release of huge amounts of greenhouse gases. Indeed, this form of travel has the greatest output of greenhouse gases per mile travelled. It is estimated that air travel accounts for roughly 2% of our current greenhouse gas emissions. This number is expected to increase since it is estimated that air travel will continue to expand.

While we have developed electric cars and green fuels for automobiles, there are few alternatives for air transportation. Some biofuels have been developed for aircraft, although they are not widely used. Electric airplanes are on the market, but they are largely small, experimental craft. Solar aircraft have also been built and flown, but they too are not widely used. It is unlikely that solar or electric passenger airplanes will be on the market any time in the near future.

It is much more likely that we will see a strong focus on energy efficiency of aircraft as a means to reduce fuel consumption. However, aircraft last for decades and any innovation in efficiency will be slow to make it into the world's airplane fleet.

Some are strongly urging people to reduce air travel overall. In my profession, it is expected that I attend at least one professional convention a year. Some professors are questioning the need for regular conference attendance, particularly with the growth of the Internet and widely available technology for face-to-face online meetings.

Some businesses that require travel to operate are examining alternatives to air travel to try to reduce the impact of travel on the environment and to save money.

Space travel

Space travel was once the realm of science fiction, but with the success of the International Space Station and the US Space Shuttle Program, more and more are looking to space as the "final frontier" of travel.

Yet the costs of escaping the relentless pull of our planet's gravity are staggering. The cost of a single space shuttle flight was approximately 1.5 billion dollars. However, new technologies are making space travel less expensive.

A number of new companies have emerged in recent years that focus their attention on developing space flights for tourists and to provide private and public organizations the opportunity to conduct research or launch satellites. Perhaps the most well-known of these is the company, Virgin Galactic, owned by Richard Branson, the British entrepreneur and explorer.

He is booking flights for the space trip at $200 000 each. The passengers will experience six minutes of weightlessness during the two-hour flight. The cost is clearly far out of reach of most people.

Space travel, both government space travel and private space travel, has been criticized as unrealistic, expensive, and environmentally dangerous. The film *Gravity* frighteningly showed the impacts of lots of space junk within our planet's orbit. There are over half a million particles over one centimeter in diameter orbiting our planet, and millions of smaller particles.

Of course, the exploration of space comes at a cost on the ground. Many of the institutions that were involved with research on space exploration in the 1900s are now contaminated with a variety of hazardous materials. Clean up at these sites, where is has occurred, costs millions of dollars. Many are concerned about the emissions from rocket launches as well. It is estimated that rocket launches have been responsible for about 1% of the ozone loss we have seen over the last 100 years. Plus the emissions from the rocket fuel in the upper atmosphere influences the heat balance of the planet in ways that we cannot clearly predict.

Plus, given the history of accidents of space craft after launch, there is concern over what would happen if a payload of hazardous material exploded in the atmosphere. Some nuclear powered equipment has been deployed and some have advocated the use of nuclear powered engines in space flight. What would happen if a rocket carrying these materials exploded during takeoff?

Obviously space flight isn't going to impact most of us in the near future. However, it is worth thinking about the impacts of space flight on our atmosphere and our environment.

Do you think it is worth exploring space given all the risks? Some believe that finding or extracting resources from other planets is the only way our species can survive into the future. Others believe that space is a cold dead place and that we are wasting our time with space exploration. What do you think?

Would you pay extra to mitigate greenhouse gases released from your air travel?

If you've booked air travel lately, you will find that some of the main booking sites offer you the option to pay into a fund that will help to mitigate the greenhouse gases from your trip. The fee is anywhere from a few dollars to a few tens of dollars, depending on the trip. The fee goes to support alternative energy projects, greenhouse gas sequestration efforts, and other initiatives focused on reducing greenhouse gas emissions.

Would you pay this if you were buying a ticket? This is essentially a voluntary tax. Some have advocated a mandatory tax on air travel to support greenhouse gas reduction initiatives. However, others have argued that a tax would put an undue burden on consumers who are already dealing with high cost of air travel.

This issue gets at the heart of one of the key policy issues associated with greenhouse gas reduction and global climate change. Who will pay for the mitigation?

Emissions from cars and trucks

There are a number of different types of emissions produced by cars and trucks that can cause environmental or health problems:

NO_x. NO_x refers to nitrous oxides and nitrogen dioxide (NO_2) compounds that are released in tailpipe emissions. The particles do not last long in the atmosphere and combine quickly to form nitric acids and other harmful chemicals that can do great environmental damage and that enter deep within the respiratory system.

Volatile organic compounds (VOCs). These are light (thus volatile) organic compounds that react in sunlight to form ozone, a major contributor to smog.

Ozone. Low-level ozone (as opposed to the naturally occurring and beneficial upper atmosphere ozone) contributes to smog formation and leads to a variety of respiratory irritations.

Carbon dioxide. Exposure to carbon dioxide can lead to health problems and even death. However, most are concerned with carbon dioxide in the current era due to its contribution to global climate change.

Carbon monoxide. Carbon monoxide is another toxic gas produced from the emission of vehicles. Exposure to this gas in the absence of oxygen leads to death

Toxic air pollutants. Along with the mix of gaseous pollutants noted above, a number of toxic pollutants are released in vehicle emissions. They include chemicals like heavy metals and some organic compounds like benzene. They are usually present in low amounts.

Particulate matter. Small particles of materials are released when fossil fuels are burned. These fine-grained materials are part of the urban dust we can feel if we run our hand across the roof of a car that hasn't been washed in a few days. The fine-grained particulate matter is especially dangerous because it can enter deep within the respiratory system.

Intermodal transportation systems

In the past, goods were transported from one port to another without much consideration about where the product was being shipped after arrival. Materials were shipped in crates, bins, and bags. Stevedores hauled these goods from ship to warehouse where they would be sorted for shipping via truck or rail.

However, in the last few decades, new ways of thinking about shipping goods has evolved to make the transit of goods much more efficient. This system is called the intermodal transportation system. This system uses uniform boxes or containers, sometimes called sea chests, that can easily move goods from their point of origin in a remote setting to another remote setting in another continent without being unpacked or removed from the crate.

This is accomplished because container ships, rail lines, and tractor trailer truck systems have coordinated to accommodate the uniform shipping containers. Due to uniformity, sea chests can be unloaded from container ships and placed within a rail system within a few hours. Once the rail reaches its destination, the sea chests can be quickly placed on a semi-tractor trailer truck. At this point, the truck can take the goods directly to a store or warehouse for delivery—all without being unpacked or disturbed on its journey.

The intermodal system is aided by high-tech port technology that partially automates the packing and unpacking of the vehicles. In addition, GPS tracking and automated inventory systems enhance the ability to move goods quickly from one place to another.

The intermodal transportation system saves considerable money and energy in our modern world.

Roads

Roads also have a significant impact on the environment. They utilize a tremendous amount of space and cost a fortune to build. For example, a basic limited access freeway mile costs over 1 million dollars to build. There are a number of different classes of road around the world. However, roads can be broken into two broad categories: open access roads and controlled access highways.

Open access roads are those roadways that allow access to any vehicles. They can range from small roads like alleys, lanes, or cul de sacs, to larger roads like highways, avenues or boulevards. Trucks, cars, pedestrians, and bikes share these open access roads. They pass through cities, towns, villages, and rural landscapes; many service residential areas. They are often managed at the local level and supported by local taxes. However, there are state and national open access roads that are supported by state and national taxes. Each local, state, or national government has particular standards by which they build and manage these open access roads, so there is great variety in the look and use of roads. In dense urban areas roads tend to have multiple purposes. They provide parking, bike lanes, pedestrian crosswalks, and room for cars. In contrast, roads in rural areas focus on moving people from one place to the next at great speed. In some older cities, the roads predate the automobile. In such settings, some cities have made room for cars, where in others, they have banned cars to allow room for pedestrian travel.

The open access roads are interconnected with each other and link different parts of cities with suburban and rural landscapes. It is easy to travel seamlessly on these roads to different places. International signage conventions make travel on these roads easy from one nation to another or from one city to another. Thus, while the character of the roads may change from one locality to the next, their signage provides a linking infrastructure element to allow users to use the roads with ease.

Controlled access highways are different from open access roadways in that they have limited entrances and exits and have particular rules (such as minimum and maximum speeds) that allow very quick traffic flow over great distances for cars and trucks. Pedestrians and bicycles are banned from these roadways. Around the world, these roadways are known by different names like autobahn, auto-estrada, autopista, expressways, highways, and parkways. Some of these roadways are supported by state or federal governments and some are toll roads managed by some type of highway authority.

These roadways, unlike the open access roads, have a distinct similarity of "look" around the world. They are typically multi-lane roads with infrequent exits. They are used heavily for transit of goods locally and over great distances by trucks. They are also used for long-distance travel. One side effect of the growth of these roadways in the late twentieth century was the expansion of the footprint of urban areas as suburbs expanded along highway routes. Many urban areas around the world, most famously Detroit, saw an exodus of people from the city to the suburbs, leading to an overall decline of some cities.

Environmental issues with roads

There are several environmental issues associated with the management and maintenance of roadways. They include storm water pollution management, street sweeping, and ground stability.

Storm water pollution management

Paved roadways inhibit infiltration of rainwater into the ground. When rains hit roads, they pool and form small ephemeral streams along the gutters or road edges. These streams are sometimes transferred from the roadways into ditches, storage ponds, or storm water sewers.

There are two main problems associated with managing storm water along roadways: the huge volume of water involved, and the pollution that the water collects as it moves across the road surface. Let's discuss each of these issues separately.

Prior to the development of cities and roads, the soils at the surface of the Earth could accommodate the majority of rain events that occurred. Certainly flooding did occur during extreme events. However, most of the rainfall could be absorbed into soil where it would filter into groundwater systems. With the advent of cities and roads, much more land became

Figure 9.7 Storm water ponds are used to divert water from roadways to avoid flooding and surface water pollution.

impermeable. The water couldn't be absorbed into the roadway and flooding became much more common.

For this reason, widespread infrastructure was developed to manage the waters flowing off of cities and roads (Figure 9.7). Large above- or below-ground storm water sewer systems were developed to move water away from places where it could do damage to roads or cities.

The sewer systems typically take water from where it is not wanted to rivers, lakes, or ponds where it can do little damage. The movement of storm water quickly across roads and cities to rivers has changed the normal flow of streams in response to rainfall events.

Figure 9.8 shows a normal urban drainage basin. As you can see, the discharge of water in a normal stream system increases slowly and it takes a significant amount of time for the water to make its way into the stream channel. In an urbanized environment, where storm water sewers and ditches enhance the collection of rainfall from impermeable surfaces, the discharge of a stream increases rapidly and most of the water moves into the channel over a shorter time period than a natural stream.

Because of this phenomenon of rapid storm water input in urban systems, rivers in cities are more likely to flood after development than they were prior to urbanization. Plus, less of this water is diverted into groundwater systems, thereby leaving these urbanized areas more vulnerable to groundwater depletion.

In areas where streams are not present, water must be diverted into man-made channels, lakes, or ponds where it must be stored for some time (Figure 9.8). These water retention ponds are common features in places like Florida where there are few streams to divert water across the landscape.

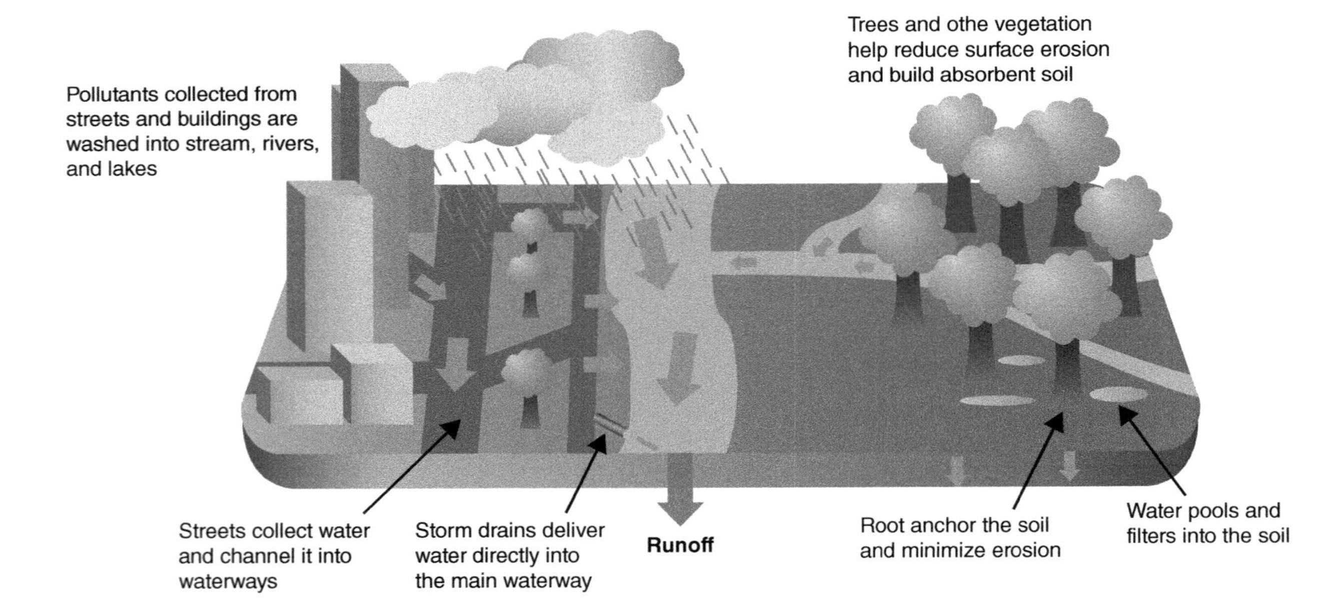

Figure 9.8 This diagram shows the difference between urban and rural drainage basins. Urban drainage basins cause water to reach streams much faster than rural basins.

Pollution is a special storm water challenge. Roads contain a number of pollutants, including oil, nutrients, heavy metals, and sediment. Oils and grease are dangerous because they can cause serious health problems upon exposure and they play havoc with ecosystems. Nutrients pose unique problems because their addition to surface water bodies can cause eutrophication of lakes and streams and cause the formation of "dead zones." Many heavy metals are poisonous not just to humans, but to other life forms. Excess sediment in storm water can clog drainage systems or lead to infilling of rivers, lakes, or ponds.

The challenge with storm water pollution is that it is a form of non-point pollution. It does not come from a single source. There are many sources of this pollution within the drainage system of a road; it can come from cars, lawns, buildings, or from waste lying on the ground. When it rains, water can pick up this pollution as it makes its way to a road.

Managing the pollution is difficult. Some places require that storm water go through some form of treatment processing before it is released into the environment. What this means is that it is collected in storm water treatment facilities where some of the pollution is removed. However, the storm water systems are more commonly managed by trying to remove sediment or litter in some traps that are designed to collect this material. It is just too difficult to try to remove nutrients, metals, oils, and other non-point pollution in most cases.

These pollution problems can be compounded in cold climates. During the spring melt, pollution that accumulates throughout the winter months is flushed suddenly into storm water via roads (Figure 9.9). Huge amounts of oils, metals, and nutrients are added during melting events when the weather warms. This is exacerbated by the inclusion of road salt and sand that were added during winter ice or snow storms. A toxic stew of nutrients, sediment, salt, metals, and oils is released during the late winter and early spring in colder climates.

Many communities are attempting to reduce the impact of non-point pollution on natural surface streams by trying to store the water in holding ponds where it can filter slowly

Figure 9.9 During the spring, pollution and salt that is stored in snow and ice is released into storm water systems.

into the subsurface. The soils and sediments help to filter out pollution prior to the water entering the groundwater system.

Yet most work on trying to reduce storm water pollution on roadways focuses on trying to prevent the pollution from happening in the first place. Many communities focus attention on reducing fertilizer use, encouraging proper vehicle maintenance, and regular street sweeping as ways to reduce storm water pollution.

Street sweeping

Street sweeping is done regularly on most paved streets daily, weekly, or monthly using vacuum and rotary brush truck sweepers. Originally, street sweeping was done for aesthetic reasons in order to remove litter and sediment from roadways. Now, however, street sweeping is done, in part, to remove pollutants to try to reduce storm water pollution.

A list of the types of things that can be found in street sweeping debris is listed in Table 9.3.

Most people do not realize how often their streets are swept. The busiest streets in cities are typically swept daily or at least twice a week. Most residential streets are swept weekly or monthly. This schedule of regular sweeping pulls tremendous amounts of pollution from roadways.

But where does this waste end up? Most of the waste taken from streets ends up in landfills and is used as topping layers on top of garbage. However, some places list street sweeping as a hazardous waste and it cannot be put into regular municipal landfills. A list of the types of materials that can end up in street sweeping waste is listed in Table 9.3. Given the heterogeneous nature of street sweeping debris, more attention is being given to its handling. In the past, many communities used the material to fill in low spots in communities. However, given the pollution potential, most communities now take the sediment to a sanitary landfill. Given the millions of miles of roads that are swept around the world every year, this waste issue is a tremendous problem.

Some have looked to find ways to recycle street sweeping debris. This is because even though sweepings do have the potential to contain some nasty pollutants, they largely contain natural sediment and organic matter. Some communities sieve the street sweeping to pull out natural sediment to use as fill and utilize the organic matter for composting. Minneapolis, Minnesota, for example, takes the sediment it pulls from street sweeping debris and uses it to sand the streets to prevent sliding on ice during the winter months. Other communities pull out the litter, such as cans or bottles, for recycling. Some have suggested using street sweeping debris as a growing medium to take advantage of the high nutrient content present in the waste.

Ground stability

One other problem with roadways that has gotten a great deal of attention in the last few years is the issue of ground stability under roads. A number of well-publicized sinkholes have formed under roadways in places like Paris, Winnipeg, and New York. Most of these sinkholes occur when water mains under roadways break for some reason (often because of aging infrastructure). When the water mains break, they wash out sediment under the roadway and cause a collapse of the pavement. Some people have been injured when these kinds of collapses occur suddenly.

Table 9.3 Types of materials found in street sweeping debris.

Type of street sweeping debris	Characteristics
Sediment	Can be local natural parent material, fill used for construction or road building, sediment (such as pieces of brick or concrete) from other human activities such as building demolition, or road materials. The sediment can range in size from coarse to very fine. The finest sediments usually contain the greatest amount of pollution. This fine-grained sediment includes urban dust that contains a number of pollutants of concern
Organic matter	Can be leaf and tree litter, grass clippings, and animal waste and corpses. Leaf litter and grass clippings are highly seasonal. During some times of the year, organic matter can make up the bulk of material collected by street sweepers. Some communities have developed composting systems to re-use the nutrients stored within the waste
Litter	Can be roadway litter associated with cars (such as spark plugs), cigarette butts, street trash (such as fast food containers), and a wide array of miscellaneous materials from condoms to old furniture.
Metals	Can be in the form of coarse pieces to microscopic chemical compounds. Includes non-toxic metals (such as aluminum, calcium, and iron), toxic and common metals (such as cadmium, copper, lead, arsenic, and mercury), and toxic and rare metals (such as gallium, tungsten, and titanium)
Nutrients	The nutrient content of street sweeping is very high due to the common use of fertilizers in cities. Of particular concern are nitrogen and phosphorus. These elements drive eutrophication in surface water bodies
Pesticides, herbicides, and other organic chemicals	A wide variety of pesticides and herbicides make their way into street sweeping waste. Anything that can be applied to lawns or gardens can find its way into the roadways. Plus, gasoline, oil, and a number of other organic fluids associated with the automobile can also be present

These vexing problems demonstrate the difficulty of maintaining roads in our modern era. These collapses are not natural sinkholes that form in limestone landscapes, but urban sinkholes that occur as results of a failure of underground urban infrastructure.

In May 2014, a water main broke in lower Manhattan near Katz's Delicatessen, a well-loved and venerable restaurant in New York City. The event was problematic due to the complexity of Manhattan's underground infrastructure of subways, storm water sewers, water mains, and sanitary sewer systems. These sinkholes are easily cleaned up and repaired. However, they demonstrate the fragility of the buried pipes just out of sight under our roadways.

Mass transit

Mass transit is designed to move many people to popular locations as efficiently as possible. There are various forms of rapid transit, such as rail, bus, and ferry. One form of transit, called rapid transit, focuses on getting people from one place to another very quickly. There are also regional and local approaches to mass transit in large urban areas. There are also a number of new emerging mass transit technologies that seek to improve efficiency while trying to get people as close to their destination as possible. Each of these themes will be discussed below.

Forms of mass transit

Railways

Railways are one of the oldest forms of mass transit in the world. Early rail systems were powered by draft animals. However, now most systems run on electricity. The most common rail systems are subways or metro systems (Figure 9.10). These railways are usually underground (or partially underground) and connect different parts of cities along set routes. Currently, there are 180 subway systems around the world. Shangai has the most extensive subway systems with about 680 kilometers of track. The Beijing and Guangzhou have the second and third most extensive systems with about 670 and 478 kilometers of track. London, Moscow, and New York round out the top six. In terms of ridership, Beijing has the most riders of any subway system (3200 million per year) with Shanghai,

Figure 9.10 Subways are integral to the economy of many cities.

Guangzhou, Seoul, Tokyo, Moscow, Shenzhen, Hong Kong, New York, and Mexico City rounding up the top 10 cities by ridership.

There are many environmental benefits to a subway system. They are very effective at quickly moving lots of people within urban settings. This reduces the need for cars, which of course reduces air pollution and the need for building car infrastructure (roads, parking lots/garages, filling stations, etc.). In addition, placing the rail system underground improves urban aesthetics by reducing noise and pollution, and by maintaining a more pleasant landscape. Cities with suburbs tend to have fewer cars per person as well, thereby limiting the overall number of cars used in cities—always a good thing for the environment.

Light rail

Light rail systems, sometimes called streetcars or trams, utilize dedicated tracks above ground to provide limited mass transit options in cities. Heavy rail systems, like subways, have greater capacity and run on more robust rail lines. However, light rail and tram systems typically utilize electricity as the main form of energy. Often, the light rail systems utilize overhead electrical lines. The rail lines are often integrated into roadway systems. In San Francisco, for example, the streetcars integrate into roadways and share the space with cars.

Light rail systems are used where ridership is lighter than could be used for subway systems. They are common in lower density cities like Miami where they try to reach all areas of the city similar to an urban subway system. They are also used to provide transit options in limited areas such as the downtown Denver light rail and the circle line in downtown Vienna.

Light rail streetcars used to be much more common in cities around the world. They provided the main form of transit from one location to another as cities expanded outside of the urban cores. However, the advent of the automobile and the expansion of suburbs made light rail less popular and many streetcar lines were abandoned. In the United States, a scandal known as the General Motors Streetcar Conspiracy, is believed to be responsible for the decline of many of the US streetcar systems. General Motors and many other companies invested in a firm that bought out dozens of streetcar systems. It was alleged that they did this to create a reliance on the automobile. Several companies were convicted of conspiring to monopolize interstate commerce via this purchase. Many of the streetcar lines were destroyed or converted to roadways. Today, many cities are trying to revive streetcar use as a viable form of mass transit.

The largest light rail systems are in Melbourne, Australia; St. Petersburg, Russia; Berlin, Germany; Moscow, Russia; and Vienna, Austria. They are cheaper to build than conventional heavy rail subway systems and tend not to need the infrastructure of the big rail lines. They work particularly well in small to mid-sized cities with low population density.

Buses

Bus mass transit is perhaps the most common form of mass transit available to the public. It is a highly flexible system in that buses can travel on almost any roadway and routes can change as the need arises. Almost every city has some form of bus system. Most are public systems run by a city or a non-profit agency tied to the city. Some for-profit systems exist in some areas.

Buses typically travel set routes within a particular timetable. The buses often connect with other mass transit options such as heavy or light rail lines. The transit buses run on a variety of fuels, although there has been a considerable effort to change vehicles over to greener fuels, particularly natural gas and hydrogen.

Buses are among the most popular transit options in the world because they are relatively inexpensive and highly flexible systems that work well in different sized communities.

Bus rapid transit

In recent years, there has been an effort to convert standard bus services into rapid transit bus systems. These systems focus heavily on using buses within a separate roadway infrastructure to allow quick movement of people without the need of developing a rail line. The bus rapid transit uses technology to synch lights so that they are not stopped by standard traffic lights. This gives priority to buses in urban systems for the rapid movement of people along existing roadways (see text box).

Ferries

A ferry system is a mass transit system that utilizes large boats to transport people along waterways. Most ferries go from one point to another with limited stops. However, some ferries make multiple stops. These are often called water taxis. They are most common in cities built close to the water. Venice and New York City are examples of places with water taxis.

However, as noted, most ferries have limited stops, usually going from one point to another. There are a number of different types of ferries that operate around the world. Some of them transport just people and some transport people and cars. Some go great distances, such as the regular ferry services between England and France or between islands in the Philippines. These ferries often connect different cities or countries. However, other ferries operate within cities. Perhaps the most well-known urban ferry is the Staten Island Ferry that operates between the boroughs of Manhattan and Staten Island. The ferry is a free service that offers regular service between the boroughs every day of the year. It is the largest ferry carrier in North America.

Transit hubs and transit-oriented development

Mass transit functions best when it is linked with other forms of transit and walkable communities. Transit hubs are places where one can link with a number of forms of transit. For example, at Union Station in Washington DC, one can catch regional rail lines on Amtrak, grab a subway, rent a car, or catch a bus. You can even rent a bike! The subway links the train station to the airport making Union Station a gateway to the world. These transit hubs allow flexibility in the options that are available for travelers. Most major cities have transit hubs like Union Station.

Even smaller cities try to create places where there are mass transit options. Where I live in suburban Long Island, New York, there are several commuter train stations of the Long Island Railroad where bus lines converge. I can catch a bus from where I live, take it to one of the transit hub train stations, and be in Penn Station in Manhattan in a relatively short time.

How easy is it to use mass transit where you live? Where are your transit hubs? Can you take mass transit to your nearest airport from your house?

If your answers to the questions above allow you access to mass transit for important trips like work or airport, you should thank a transit planner. Across the world, experts in mass transit have worked to try to provide transportation alternatives to the personal car. They have designed means of getting people to transit hubs where they can catch other forms of transit to their final destinations. This effort is aided when communities are designed around mass transit. This form of development, called transit-oriented development, is an urban design that focuses on building dense, walkable communities with access to mass transit.

Transit-oriented development also features mixed use development. What this means is that residential and commercial land uses are mixed within a single place. Shopping options are mixed within residential spaces. In many ways, transit-oriented development re-creates the dense downtowns of the past that we lost as we developed suburbs in the middle of the twentieth century.

Mid-century suburban landscapes featured residential neighborhoods where each house had a driveway and garage. In those places, one had to drive to commercial districts where one found strip malls, big box stores, and many parking lots.

Transit-oriented development, in contrast, merges the commercial district with the residential landscape to create places where one can live without a car. Shopping is within walking distance and transit options are available for people to go to work or school.

This form of development is gaining in popularity, particularly among young people who wish to live in denser, pedestrian and bike friendly places. They are abandoning the suburbs of the past and choosing to live in older dense downtown areas or in newly created communities built around transit options.

Many older suburban areas have seen some decline as a result of this demographic shift. As a result, many suburbs are creating new "downtown" areas to create transit hubs and mixed-use development. This helps to give these suburbs a new sense of place that helps to re-define them for a new generation seeking walkability and access to mass transit.

The future

It is always fun to ponder the future of transportation. If you've ever seen The Jetson's cartoon, you know back in the middle of the twentieth century many thought our cities would have flying cars! We don't have flying cars yet, but we are moving into a world where automatic driverless cars will transform the way we drive from place to place. The automation will certainly save energy and help to reduce the use of fossil fuels. We are also seeing advances in electric cars. New electric charging stations are popping up across most urban landscapes of the world.

We are also seeing more high-speed rail lines and other innovative forms of mass transit like rapid bus transit. Ships are becoming bigger and more efficient at moving goods from place to place. While modifications in airplane efficiency are slow to make their way into the market, the air travel industry is very concerned about the impact of airplanes on the environment.

Our transportation infrastructure is also changing. We are developing greener ways of designing roadways and some have advocated for transforming our paved roads into solar

Figure 9.11 Many people are interested in walking or biking to work or school.

panels. We are also seeing advances in rail and bus infrastructure. In shipping, enhanced ports and the expansion of the Panama Canal allow the more efficient transport of goods.

We are also seeing cities re-thinking transportation infrastructure. Many are striving to promote transit-oriented development by creating denser, walkable communities. Some communities are re-thinking zoning rules and are creating places that have commercial and residential space together to promote living, working, and shopping options in one place.

Even with all of these innovations, there is growing concern over pollution caused by our transportation options. By far the most significant problem is greenhouse gas pollution. All forms of transit produce these harmful chemicals that are responsible for global climate change. Other pollutants associated with transit, including oil spills, are also a concern. So, while we are making progress, it is worth noting that all of our transit options come at a measurable cost to the environment.

That is why so many people are looking toward biking and walking. More people are interested in living where they can bike or walk to work (Figure 9.11).

Long Island: America's first suburban landscape built around the automobile

When Robert Moses looked outside of New York City for new parkland for the burgeoning 1920s residents of New York City to enjoy, he looked directly at the beaches and coastlines of Long Island. Robert Moses was the "master builder" of many of New York's roadways and parks in the New York City region. He is often credited as the father of modern suburbia.

The New York City and Long Island regions were settled by westerners in colonial times and the urban footprint of New York City had expanded into the Long Island communities of Brooklyn and Queens. However, much of the rest of the vast island remained wilderness and farmland. Prior to the advent of the automobile, Queens was quite distant from the heart of the city in downtown lower Manhattan.

In the first part of the twentieth century, Manhattan was bursting at the seams. Immigrants from all over the world came to the island where population density exceeded 330 000 people per square mile!

When Robert Moses looked to find recreational opportunities for the crowds in New York City in the 1920s and 1930s, he found an abundance of natural resources in Long Island. But he needed a way to get the crowds to them. Up to that point, interstate highway systems were not imagined. So, he developed a network of roadways called parkways that took people in the newly mass-produced automobile to the beaches.

He designed the parkways as parks for motorists. He wanted drivers to feel as if they were driving in a beautiful wilderness or garden. He designed the roads with low bridges to limit access to cars. Buses, trucks, and other high vehicles were not able to travel on these roadways.

The parkways were highly efficient at bringing commuters in and out of New York City to the seashore. However, some decided to stay and put down roots to get out of the cramped city. The population of Long Island soared as more and more people came to the region. Eventually, Levittown, considered America's first suburb, was built between 1947 and 1951 as a mass-produced neighborhood built around the automobile.

Levittown served as a model for suburban development in North American and much of the rest of the world. No longer did we desire to live in the densely populated, but transit efficient, cities. Instead, many of us strove to escape to suburban landscapes. In order to do that, we needed cars and roads.

Golf cart communities

One of the newer developments in urban planning is the advent of "golf cart communities." These are places where golf carts are the preferred means of transport over cars. While there is some limited car access in most cases, golf cart travel is the main mode of transportation within the communities. What this means is that there is less of a need for roadways and associated car infrastructure. Land is used more efficiently and pollution is reduced.

Many of the golf cart communities are built in warmer areas of the world. They are common in the Sunbelt of the southern United States from Florida to California where warm temperatures make the casual outdoor golf cart experience pleasant. They are often found in resort or island settings or in retirement communities where the leisure lifestyle limits the need for cars.

(Continued)

The Villages of Florida are examples of Golf Car Communities associated with a retirement or age-restricted community. Here, it is quite common to see golf cart parking spots at grocery stores or shopping malls. While roads are present and many own cars in this community, golf carts are more commonly used for short trips. The infrastructure of the community is built around cars and golf carts.

In contrast, Bald Head Island in North Carolina, does not allow cars. The only means of transit is by biking, walking, or golf cart. The village has over 1100 private homes and is a mecca for tourists seeking quiet beaches and a simple, secluded vacation experience. Residents seek to maintain their community as a car-free environment. They need fewer streets and require less of the associated expensive infrastructure. Plus, they prize their green space and recognize that its preservation is key to the long-term economic survival of tourism on their island.

While these two examples are distinctly unique situations, they demonstrate that low-density communities can survive without the car and without extensive infrastructure to support our modern car culture.

Bus rapid transit in the Bronx

One of the most successful bus rapid transit systems in the world is in the Bronx in New York City. The Bronx is a densely populated borough of New York City where many people live in apartment buildings and multi-family homes. Many complained for years about the length of time it took to get from one place to another in the Bronx due to the lack of subway systems and the slow pace of the bus service. However, in 2008 the city opened a bus rapid transit line called "Select Bus Service."

The advent of the service cut the time of travel from one end of the line to the other in half. Riders were required to purchase fares in advance. Drivers used a transit signal priority system that kept green lights green for the buses. Since fares were purchased in advance, riders could board at front and back doors. Plus, special lanes were given over to the service.

One of my students noted that her commute on this line used to take forever. Now, she finds that the system is very quick and efficient.

10

Pollution and Waste

In our technological age of wonder, pollution is a growing problem. We are using natural and manmade materials that have never been utilized in all of human history. Their production, use, and disposal are of great concern. Plus, we are concentrating some normally benign materials, like phosphorus, in environmentally damaging ways. At the same time, our consumerist culture is creating huge amounts of waste. How we manage and handle this waste varies considerably around the world. Finally, the management (or lack thereof) of sewage waste has tremendous implications for public health.

Pollution

Pollution is a growing problem on our planet as more of us are exposed to materials that can cause personal or environmental harm. This section reviews the main types of pollution and provides management strategies for dealing with pollution problems.

Chemical pollution

Chemical pollutants are the most problematic of all pollution types. This is largely because the chemicals are unseen and difficult to determine. We are unable to assess whether or not they are present in our environment without some type of testing. A listing of some of the major types of pollutants are listed in Table 10.1, broken down into categories of metals and metalloids, organic chemicals, nutrients, and radioactive particles.

Metals

A number of metals and metalloids are considered important pollutants, particularly arsenic, cadmium, chromium, lead, and mercury. These materials are problematic because they are elemental—meaning that they do not break down in nature. These materials are toxic to humans and other organisms at high levels and can cause health problems at lower levels. One of the challenges with metals is that they can bio-magnify, which means that they can become more concentrated as they move up into the food chain. Metals can be remediated in the environment by transforming them into insoluble compounds or by removing them to a managed waste facility.

Introduction to Sustainability, Second Edition. Robert Brinkmann.

Companion website: www.wiley.com/go/Brinkmann/IntroductiontoSustainability

Table 10.1 Major chemical pollutants.

Metals and metalloids	Organic chemicals	Nutrients	Radioactive particles (isotopes of the elements listed)
Antimony	Aldrin	Nitrogen	Americium
Arsenic	Chlordane	Phosphorus	Barium
Beryllium	Chlordecone	Variety of trace elements	Californium
Bismuth	Dichlorodiphenyl trichloroethane (DDT)		Cesium
Cadmium	Dioxin		Cobalt
Chromium	Dieldrin		Iodine
Cobalt	Endosulfans		Krypton
Copper	Endrin		Neptunium
Gold	Heptachlor		Potassium
Lead	Hexabromocyclodecane		Plutonium
Mercury	Hexabromodiphenyl ether		Protactinium
Nickel	Hexachlorobenzene		Radium
Palladium	Hexachlorocyclohexane		Radon
Selenium	Lindane		Selenium
Silver	Mirex		Strontium
Tellurium	Pentachlorobenzene		Technetium
Thallium	Perfluoroctanesulfonic acid		Thorium
Tin	Polychlorinated biphenyls		Uranium
Zinc	Polychlorinated dibenzofurans		Yttrium
	Polycyclic aromatic hydrocarbons (PAHs)		
	Tetrabromodiphenyl ether		
	Toxaphene		
	Tributyltin (TBT)		

Organic compounds

In our technological era, we are producing a huge variety of organic chemicals. Many of them are benign, but many others cause significant health problems upon exposure. The list in Table 10.1 only includes organic chemicals that are persistent in the environment. What this means is that they do not readily break down through normal environmental processes. In many ways, they become like metals in that they are difficult to clean up without removing them or chemically altering them in some way. The persistent organic chemicals listed in Table 10.1 are problematic because they can disrupt reproductive and endocrine systems as well as cause heart disease, cancer, and diabetes. In addition, children born of women who are exposed to some of these chemicals can have developmental disorders and learning disabilities. There are a number of less persistent organic chemicals that also are pollutants of concern.

Nutrients

The natural nutrient content of soils varies with local geology. As a result of this natural variability, farmers and gardeners have developed ways to add nutrients to the soil. For centuries, growers used natural fertilizers like manure and compost. However, in the last century, a tremendous variety of fertilizers have made their way to market for farmers, gardeners, and those interested in lawn care. The improvement in fertilization technology led to huge increases in agricultural output throughout the world (Figure 10.1). Unfortunately, the use of fertilizers has also led to widespread nutrient pollution.

The main plant nutrients are broken down into macronutrients and micronutrients (Table 10.2). The macronutrients are needed in great abundance and the micronutrients are needed in small quantities. When any of these elements are not present in the soil, it limits the ability of plants to grow to their fullest potential. In most cases, there is an abundance of several of the macronutrients and micronutrients in soil, particularly carbon, oxygen, and iron. Importantly, low levels of nitrogen and phosphorus are the main limiting factors for strong plant growth in most soils. These two elements make up

Figure 10.1 Fertilizers are great for producing healthy gardens and farms, but their over-use or mismanagement is a serious pollution problem.

Table 10.2 Macronutrients and micronutrients needed for productive plant growth.

Macronutrients	Micronutrients
Calcium	Boron
Carbon	Chlorine
Hydrogen	Cobalt
Magnesium	Copper
Nitrogen	Iron
Oxygen	Manganese
Phosphorus	Molybdenum
Potassium	Nickel
Sulfur	Zinc

a great proportion of fertilizers. These are also elements that cause significant nutrient pollution problems because they are present in fertilizers in soluble compounds that can enter surface water and groundwater systems.

The widespread use of fertilizers in agriculture is problematic for rural environments. However, fertilizers are added, often in great quantities, in many urban settings. Suburbanites love their green lawns and abundant gardens. Golf courses carefully tend their grounds to ensure luxurious grass growth (Figure 10.2). Many playgrounds, ball fields, and parks depend on fertilizers to maintain a particular look. The application of fertilizers in both rural and urban settings makes them among the most common pollutants on the planet. They can be found in water, soil, and street sediment in almost any humanly altered landscape on Earth.

Nutrients are found naturally in surface and groundwater systems, just like soils. Also, just like soils, nutrients are extremely variable in water. Some waters are naturally nutrient rich. Rivers running off of nutrient-rich soil, for example, often contain high levels of nutrients. These rivers often have abundant water plants like cattails and water lilies. They may also have a great deal of algae present. Nutrient-poor water bodies tend to be clear systems with very low aquatic plant growth. Rivers running off of steep mountains with little soil often are low-nutrient environments.

Entire aquatic ecosystems have evolved for thousands of years around the natural occurring nutrient content entering the system through groundwater or surface water. When we add nutrients to these systems from fertilizer runoff, we throw them out of their natural balance that has been in place for generations.

Radioactive Pollutants

There is great concern over pollution released after the Fukushima nuclear power plant disaster that occurred in 2011. However, radioactive pollutants are present in many areas besides Fukushima. Nuclear waste and pollution are produced through nuclear power

Figure 10.2 Many who live near the water try to limit the use of fertilizers and pesticides in their lawn and garden.

generation, medical applications, atomic bomb testing, and research. Table 10.1 lists elements that can be nuclear pollutants when they are in particular isotopic forms. An isotope is a form of an element that has a different number of neutrons from the standard atomic number. The atomic number is the number of protons in the nucleus of the atom. Many isotopes are unstable and break down into more stable forms within a set amount of time called a half-life.

Pharmaceutical pollutants

The proliferation of readily available pharmaceutical products around the world has led to concerns over their pollution of water. Testing has found elevated levels of hormones, anti-depressants, and a range of other products in surface waters in many areas. Some of these materials pass through the human waste stream and enter surface waters via releases from sewage treatment plants. Others find their way into surface waters from animal farms that inject or feed animals with pharmaceutical products. In the environment, it is believed that some pharmaceutical products, particularly hormones, are disrupting reproductive systems and causing birth defects in some animals. The challenge for managing this pollutant class is that pharmaceuticals are difficult to remove from wastewater. Most treatment regimens for drinking water do not remove pharmaceuticals, much less test for them.

Heat pollution

Many of our industrial processes create excess heat that is released into the air, water, or ground. This heat is considered a pollutant when it causes environmental problems.

Because water is used so frequently for cooling in industrial processes, there are many examples of heat pollution in lakes, rivers, and estuaries throughout the world. Ecosystems that evolved within particular thermal regimes are harmed when they heat up. Often, non-native species are drawn to the warmer waters, thereby disrupting normal ecosystem processes. In addition, the warmer waters can change normal migratory patterns, thereby putting some species at risk.

Local heating of air in cities from the burning of fossil fuels and the release of long wave radiation stored in concrete and other materials has been well documented in a phenomenon called an urban heat island. These islands act like bubbles over cities throughout the world. The differences in heat are most noted at night. When surrounding rural areas cool after sunset, urban areas remain warm due to the slow release of long wave radiation (heat) that was stored in roads and other non-vegetated areas during the day, when they received solar radiation.

Heat islands create their own weather patterns. Researchers throughout the last century have documented changes in wind, cloud cover, precipitation, and humidity. The added heat can lead to extreme health problems during heat waves. Of course, concerns over global climate change have highlighted other health and environmental problems that are emerging as a result of global environmental change.

Light pollution

Widespread electrification over the last century has transformed the night sky. For millennia, plants and animals evolved with the subtle changes in lunar and stellar light that is so noticeable in dark skies. Today, throughout the world, artificial lighting makes seeing the Moon and the stars difficult. The outdoor environment is different (Figure 10.3).

Figure 10.3 The night sky throughout the world has changed. What kind of impact do your actions have on the night environment?

Because so many animals and plants are impacted by night light, the pollution produced by lighting greatly impacts their behavior and can disorient them. Breeding of some species is timed with phases of the moon. Lighting of coastlines in developed areas can impact sea life. Sea turtles lay their eggs on sandy beaches in many parts of the world. When the eggs hatch, the young turtles are drawn via evolution to the naturally lighter reflective ocean. When coastlines are lit artificially, the turtles turn away from the water and are compelled to move toward the land.

Light pollution is not just a broad problem of the overall sky. Individual lights can also have a direct impact on ecosystems. For example, adding a light to a property significantly alters the site. Just think about your own experiences with lighting. Insects are drawn to light, but many animals (including insects) shun light. Lighting also has an impact on human health. Changes in lighting impacts our natural rhythms and can create stress.

Noise pollution

Loud noises create biological stress in many organisms (Figure 10.4). Prolonged exposure to noise can impact hearing and lead to high blood pressure and other illnesses. In the animal world, noisy environments make it difficult for animals to hear predators. Many airports and roadways have built noise-reducing structures to mitigate the impact on residents.

The use of sonar in oceans has elevated concerns about noise pollution in water. Whale songs change in environments that are particularly loud (Figure 10.4).

Visual pollution

While not often classed as a traditional pollutant, visual blight is of growing concern to many around the world. The problem of visual pollution is not new. However, it has grown in prominence in recent decades as we have seen expansive development of roadways, power lines, and suburban sprawl. Many beautiful views have been destroyed due to

Figure 10.4 Noise is a problem for many of us who live in big cities. However, noise is also a problem for marine animals.

over-development and visual clutter. As we go about our daily lives, we are exposed to a number of unpleasant views that include advertising, litter, ugly buildings, electrical infrastructure, and poorly maintained properties.

Many use the term "viewshed" to describe the space one sees in one's daily life. Many communities try to manage community viewsheds to eliminate visual pollution. This is done by very strict zoning and planning, as well as by encouraging property maintenance and aesthetic landscaping.

Littering

Littering is improper disposal of waste. While most litter is not harmful to the environment, it reduces the overall quality of ecosystems when out of control. The problem is especially harmful in areas of the world without regular waste collection and recycling. Geologists have coined the term plastiglomerate for the solid rock-like material that forms in some areas where large amounts of plastic waste are deposited. The most littered item in the world? Cigarette butts.

Some litter poses special problems. For example, fishing line is a serious problem for water birds and fish. They can get trapped in the lines. Six-pack containers are notorious for trapping animal necks and constricting growth. Necropsies of animals have found cigarette butts and other litter impacting digestive systems. Extensive littering can cover soil or aquatic substrate, thereby limiting plant growth. Large accumulations of litter, such as tires, can be a fire hazard.

Understanding pollution distribution

Pollution can be found in air, water, soil, and biological organisms. Regardless of media, the pollution originates at a source. Sources can be point, line, or areal in nature.

Point sources of pollution are single points where pollution emerges into the environment. Point sources include smokestacks, tailpipes, or an individual drainage pipe. Their effect is most significant at the point of the source with dispersion away from the emission. Line sources, in contrast, are linear segments of the landscape where pollution is emitted. Line sources could be things like roadways, leaking pipes, or lines of light. The impact can be seen in a linear pattern on the landscape where the impacts are found along a line with dispersion of the material radiating away from the line. Finally, areal pollution occurs when there is an aerial or regional source of pollution. Nutrient-rich runoff from agricultural fields or suburban lawns is a good example of areal pollution sources.

Managing point or line pollution is far easier than managing areal pollution sources. Single sources can be managed at the point of emissions. Individual factories, power plants, or cars can be regulated to limit pollution to acceptable levels. They can be managed in partnership with the organizations responsible for producing the pollution.

Areal pollution, in contrast, is much harder to manage. The pollution problems associated with areal pollution are much more dispersed. It is not a matter of regulating and working with one single property owner. Instead, many property owners, indeed, whole communities, are responsible for the pollution problems.

This has led to the classification of pollutant sources as point and non-point pollution. Because point pollution is easily regulated and managed, many guidelines have been established for emissions for point solution sources like smokestacks. It is far harder to regulate non-point pollution.

According the US EPA, sources of nonpoint pollution can include (Quoted directly from http://water.epa.gov/ polwaste/nps/whatis.cfm):

- Excess fertilizers, herbicides, and insecticides from agricultural lands and residential areas
- Oil, grease, and toxic chemicals from urban runoff and energy production
- Sediment from improperly managed construction-sites, crop and forest lands, and eroding stream banks
- Salt from irrigation practices and acid drainage from abandoned mines
- Bacteria and nutrients from livestock, pet wastes, and faulty septic systems
- Atmospheric deposition and hydromodification.

Given that most of the challenges associated with non-point pollution are caused by runoff from polluted lands, a significant amount of attention has been given to reducing pollution in storm water runoff.

Storm water is water that flows overland during rainfall events. In rural areas it enters streams after flowing off of agricultural or natural lands. In recent years, there has been significant attention given to trying to reduce runoff from farms, particularly lands that hold animals. Fields, pastures, or pens that contain large numbers of animals also concentrate animal waste. During rainy seasons, this waste can enter streams to cause significant nutrient pollution. Some areas with large numbers of animals have even started using mini sewage treatment plants to try to lessen the impact of animal wastes on non-point pollution.

In urban areas, storm water is diverted to roadways where it is moved to storm water sewers or ditches that divert the water to surface water bodies like lakes, rivers, or ponds. Cities with significant surface water pollution problems divert storm water to sewage treatment plants in order to remove pollutants like nutrients.

More and more attention is being given to storm water pollution in cities. This is largely because urban storm water contains high amounts of nutrients that can lead to eutrophication of surface waters. Urban storm water also contains metals, litter, organic chemicals, and *E. coli* bacteria from pet feces.

One of the most important developments in point and non-point pollution management in recent years is the development of something called total maximum daily loads, or TMDLs. TMDLs are set at the watershed, or drainage basin level for specific pollutants. Thus, managers must work with point and non-point pollution issues to achieve compliance to pollution standards.

The management of TMDLs is complex. Prior to setting TMDLs, government organizations must study the watershed in order to understand its overall hydrology, ecology, and pollution problems. Watershed land use is highly variable and thus it is important to understand the overall range of pollution sources within the watershed in order to best design a strategy for managing pollution. Once this study is complete, maximum pollutant loads can be designed in order to reduce the impact of pollution on the region. If a drainage basin is out of compliance, strategies are developed to reduce pollution.

The benefit of using TMDL management strategies is that they focus on all pollution sources in designing a regional approach to pollution management. Instead of targeting a single industry or source, a regional approach is employed to make improvements. TMDL management is typically used in large impaired drainage basins (see text box, Manhasset Bay Protection Committee).

The Manhasset Bay Protection Committee

Manhasset Bay is a small embayment in Long Island Sound several kilometers east of New York City. In the late twentieth century, the waters were severely impaired. Shellfishing, which was once a staple of the economy, disappeared due to the pollution in the bay. Eutrophication was a regular problem. A wide variety of pollutants, including metals, nutrients, and litter, were a severe problem in the bay. To try to combat the pollution problems, several local governments that surround the bay formed The Manhasset Bay Protection Committee in 1998. This group is responsible, in part, for managing the water quality of Manhasset Bay at the watershed level.

The Committee consists of representatives of twelve villages, one town, and one county. Collectively, they are seeking to reduce storm water pollution via education of citizens with education programs and through partnerships on projects with local governments. Since its formation, the water quality of the bay has improved tremendously. While shellfishing is still prohibited in the bay due to pollution, beaches have opened and many are enjoying the improvements in water quality that have been made due to the regional cooperation of this organization.

The US approach to pollution

In the United States, pollution is regulated using a variety of laws that were passed by congress in the 1960s and 1970s. These include the Clean Air Act, Clean Water Act, National Environmental Policy Act, and the Comprehensive Environmental Response, Compensation and Liability Act. Each of these will be discussed below.

Clean Air Act

This law was established in the United States via legislative processes in 1963 (Figure 10.5). It was amended and expanded in 1967, 1970, 1977, and 1990. Today, the law is widely applied nationally and at the state and local level to regulate air pollution from stationary and non-stationary (cars, trucks, airplanes, etc.) sources.

The Act consists of six parts, each called titles. Title I defines the programs and activities associated with air pollution that the EPA is required to regulate. It includes very distinct guidelines on air quality and lists national standards that are required for attainment of clean air. It requires permits for all new significant sources of air pollution. Importantly, it also spatially defines regions as attaining or not-attaining particular air quality standards via the assessment of air quality guidelines. There are consequences to non-attainment.

Figure 10.5 The clear air that is enjoyed in many parts of the world is due to sound management and regulation.

Plans must be developed to inventory and quantify all emissions from stationary and non-stationary sources, and develop a timetable for ensuring compliance. Specific policies and measures must be outlined to achieve compliance. In addition, contingency plans must be made in case the region is not compliant in the future.

Title II focuses on emissions from mobile sources. This act focuses on air pollution associated with moving vehicles, especially cars and trucks. It sets specific standards on air pollution emissions from individual cars. These pollutants include hydrocarbons, carbon monoxide, nitrogen oxides, and particulate matter. It also sets rules for the materials that may or may not be present in fuels.

Title III reviews general provisions in the law and includes a number of different issues. Perhaps the most important aspect of Title III is that it allows citizens to sue for air quality compliance. What this means is that individuals can sue to require a non-compliant entity to become compliant with air quality standards. This provides a unique approach to enforcement of air quality standards. Title III also better defines air quality monitoring guidelines.

Title IV stresses acid rain reduction by setting standards for acid-producing pollution, particularly sulfur dioxide and nitrogen oxides. It also provides incentives for the development of clean coal technology. This makes sense since coal-burning power plants contribute the acid-producing chemicals into the atmosphere.

Title V requires permits for any major air pollution source. The permitting process makes emitters develop a report that summarizes the emissions quantity of regulated air pollutants and that outlines plans for pollution reduction. The permitting process is largely managed by state and local governments.

Title VI provides rules for the protection of the ozone layer of the stratosphere. This is the layer of air that protects the Earth from harmful UV radiation. Title VI lists the chemicals that are regulated, provides a timetable for their removal from use, and encourages research on alternative products. The law also provides a framework for international cooperation on removing these products from international production.

Together, the compilation of rules set by the Clean Air Act provides a framework for the nation to regulate air quality. Many of the rules are managed in partnership with state and local governments. What is important in the context of the Act is that it comprehensively addresses many of the key issues associated with air pollution: individual large polluters like factories, many individual small polluters in transportation, and regional air quality problems.

Clean Water Act

The Clean Water Act was passed by the US Congress in 1972 in response to serious water pollution problems in the country. It was modified twice in 1977 and in 1987. There have also been some modifications to the interpretation of the law by the courts. Originally, the law was intended to focus only on single point pollution. However, in recent years, it has been used to regulate non-point pollution like storm water pollution.

The law is divided into six main parts, called Titles. Titles I and II focus on providing funding for research for technology, pollution control, and improved sewage treatment plants. This section of the law was intended to assist with identifying the best ways to manage water pollution and deal with a number of particular pollution sources, while also funding improvements in local sewage treatment plants to improve surface water quality.

Title III of the Clean Water Act focuses on designing pollution standards and providing a mechanism for enforcement for the law and requires permits for discharging into surface water. Specifically, it requires science-based standards for water quality of sewage treatment plants, and industrial pollution sources. It also sets parameters of pre-treatment of waste products prior to their release into the environment for a number of industrial activities such as iron and steel manufacturing. Title III also initiated the development of a Water Quality Standards Program that sets standards for pollution levels in water based on risk to public health and the environment based on the use of the water (recreation, drinking water reservoir, etc.). When waters are impaired, they are placed on a list of impaired water bodies and plans must be made to make improvements. In addition, they are required to assess the TMDLs of pollutants the water can receive.

Along with the development of standards, Title III also sets guidelines for the development of a national water quality inventory, conducted every two years. Each state and jurisdictional entity must submit a report summarizing water quality and defining issues within their borders that are of national significance.

The enforcement aspect of Title III focuses on the number of fines and amount of jail time that can be given to violators. First or second time offenses can be up to $50 000 a day for water quality offences. Intentional violations that put people at risk can lead to up to 15 years in prison and fines up to $1 000 000 for an organization.

Title IV details the rules associated with managing discharge pollutants. It sets guidelines that states must follow in managing pollution permits. Title V provides guidelines that protect employees who disclose pollution problems within their companies (whistleblower protection) and allows individuals to sue an individual or organization that violates the Clean Water Act. It also allows individuals to sue the EPA for failure to regulate an offense. Finally, Title VI provides funding for state and local governments to develop large projects to improve water quality in areas such as sewage treatment and non-point pollution.

The Clean Water Act has been very effective at improving water quality in the United States. While there are many remaining problems, particularly nutrient pollution from non-point storm water pollution from cities and agricultural fields, many of the surface waters in the nation are much cleaner today than they were when the Clean Water Act was passed.

National Environmental Policy Act

The National Environmental Policy Act, or NEPA, was passed by the US Congress in 1969 to protect the country from pollution and other environmental damage as a result of large projects regulated by the government. NEPA is considered one of the most important environmental rules established by the government because it is focused on preventing environmental problems before they occur.

NEPA rules require all major projects to undergo a review so that environmental considerations are as important as other aspects of a project. What this means is that it places the environment on equal footing as need, economics, or other issues that can move an organization to undertake a large project.

Prior to even beginning a project, an environmental assessment must be made in order to evaluate whether or not the project will have any major environmental impact and whether or not there are alternatives to the project that could be explored in order to prevent any environmental damage. The environmental assessment may conclude that there is no significant environmental impact as a result of the implementation of the project. If that is the conclusion, NEPA rules have been satisfied and the project can proceed. If they have not been satisfied, an environmental impact statement must be prepared that seeks input from the public and experts.

The environmental impact statement has several goals. It has to describe the environmental problems that will occur if the project moves forward, especially those that cannot be avoided. It also has to evaluate alternatives to the project and the resources that will be needed to maintain the project over the long term.

While environmental impact statements do not have any rule-making power, they do influence decisions. They are largely a decision-making tool that those in authority use to balance whether or not a project can go forward. In other words, the statements do not necessarily recommend a project or reject it. The documents review the impacts and suggest alternatives.

Perhaps the best example of how environmental impact statements are used is the controversial Keystone XL Pipeline. This is the pipeline that developers built to bring raw petroleum products from the Athabasca Tar Sands region of Canada to refineries in the United States. The pipeline, once complete, will be over 3000 kilometers long.

In order to build the pipeline, an extensive environmental impact statement was written that summarized the issues associated with the development of the project, especially those that relate to global climate change. Concerns have been raised about using the pipeline to develop the tar sand reserves of Alberta and how their use will impact global climate change.

The environmental impact statement was released by the State Department. They oversaw its development because the project crosses the border between Canada and the United States. The report largely focuses on whether or not the project will impact global climate change. It reviews the environmental impacts of the development of the pipeline and provides some suggestions for alternatives. However, the report is not linked with a final decision. In this case, the President must authorize the development of the pipeline and he will use the statement to better understand the implications of his decision. As of this writing, the President has not made it clear whether or not he will approve the project. He has been aggressively using the Presidential authority to set rules for climate change, but has not signaled his intent on the pipeline.

Some have argued that the NEPA process slows development. Perhaps that is true. In the case of the Keystone XL Pipeline, it is worth noting that the project was first proposed in 2008. Environmental impact statements were prepared by the State Department in 2010 and 2011. The final report was released in 2014. The project was initially rejected in 2012 by President Obama as a result of his concerns over the potential for environmental damage to the Sand Hill district of Nebraska. Since then, there have been numerous attempts by Congress and President Trump to permit the extended pipeline, but it has not yet been built.

Nevertheless, NEPA is a very powerful tool to try to avoid pollution problems. Indeed, the program is so successful that many states use NEPA as a model. Many state governments require environmental assessments and subsequent environmental impact statements for state permitted projects. Some local governments also have this requirement.

Superfund

One of the newest laws that impact federal environmental policy is the Comprehensive Environmental Response, Compensation, and Liability Act of 1980. More commonly called Superfund, or CERCLA, the law was established to clean up and restore properties that were impacted by toxic pollution.

The idea of a "superfund" was established after evidence emerged of widespread environmental contamination of some difficult to manage properties. In some cases, the contamination was in the past and new land uses were now on top of the old contamination. In other cases, the property was abandoned or the current owner had nothing to do with the contamination. As a result of these complex problems, guidance was needed at the federal level to try to deal with the complicated matters that arose as a result of these challenges. Who was responsible for clean up? Who was responsible for paying for it? Who would ensure that the properties were safe?

Perhaps the most notorious problem that provided the impetus for the development of CERCLA was the widespread contamination of the Love Canal area near Niagara Falls, New York. Love Canal was used as a dump site for the city of Niagara Falls and then the

Hooker Electrochemical Company. The company dumped hundreds of barrels of waste that included everything from solvents and caustic materials to dyes and perfumes. The company sold the dump property in 1953 to the Niagara Falls School District. In the sale, the company advised closing off the development of the dump and fencing it in to prevent exposure.

However, the District built a school on the site and developed the property for homes, even though barrels of waste were exposed in the construction process. By the mid 1970s residents noted a variety of common health problems. Besides unusually high rates of miscarriages, more than half of the children born to families living in Love Canal had at least one birth defect and chromosomal damage was found. Eventually, a national environmental emergency was declared and residents were relocated and compensated. The site is largely abandoned today. The Love Canal disaster is often seen as the event that triggered the birth of CERCLA.

When it was established, Superfund was designed to receive funding from polluters to fund clean-up projects. To date, about 75% of the funding for the clean-up was received by the responsible parties. When the law started, a petroleum tax was used to pay for funds for projects where the responsible parties could not be found. That tax was abolished in 1995. Today, the US Congress must authorize funds for clean-ups where responsible parties are not found.

It can often be difficult to gauge who is responsible for contamination of a property. This is especially true since some of the worst contamination problems pre-date the development of the environmental laws that we now use to regulate pollution. Responsible parties include current owners or operators of the property, the owners or operators at the time of the contamination, the parties that arranged for disposal of the material, or the parties that transported the material. In many cases, an environmental historian is needed to reconstruct ownership and activities on a property to evaluate overall responsibility for the contamination.

CERCLA is managed by the EPA. As such, it maintains a listing of all of the potentially hazardous sites in the United States. After conducting an initial review, it ranks the site to determine whether or not it should be elevated to Superfund status. Many contaminated sites can be managed in partnership with state and local officials in order to remove contaminants and reduce risk. However, if the contamination is serious enough, the site can be given Superfund status by placing it on a National Priorities List, sometimes called the Superfund list.

The name Superfund is unfortunate. It implies that the government has a deep reservoir of funds that can be used to implement clean-up of listed sites. Nothing could be further from the truth. While the government did raise funds for clean-up initially in the early 1980s, there is no source of revenue set aside for cleaning up Superfund sites. The EPA works with local authorities to identify the responsible parties who are required to pay for the remediation.

Of course, the clean-up costs are prohibitive for some. The EPA, in partnership with state and local governments, will work with responsible parties to help, particularly in providing expertise and technical assistance. Some of the clean-up operations are complex and beyond the scope of some local environmental consulting firms or local agencies. CERCLA officials will help and advise on complex clean-up initiatives.

To date, there are approximately 1300 properties that are on the Superfund list. Just over 400 properties have been taken off the list. This means that in the 40 years of operation, CERCLA has been able to remediate approximately 10 properties a year. Some have criticized the rate of clean-up of these sites. However, given the limited funding and the challenges associated with the size and complexity of some of the sites, the rate of clean-up is not all that surprising. Indeed, many properties provide significant scientific challenges. A number of innovative clean-up techniques and technical discoveries emerged from the challenges of mitigating some of these sites. Perhaps with the advances in technology the list will decrease faster in the coming decades. Plus, we must remember that we are not creating as many toxic dump sites as we did in the past due to improvements in rules. We are hoping we are avoiding the Love Canals of the future through better environmental management.

It is important to note that the Agency for Toxic Substances and Disease Registry (ATSDR) emerged out of the 1980 CERCLA legislation. It was created in 1983 to assess the public health risks of people living near Superfund sites and to expand the knowledge about health effects caused by exposure to hazardous materials. It was also charged with finding ways to prevent exposures. Since it formed, the mission of ATSDR has increased dramatically and focuses on a number of areas of public health as it relates to toxic substances. It is managed by United States Department of Health and Human Services in consultation and partnership with the Centers for Disease Control.

ATSDR also manages several databases that are important indicators of national public health risk. These include a listing of minimal risk levels of chemicals that are likely to be without a risk of adverse non-cancerous health risk, a list of priority substances that pose a potential threat to human health, and a national toxic substances incidents listing.

The EPA also manages a toxic release inventory database that can be searched (https://www.epa.gov/toxics-release-inventory-tri-program) by address to find the type of toxic chemicals that are released in any place in the United States.

Sewage treatment

One of the most problematic waste issues in the world is sewage. It is so problematic because it is a vexing public health issue due to the potential for disease transfer and because it contains a large volume of nutrients. Thus, sewage treatment is important for public health and for the protection of the environment.

Places in the world with limited sewage treatment infrastructure have distinct public health and environmental challenges. Diseases like cholera can be carried in sewage. When sewage is not treated appropriately, it can enter drinking water systems. When it enters surface water systems, the excess nutrient content can lead to eutrophication and the solids present in the waste disrupts natural systems in the bottom environments of the aquatic ecosystems.

There are several forms of sewage treatment that can be divided into two groups: on-site and off-site treatment. On-site treatment focuses on treating sewage where it is produced. Off-site treatment collects sewage from multiple sites via sewer piles to a single processing facility.

The most common form of on-site sewage treatment is a septic system. Septic systems are used most commonly in places where off-site systems are not available. On-site treatment is often considered more desirable than septic systems since sewage treatment plants are able to limit local impacts. Septic systems require two types of infrastructure: a septic tank and a drain field.

When sewage waste leaves a home, it is transported via a pipe to a septic tank. In the tank, heavier solids filter to the bottom where they partially decompose via anaerobic decomposition. Liquids are diverted into the drain field. Septic tanks have to be pumped regularly to remove solids that do not decompose or else the septic system will become unusable. High-use small septic systems must be pumped every few years. Other systems can be pumped at much longer intervals. Failure to pump tanks will cause clogging. In some cases, solids will overflow into the drain field where they can clog drain field piping.

Drain fields are connected to the septic tank by a pipe. From this pipe, several pipes radiate out into the drain field to distribute liquids over a broad area. When the drain field is constructed, gravel is often laid down to create the drain field to encourage gravity flow of the liquids into the subsurface.

There is growing concern over the environmental impact of septic systems. The liquids that enter drainage fields can enter groundwater and surface groundwater systems. These liquids carry high amounts of nutrients. Some areas have seen significant nutrient water pollution problems. As a result, many communities are trying to get rid of old septic tank infrastructure. They are putting in sewer lines that will transport sewage to treatment plants. However, putting in this sewage infrastructure is very expensive. Since most septic systems are in areas with relatively low density housing, lots of piping must be put in to service very few families. Because many people move to low density communities to take advantage of low tax rates, the expense of putting in sewer lines is often politically unpalatable due to significant increases in taxes that are needed to pay for construction of the sewer pipes.

Off-site sewage treatment is used to process sewage prior to its release into the environment. The treatment is done, in part, to reduce the biological oxygen demand (BOD) of the sewage. If it were released without treatment into surface water bodies, organisms would use a tremendous amount of oxygen to decompose the sewage. This would cause eutrophication and environmental problems.

There are three types of sewage treatment: primary, secondary, and tertiary. Each type of treatment cleans the sewage and progressively removes harmful materials prior to release into the environment.

At the most basic level, primary treatment separates solids, oils, and floating materials from liquids. This is done in settling tanks that allows heavy sludge solids to settle, while also allowing floating materials and grease to rise to the surface. Each of the separate materials can be further processed, collected for practical uses, or released into the environment. The oils and greases are sometimes collected to make soap. The solid sludge can be gathered for use on agricultural fields or it can be taken to a landfill. It can be applied to agricultural fields while wet or after drying. The City of Milwaukee uses its dried sewage sludge to make a product called Milorganite which is sold throughout the United States as a nitrogen-rich soil amendment (Figure 10.6).

Figure 10.6 Many playing fields are fertilized with nitrogen-rich Milorganite to keep them green. The fertilizer is a by-product of sewage treatment from the City of Milwaukee.

Secondary treatment is used to process the material that remains after the heavy sludge and floating materials have been removed. While there are a number of ways that secondary treatment is done, the process focuses on using bacteria to further break down the sewage and reduce the biological oxygen demand. In some cases, the sewage is processed through constructed wetlands where natural processes reduce the oxygen demand. At the close of the secondary treatment process, solids are allowed to settle. Once this occurs, the sludge is collected and processed with the sludge collected in primary treatment. Many communities around the world release the remaining liquids to surface water bodies at this point. It must be noted that the released liquid still contains nutrients and often contains a relatively high, but certainly reduced, BOD.

In some places, expensive tertiary treatment is used to remove nutrients and other harmful materials from the remaining liquid prior to releasing it into the environment. In some situations, the liquids are disinfected with chlorine, ultraviolet light, or ozone to remove microorganisms. This usually turns the sewage into water that meets local drinking water standards. Tertiary treatment is often called a "toilet to tap" process because the resulting material is so clean. It is expensive, but crucial in some key areas.

Tertiary treatment is used in many places around the world to try to protect fragile or damaged ecosystems that would be harmed by the release of nutrients into the environment. In Great Britain, for example, tertiary treatment is used for areas that are designated "sensitive."

Sewage and sustainability

Obviously, the most important aspect about managing sewage is to try to manage it so that it does not do any damage to the environment. However, thinking about sewage as a potential resource is also important. In the past, human waste was used as fertilizer, in tanning leather, and as a fuel. Today, many are looking at ways to transform the sewage we collect into usable products.

As noted earlier, fertilizers are commonly made from treated sewage. However, there is growing interest in using sewage for energy production. Methane and other burnable gases can be collected from sewage to run power plants. Plus, dried bricks made from sewage sludge can be burned as a biofuel. In addition, sewage provides a source of precious water in places that are susceptible to drought. Many communities pump the liquids remaining from secondary or tertiary sewage treatment into surface water bodies or groundwater systems to try to mitigate the impacts of human water withdrawals.

Garbage and recycling

What we do with our waste varies all over the world. Many parts of the world do not have regular garbage collection. Waste is managed by individual families. In other areas of the world, waste management is a highly organized process that is undertaken by governments with strict regulations. This section will review waste management and recycling options in areas of the world that manage waste.

Before delving into waste management in more developed settings, it is worth noting a few points about the issues of waste in places with limited collection (Figure 10.7):

1. When there is limited garbage collection, waste must be handled at the family level.

Figure 10.7 In some places, such as here in Guyana, the infrastructure is not in place to deal with municipal waste (photo courtesy of Chontelle Sewett).

2. Individuals are likely to produce less waste in such settings or re-use materials that remain.
3. Individuals are likely to mishandle hazardous materials.

While there are some benefits to managing one's own trash, it is easy to mishandle the materials. One can cause environmental or health problems due to bad handling. For example, disposal of waste oils can pollute groundwater systems or burning hazardous chemicals can cause health problems to those exposed to the smoke.

Garbage composition

According to the US EPA, the composition of garbage, which they call municipal solid waste, consists of many materials that could be recycled. The most common materials in municipal garbage (prior to recycling) are paper and paperboard, food waste, yard trimmings, and plastics. A full summary of the composition of municipal garbage is listed in Table 10.3. There are many resources present in the waste that can be re-used or recycled in some way.

Managing garbage

Managing garbage within communities involves several steps. It must be collected, transported, sorted, and disposed. Each step involves a degree of sustainability.

Collection

Many communities that collect garbage have rules for what can and cannot be collected. For example, communities may collect household garbage one day and yard waste another. Some may have special collection days for electronic waste or pharmaceuticals. Some communities also require that the garbage be sorted into different bins for the collection of recyclable materials. Others recycle materials in a factory-like environment where workers sort through trash after it is brought to a transfer station. In other words, producers of garbage have to abide by distinct rules set by their communities and those rules vary considerably from place to place.

Table 10.3 Summary of the composition of municipal solid waste (http://www.epa.gov/epawaste/nonhaz/municipal/)

Type of material	Percentage by weight
Paper and paperboard	27.4
Food waste	14.5
Yard trimmings	13.5
Plastics	12.7
Metals	8.9
Rubber, leather, and textiles	8.7
Wood	6.3
Glass	4.6
Other	3.4

Transportation

There are many different types of garbage collection truck. Some have mechanized systems that will transfer garbage from bins via a hydraulic lift. Others require labor to transfer the waste into the loader. A hydraulic paddle bar moves trash forward to compact it. Garbage trucks can carry 20–40 cubic yards of trash prior to being offloaded. Recycling trucks have a different design from garbage trucks. These vehicles often have multiple bins to collect paper, bottles, cans, and other materials. Some communities have different days for recycling pickup. Once the material is collected, it has to be taken to a transfer station or disposal site.

Transfer stations

Transfer stations are places where the garbage is transferred to larger vehicles for transfer to other locations for disposal. Sometimes, the garbage is sorted at this point to remove recyclable materials. Many parts of the world have run out of landfill options and must transfer garbage long distances. In New York City, for example, garbage is hauled by truck, barge, and train to landfills as far away as Virginia and Ohio.

Disposal sites

The garbage eventually makes its way to disposal sites. The most common form of disposal used today is the sanitary landfill. However, about 12% of the waste is diverted to produce electricity in waste to energy facilities.

Landfills

When I was a little boy in rural Wisconsin, we would take some stuff to our local dump and bury the remainder (largely household food scraps) in trash pits in the woods on our property. The local dump was managed by a dump manager who would burn some of the waste and bury the rest. Organic waste was left to rot in the open air. Bears, birds, and other creatures would come to scavenge from the dump. Dumps like this in many parts of the world are still common. However, many other parts of the world have set stringent guidelines on waste facilities. These more carefully managed dumps are called landfills.

In the European Union, for example, countries are required to place their municipal garbage in landfills that have distinct management rules for construction. In addition, they cannot contain any hazardous materials, liquid waste, flammable waste, medical waste, or tires.

Most of the developed countries of the world have rules about landfill construction. The EPA summarizes US guidelines here: http://www.epa.gov/solidwaste/nonhaz/municipal/landfill.htm.

Location restrictions – ensure that landfills are built in suitable geological areas away from faults, wetlands, flood plains, or other restricted areas.

Composite liners requirements – include a flexible membrane (geomembrane) overlaying two feet of compacted clay soil lining the bottom and sides of the landfill, protect groundwater and the underlying soil from leachate releases.

Leachate collection and removal systems – sit on top of the composite liner and removes leachate from the landfill for treatment and disposal.

Operating practices – include compacting and covering waste frequently with several inches of soil help reduce odor; control litter, insects, and rodents; and protect public health.
Groundwater monitoring requirements – requires testing groundwater wells to determine whether waste materials have escaped from the landfill.
Closure and post closure care requirements – include covering landfills and providing long-term care of closed landfills.
Corrective action provisions – control and clean up landfill releases and achieves groundwater protection standards.
Financial assurance – provides funding for environmental protection during and after landfill closure (i.e., closure and post closure care).

Based on the guidelines set by the EPA, most landfills are hi-tech compared to the dumps of a generation ago.

Greater attention has been given to the energy generating power of landfills in recent years. As refuse decomposes it produces heat and methane gas. There are many communities taking advantage of these resources from some of their local landfills.

Reducing waste

While we have found ways to deal with waste, it is much better to reduce waste in general. Many individuals have changed their lifestyles to significantly reduce the amount of waste they produce. Bea Johnson is perhaps the world's leading expert on a zero-waste lifestyle. Check out her Website at www.zerowastehomecom. She provides lots of practical advice for those seeking to live a simpler lifestyle. The motto on her blog is refuse, reduce, re-use, recycle, rot. What she means by refuse is that we need to start refusing things that are given to us in our day to day lives. We don't need plastic bags, bottled waters, swag, packaging, disposable drink cups, junk mail, and other junk we usually throw away. We also should start to refuse to buy so much stuff and at the very least reduce our consumption. Once you start asking yourself if you really need something you will find that you likely don't. Living with less is a big part of reducing one's annual waste.

Anything that we do own and no longer need, we should try to re-use it or recycle it in some way. By re-using it, we avoid the energy costs associated with recycling it. So, if something you own can be re-used by someone else or re-purposed by you or someone else you will have a smaller footprint on the planet. Many things in our lives can be substituted for re-usable materials. Paper towels, diapers, and pens are a few examples.

Johnson's final suggestion—rot—is that we compost materials from our kitchen that can decompose. By throwing out kitchen waste, we are throwing away nutrients that can be added to our local soil to promote plant growth.

Johnson's annual household garbage that cannot be re-used, recycled, or composted can fit in a jar. Think about the waste that you produced today? Is it more than the size of a jar? How much waste did you produce this year compared to Johnson? What can you do to try to reduce the waste that you produce each year?

Minimalism is another approach that is gaining considerable attention. The idea of minimalism focuses on a broad lifestyle change to simplify one's life. The reduction of waste is just a side benefit of minimalism. Minimalists focus on living simply in many ways including reducing technology, buying and owning less stuff, and reducing consumption of popular culture.

Composting

Bea Johnson, like millions of people all over the world, composts her kitchen waste. There are a number of ways that composting can be done. Some are highly scientific systems that focus on creating effective aeration and temperature control within a compost bin to ensure that materials decompose quickly. Other systems are much simpler and do not require much work. As we all know, food will rot with or without our help. But by using composting methods, the food will rot quickly.

While it is easy to set up a home composting system, many people do not have access to composting facilities. They may not have the yard or kitchen space to effectively run a home composting system. They may be in apartments or live in assisted living facilities where they do not have the ability to individually manage their kitchen scraps. Plus, large food processing facilities, restaurants, and hospitals, often cannot compost the volume of materials they produce easily.

To deal with these situations, some organizations install industrial-scale composting systems that work to aerate and heat the waste to quickly decompose the material. In addition, some communities have collection sites where individuals can bring their food waste. Some cities, like New York City, have instituted special collections of compostable materials. Given that food waste and yard trimmings, both compostable wastes, make up over 25% of the waste stream, the potential to reduce waste via the introduction of a composting collection system is quite high.

Recycling

Recycling, of course, is using resources found in garbage to make new items (Figure 10.8) and has been in use for centuries. Metals, particularly precious metals like gold and silver, have been melted and re-used to make new goods for millennia. However, given the significant amount of waste produced in our current era, there is great interest in recycling a variety of materials, particularly the high volume materials: paper, plastic, glass, and metals.

One of the challenges of recycling is that there must be a market for waste. While the recycled metal market has always been robust (indeed, there have been problems of theft of metal that is in use to sell as recycled scrap), the market for plastics, glass, and paper is less strong since in many cases it is cheaper to use raw unrecycled materials.

Plastics are a special problem. There are many different types of plastics. When they are gathered together and melted as a group, there are challenges with re-framing the plastics into usable materials because the chemistry among the plastics is so variable. As a result, plastic consumer products have been given a plastic identification code that allows better sorting of material. This allows greater separation of the plastics among type to ensure appropriate recycling strategies are employed.

While only a small percentage of plastics that are thrown away are recycled (about 7% in the US), there is a market for the plastic. It is used to make fabric, take out clam shell boxes, and plastic lumber.

Figure 10.8 Managing garbage and recycling is a major undertaking in many cities throughout the world, including in Oxford, England.

Glass recycling is widely employed around the world. However, to be effective, the glass must be separated by color. Manufacturers tend to use particular colors of glass for their products. Wine bottles, for instance, tend to be dark in color, whereas soft drink glass bottles are clear. Most of the glass that is collected for recycling is melted and turned into other glass containers. However, there is increased interest in using glass for a wide variety of secondary uses. For example, recycled glass is now used as decorative elements in countertops and tiles. It is also used as an abrasive and as an aggregate in cement. Because of the versatility of glass, manufacturers are able to use it as a resource in a variety of manufacturing processes.

Many communities encourage glass recycling by putting a deposit on the bottles. At check out, one pays a small fee called a bottle fee. If you return the bottle to a recycling center, the fee is returned to you. Obviously, not all bottles are returned. The money that remains in the bottle fund can be used to advance recycling programs.

Paper

Paper is commonly recycled throughout the world. Waste paper is a very good resource that can be used to make a variety of other paper products. Since roughly 30% of trees cut each year are used to make paper, recycling helps to preserve forests and reduce the environmental impact of deforestation. Europe and the United States have paper recycling rates of approximately 70%. In the United States, approximately 75% of cardboard is recycled.

These high recycling rates are great, but it must be noted that about half of the recycled paper in the United States and other developed countries is shipped overseas to developing countries for recycling. There are complex economic reasons why this happens. However, many paper recycling companies outside of the developed world do not have strong environmental rules or regulations. Thus, in many ways, the recycling issue is an example of how the developed world is exporting its waste and the associated environmental problems. While it is great that the paper is getting recycled, it is important to recognize that there are many pollution problems associated with waste recycling. While we can feel good about buying recycled paper, we have to recognize that a considerable amount of recycled paper is produced in factories with limited pollution control.

In the last few years, many countries have stopped accepting paper and other recyclable materials due in part to some well publicized cases where the material ended up not getting recycled and ended up as waste. As a result, there is growing interest in domestic recycling in the United States and Europe.

To recycle paper, the used paper is mixed with water and chopped up in a mill. This creates a slurry of material that goes through a process to remove inks and impurities. This mix is then separated and dried to be used to make paper and other products. Paper can be recycled up to seven times before it becomes unusable.

The market for recycled materials

One of the tougher aspects of ensuring the success of recycling programs is ensuring that there is a market for end products of recycled materials. Some recycled materials cost more than materials that come from non-recycled resources. For example, in the United States, a case of office paper containing recycled fiber costs several dollars more than the same quality paper without recycled paper content. This obviously creates an economic challenge for recycled materials to make their way into the shopping baskets of the average consumer.

Plus, some individuals and organizations do not like products made from recycled materials. This is, of course, a matter of preference. Some may not like the "look" of recycled paper or like using plastics made from drinking bottles. However, this attitude is changing with the help of purchasing and procurement guidelines set by large organizations. Retail giants like Wal-Mart and large users of materials, like universities, are setting requirements for the use of recycled materials. Check out your own university or school. Do they require the use of recycled materials in their purchasing guidelines? What about businesses in your community? Do they buy recycled materials? What about your own preferences and those of your family? Do you prefer to buy paper that is recycled or not? What drives your preferences?

The environmental justice critique of recycling

Recycling is great! Right? However, consider the people employed in the recycling field. They are often low-paid workers who are exposed to a variety of materials as they sort trash and haul materials that may or may not contain harmful objects or chemicals.

Communities that pull recyclable materials from trash in a sorting line employ workers to pick through home garbage. This waste can contain materials that are physically harmful such as sharp objects or broken glass. It can also contain chemicals that may be harmful on exposure such as oven cleaning chemicals, pharmaceuticals, or battery acids. The waste transfer facilities are often located in low income or minority communities.

These low-income individuals then are dealing with the consumerist waste that is produced by wealthier individuals. Talk to any garbage collector and they will tell you that there is a very distinct difference in the waste produced in different communities and they will tell you that wealthier communities produce more waste than poorer communities. So, the workers involved with recycling industry are bearing the burdens of the wasteful practices of others. Plus, many developed countries send their waste overseas for poorer countries to manage. Sometimes, the waste ends up net entering the recycling stream and ends up a pollution problem for the host country.

Pollution shuts down the water supply in Toledo, Ohio

In August of 2014, the City of Toledo shut down its public drinking water supply because it was contaminated with microcystin, a byproduct of cyanobacteria, a form of blue-green algae. During the warm months, algae blooms are common in surface waters throughout the world. The blooms are natural phenomena that occur as the algae takes advantage of nutrients that run off into surface water bodies during the warmer months. However, too much algae can produce high amounts of microcystin that can contaminate drinking water sources. Ingestion of microscystin can lead to severe health problems.

Toledo receives its drinking water from Lake Erie, one of the Great Lakes that make up the border between Canada and the United States. Lake Erie became famous in 1969 when the Cuyahoga River in Cleveland caught fire near the point where it enters Lake Erie. The river was so full of contaminants that it was flammable. Events like this helped to spur action to pass the Clean Water Act. At the time, it was well known that the lake also had severe nutrient pollution from agricultural and sewage runoff. High levels of nitrogen and phosphorus in the lake caused the formation of high amounts of algae, which led to anoxic (low oxygen) conditions in the water. The low oxygen conditions, of course, caused fish to die, which caused even greater pollution problems, especially when the dead fish ended up on the shores of densely populated communities.

While there have been many successes in cleaning up industrial pollution from Lake Erie, nutrient pollution remains a significant problem. When microcystin was found in Toledo's drinking water, the Mayor of the city told residents to not drink water or even boil it for drinking. Boiling of the water doesn't remove the contaminant. For three days, hundreds of thousands of residents were unable to use the public drinking water supply. Truckloads of bottled water were shipped to the city from surrounding areas.

This may be one extreme example of water quality problems from nutrient pollution, but it is likely that unless changes are made, more of these types of problems will occur in the future. In rural Dunn County, Wisconsin, 26 cases of illnesses from algae were caused by drinking water from Tainter Lake. Microcystin was also found in Lake Winnebago, the main drinking water source for major Wisconsin urban areas like Appleton, and Oshkosh. Some communities have taken aggressive steps to improve water treatment to remove microcystin, and others are working hard to try to reduce nutrient pollution. However, problems like the one that occurred in Toledo will continue into the foreseeable future.

How sustainable is paper recycling?

Paper recycling is one of the more successful forms of recycling we have today. About 65% of all paper waste in developing countries is recycled in some way. Products made from recycled paper are making their way to the world market. The system in place is widely lauded for its ability to remove waste from landfills.

However, there are some challenges that call into question the overall sustainability of paper recycling. Specifically, many are concerned over energy costs, pollution costs, and environmental justice issues.

Paper recycling is energy intensive. While it uses somewhat less energy than producing paper, it does have a large transportation energy use. Plus, many of the world's forest reserves, and thus the world's paper mills are in places that utilize hydroelectric power, a green and renewable form of energy. These plants also use waste products from paper manufacturing for fuel. In other words, a considerable amount of paper from wood pulp is produced using renewable energy, whereas paper produced via recycling generally uses fossil fuels.

Another problem is the pollution associated with paper recycling. Inks are removed using caustic chemicals. The inks, can contain a variety of harmful organic and inorganic chemicals like plastic polymers, chromium, and cadmium. These materials can be discharged into surface waters. Finally, solid waste that consists of unusable fibers and impurities are left behind and must be landfilled. Thus, the process has distinct environmental impacts that are different from standard paper manufacturing.

Finally, paper recycling has important environmental justice issues. Roughly 50% of paper from developed countries is shipped to developing countries for recycling. The paper is sent to developing countries in container ships after they leave consumer goods in their ports. The waste is delivered to factories that have limited environmental rules and that are taking advantage of the interconnected globalized world. Once the recycled paper is produced, it is shipped back to developed countries where consumers feel good about buying recycled content paper.

While it is important to support recycling whenever we can, we have to recognize that recycling can have unintended consequences. We must examine the manufacturing stream to ensure that it is done fairly.

11

Environmental Justice

One of the issues that distinguishes the modern sustainability movement from the environmental movements that preceded it is that the modern sustainability movement focuses, in part, on issues of social equity. Indeed, the very definition of sustainability incorporates equity within the three "E"s of environment, economics, and equity. In this chapter, we will explore issues of equity within the context of the environment. As we will see, this is a new approach that is looking at problems that have accelerated in the last several decades—in part as a result of racism, classism, sexism, and globalization.

What is striking about issues of environmental justice is that the problems are all around us, yet many of us feel powerless, unable, or unwilling to address the issues. However, we will also see that progress is being made in many ways. While there are many problems to confront, efforts over the last few decades have shed light on them and many in the sustainability community are focused like a laser on trying to solve them. Indeed, issues of environmental justice have gained the attention of many in the media, government, and the public and private sectors.

Social justice

The environmental justice movement emerged from the environmental movement and the social justice movement, both of which gained great momentum in the nineteenth century as the impacts of the industrial revolution were felt around the world (Figure 11.1). As we know, great environmentalists like John Muir were concerned with preservation and protection of the land. However, many in the social justice movement were focused on issues such as how to ensure that workers earned a fair wage and had access to reasonable housing options.

In our modern era, social justice has been defined in many ways, but is broadly considered as ensuring that individuals are treated fairly and have equal access to liberties, rights, and opportunities. While many have written on social fairness for centuries, the modern concept of social justice emerged approximately 150–200 years ago. This was a time well documented by writers of the era, including Charles Dickens, who wrote eloquently about issues such as child labor, working conditions, pollution, poverty, and housing. Social organizations, many of them originating in religious groups, formed during this time to focus efforts on developing sound public policy to address societal ills.

Introduction to Sustainability, Second Edition. Robert Brinkmann.

Companion website: www.wiley.com/go/Brinkmann/IntroductiontoSustainability

Figure 11.1 Physicians protesting against nuclear war in Washington, DC (photo courtesy of Hofstra University Special Collections).

Social justice advocates often look at broad societal institutions for solving problems. Thus, social justice confronts injustice in all levels and types of societies. Thus, it is often in conflict with governments, no matter if left or right, and with other social institutions such as religious organizations, businesses, and non-profit organizations. At the same time, many of these same organizations strive to improve the social conditions of members of society.

Social justice initiatives are often focused on the poor. However, it should be noted that many other themes have emerged that are of concern to those engaged with the social justice movement, including: race and ethnicity, immigration status, religion, sexual orientation, gender, healthcare, income inequality, labor and working conditions, and overall human rights.

In recent years, many in the social justice community have been focused on broad issues of income inequality. The income disparities in countries all over the world, in both developed and less-developed countries, is increasing. As a result, many activists have been investigating how to deal with this growing social divide that has emerged in recent years. The reality of our times is that the rich are growing richer and the poor are growing poorer. Such conditions are not sustainable for a fair and equitable society and many are concerned that social equity is diminishing, even in our time of greater knowledge and awareness.

Indeed, in 2006, the United Nations published a report called Social Justice in an Open World. In it, the report notes that there has been a "... rise in poverty in all its manifestations, along with the increase in the number of refugees, displaced persons and other victims of circumstance and abuse..." In the report, six areas were identified as *trends* of inequality among people:

1 Rising inequality in the distribution of income. Certainly some in the past tried to make a world where there was an even distribution of income. Yet the reality of human nature has shown us that such a system is doomed to failure. Most understand that unequal incomes are the normal result of differences that emerge when people strive for personal and familial betterment. However, there is no doubt that concerns emerge when large income disparities occur. It is often seen as politically unsound, morally inappropriate, and economically unwise. Considering the vast differences within countries, it is also important to recognize that vast income disparities exist across regions of the planet as well.
2 Rising inequality in the distribution of assets. Assets are resources controlled by an individual or country. What has happened with assets is concomitant with what has happened with income in the last generation. Fewer individuals have control of greater numbers of assets and many nations have lost control of key assets as a result of the transfer of assets to other nations.
3 More work opportunities for a few and increased unemployment and underemployment for the majority. In the last several decades, we have moved into a highly technical and specialized economy where the highest paying jobs require advanced education and training. In part due to mechanization and efficiencies, there is greater unemployment and under-employment for many.
4 A better distribution of information and perhaps of knowledge, but a more uneven distribution of opportunities for quality education. In the last several years, we have seen information explode on the Internet. At the same time, the cost of high-quality education has risen. As a result, we have a cognitive dissonance situation occurring whereby we have great access to knowledge, but limited access to educational degrees throughout the world.
5 Growing inequalities in health care and social security and the apparent emergence of environmental inequalities. There is no doubt that throughout the world there are differences in the ability to gain access to quality affordable healthcare. At the same time, there are distinct differences in things like age of retirement and day care.
6 Ambiguous trends in the distribution of opportunities for participation in civic and political life. There is growing concern over the role of money in politics. Certainly money has always and will continue to play a role in access to power. Yet, in recent years, the amount of money in politics is increasing, thereby limiting the impact of less wealthy individuals in the political process. Indeed, there is growing concern over the disparities starting to emerge throughout the world. For example, in the United States, more than half of congressmen are millionaires. The median net worth for members of the US House of Representatives was $896,000 and the median net worth for Senators was $2.5 million (http://time.com/373/congress-is-now-mostly-a-millionaires-club/).

While this UN report is not the definitive guide to social justice, it does present many of the issues of concern in the social justice movement internationally.

In the United States, the environmental justice movement can clearly be traced to the advent of the Civil Rights movement, which itself rose out of the broad social justice movement that extends back to early anti-slavery efforts.

Civil rights and the modern environmental movement in the United States

The US environmental movement as it emerged in the 1960s focused strongly on the connections between humans and the environment (Figure 11.2). People were very concerned about issues like loss of natural lands, over-development, and pollution. The issues were very Earth-centered and not people-centered. Protests in that era helped encourage the development of major environmental laws like the Clean Air Act, Clean Water Act, and Endangered Species Act—the main federal laws that are still in place to protect the environment.

At the same time that the modern environmental movement was coming of age, the American civil rights movement was working to promote equality of the races. Throughout many areas of the United States in the middle of the twentieth century, racism was a serious problem. White privilege limited the ability of African Americans and other minorities to gain access to good jobs, excellent schools, particular neighborhoods, etc. Several well-publicized marches, especially those led by Dr. Martin Luther King educated the public and politicians about the problems associated with institutionalized racism. New federal, state, and local laws were implemented during this era that sought to remove social barriers and eliminate institutionalized racism.

Of course we all know that the environmental laws that were implemented in the 1960s and 1970s did not eliminate all of the environmental problems in the world. Likewise, the

Figure 11.2 This is the famous bus (now located at the Henry Ford Museum) where Rosa Parks refused to move the back seat. The American Civil Rights Movement ushered in a new focus on environmental justice.

new civil rights laws did not solve racism. Efforts by activists continue to strive to improve environmental conditions and civil rights of minorities.

Looking back, it is surprising that the environmental and civil rights communities did not work together. At the time, the environmental community was focused less on environmental issues in cities, and more on issues in the wilderness. This is not surprising since American environmentalism has its roots with individuals like Muir and Thoreau who found great solace in nature. Muir advocated strongly for preservation of wilderness for wilderness' sake. It wasn't until Rachel Carson published *Silent Spring* in 1962 that the mainstream environmental community started to consider environmental issues in urban and suburban areas. Thus, it is not all that unexpected that the civil rights community did not see a clear link with the environmental community and the environmental community did not see a clear link with the civil rights world.

Thus, it wasn't until the 1980s that people started to understand that issues of environmental protection and civil rights were linked within a new combined area of environmental justice. Since then, there has been a steadily growing interest in how civil rights and equity are intertwined with environmentalism. Indeed, equity has emerged as a key theme of sustainability.

Lead pollution and the growth of the urban environmental justice movement

Prior to the start of the environmental justice movement in the 1980s, many people in cities started to notice that things were not right. They started to notice that children were suffering with learning disabilities and other problems that were not common a generation before.

For centuries, lead was used in common household products like eating utensils, pipes and solder, and bullets. Occasionally, users and manufacturers of these goods would develop lead poisoning. But such cases were not common, or at least they were not commonly diagnosed.

In the last century, with the advent of the widespread use of lead-based paint and leaded gasoline (now banned in many countries), occurrences of lead poisoning increased (Figure 11.3). In fact, lead in paint and in gasoline were the most common ways that people became exposed to lead in the twentieth century. Lead is an excellent additive to paint in some cases. It makes paint very hard, durable, and resistant to weathering. Because of this, it is still used on bridges and in some industrial applications. Lead is added to gasoline as an anti-knock agent and lubricant.

When paint breaks down, it can turn into dust in or outside of the home. Plus, some children peel paint and eat it or suck or chew on windowsills. In these situations, extreme lead exposure can occur from direct ingestion. Lead present in automobile exhaust enters the atmosphere where the heavy lead particles can fall near the point of exhaust, or they can be carried some distance from the source. In most cases, higher levels of lead were found near roadways and in or near homes painted with lead. However, since these particles are not visible to the eye, it is impossible to know where contamination is present. Even today,

Figure 11.3 Children are more susceptible than adults to lead poisoning, which is why many nations have banned using lead additives in paint and gas.

decades after lead was banned in most paints and gasoline, remnants of lead are found in soils and traces of lead paint are found in older homes.

Unfortunately, the ones who are most impacted by lead ingestion are children, who absorb many more times the amount of lead in their growing bodies than adults. Plus, children, who like to play on floors, in dirty yards, and along streets and sidewalks, pick up lead through their regular activities. They have much more hand to mouth activity than adults (when is the last time that you saw an adult sucking their thumb?) and are exposed to dirt, sediment, and dust containing lead directly via eating or breathing in contaminated materials.

While the severe symptoms of lead poisoning are obvious and include things like vomiting and neurological disorders, low-level poisoning symptoms are harder to detect, but significant. Learning disabilities, behavioral problems, and listlessness are among some of the issues. Problems arise with lead exposure because there is no known metabolic use for lead in the human body. It often takes the place of calcium or magnesium in bones and brains, and other areas of the body where it does not have the positive effect of these elements.

Along with learning disabilities and other problems, many believe that lead pollution and poisoning is responsible for increased violent crime rates. Noted economist Rick Nevin has conducted research on lead poisoning and crime in the United States and other countries and found a clear link between exposure and crime. Most particularly, he finds that certain groups with high incidents of violent crime have had greater exposure to lead in the past. Interestingly, Nevin suggests that similar links exist in Latin America. In recent years, Latin America has seen very high violent crime rates that are coinciding with maturation of individuals who were exposed to high lead levels from auto exhaust. Latin America was one of the last regions of the world to ban leaded gasoline. Indeed, Venezuela, with one of the highest murder rates in the world, was the last country to ban leaded gasoline in 2004.

What does all of this lead pollution information have to do with environmental justice? In the United States, it is clear, based on research conducted by Gerald Markowitz and David Rosner, that many in the lead industry knew about the health problems associated with lead in paint and gasoline, but fought against banning their products. This was perhaps the first widespread environmental justice situation that gained the attention of the American public. Data from the 1960s and 1970s demonstrated that many of the people impacted by lead poisoning were poor children of color living in cities.

Thus, entire generations of children were exposed to lead and many involved with the lead industry knew of the impacts of their products on public health. This set in motion a whole series of laws and started people thinking about issues of social justice, health, and the environment. The advent of mapping technologies, such as computerized maps and geographic information systems, furthered the understanding of where lead pollution was located and what types of people were impacted. By the 1970s it was clear that the children most at risk of lead poisoning were poor children of color.

The environmental regulations associated with lead required the United States to test young children for lead poisoning as part of overall health screening. By far most of the cases that were identified were in poor urban areas, although pockets of lead poisoning are also found in older rural communities. New national rules for the management of lead-based houses were established that required clean-up of publicly owned housing and disclosure of the presence of lead in homes if a house is sold to a new owner. While not all of the lead in the United States is gone and while there are still cases of lead poisoning that occur, the risk of lead poisoning and the number of reported cases has declined significantly.

Environmental racism in the United States

In the 1980s and 1990s, noted US researcher Robert Bullard started to recognize that there were significant disparities in the way that individuals were exposed to waste in the United States. He and others found that there were distinct differences in where hazardous waste sites were located. Instead of these facilities being equally distributed across space, he found that African American and low-income neighborhoods were disproportionately impacted by the siting of hazardous waste sites. This was seen as a form of environment racism. While the term environmental racism is used in many ways, it is most often used in the context that Bullard identified: disproportionate exposure to environmentally hazardous materials or degraded environments based on race.

For his efforts, Robert Bullard is often called the "Father of the Environmental Justice Movement." His endeavors helped to put issues of environmental racism front and center within the social justice and environmental movement.

But if Robert Bullard is the father of environmental justice, Hazel Johnson (1935–2011) is the grandmother.

Johnson was an environmental activist in Chicago who found her calling after seeing friends and relatives fall ill due to environmental exposures in public housing. She started to document and record each problem she heard about in order to understand patterns and draw attention to the problem of environmental health and housing.

In many ways, Johnson's activism began in 1969 after her husband died of lung cancer. She began to be concerned about issues of asbestos in public housing and started a group called People for Community Recovery in 1979 to focus attention on problems in Chicago's public housing communities. A big focus of her work was asbestos and its removal.

As Johnson began documenting health problems in the neighborhood, her attention focused on the range of unregulated dumps and hundreds of underground storage tanks present. Indeed, the community was surrounded by what some called a "toxic donut." Unfortunately, many homes did not have access to treated drinking water and relied on well water. Her determined focus and community organization brought attention to these problems and eventually water and sewer lines were brought to the area.

Hazel Johnson's efforts serve as an example of the power of one individual in trying to make a difference in their community when confronted by environmental problems. She worked on these vexing environmental and social problems in a largely African American neighborhood well before the terms environmental racism or environmental justice were clearly defined. Yet, her research and advocacy changed her community and the lives of hundreds of people for the better. Late in her life, many local, national, and international leaders recognized her for her accomplishments and her many contributions to public health and environmental justice in the United States.

Hazel Johnson served as an example to many who saw problems in their community. Her efforts combined with the ideas of Robert Bullard came to fruitition in new ways in New York City. Since 1990, many examples of environmental racism have been identified in the region. For example, in the South Bronx of New York City, many residents were impacted by a number of environmental and social problems, including dumps, pollution, and housing problems. Because of all of the issues, it had crime problems and became one of the more notorious neighborhoods of New York City. It was not only polluted, but it was unhealthy and dangerous. However, because of the low housing costs, it was a mecca for immigrants and low-income families. Thus, not only were immigrants and minorities disproportionately exposed to environmental and social problems, but high numbers of children were also impacted.

One resident of the South Bronx, Majora Carter, saw all the problems and started to work to solve them. She started a community group called Sustainable South Bronx to try to address some of the environmental issues that were caused by decades of neglect and poor urban management. She felt that the neighborhood should not carry the disproportionate environmental impacts of the New York region. She and others within Sustainable South Bronx worked hard to shed light on the problems of the neighborhood. They held numerous community meetings, listened to residents and other stakeholders to find neighborhood concerns. They found that residents wanted a healthy community with a flourishing environment. They also wanted job opportunities and better community amenities. Sustainable South Bronx worked with the City of New York government, received grants, and accepted donations to make significant improvements in the community. While not everything is perfect in the South Bronx, there is no doubt that the efforts of Sustainable South Bronx made a huge difference to the community.

After the success of the Sustainable South Bronx, Carter went into the private sector where she works as an urban revitalization specialist, but the organization is still active. According to its mission statement, Sustainable South Bronx "... works to address economic and environmental issues in the South Bronx—and throughout New York City—through a combination of green job training, community greening programs, and social enterprise." It is one of the most active local environmental justice organizations in the United States and is often used as a model of success. Many who know what the south Bronx was a decade or two ago know that the organization has been successful in transforming the neighborhood into a greener, healthier community.

Carter's efforts demonstrate that environmental justice provides a framework for broader community re-development and improvement. Carter recognized that by shedding light on the problems in the community, she could work with others to make improvements for the betterment of all. By drawing a line in the sand and stating "No more!" she was able to eliminate the continuation of practices that were detrimental to her community.

Brownfields, community re-development, and environmental justice

One of the major concerns in many parts of the developed world, particularly the United States, are brownfields, which are defined as parcels of vacant or abandoned land that may or may not be contaminated (Figure 11.4). Brownfields can be commercial, industrial, or residential sites that are problematic for re-development because developers do not wish to incur the costs of the clean-up of known or potential contaminants on the property. Because environmental laws and regulations were lax in the past, many areas have properties that

Figure 11.4 Many brownfield sites are fenced and closed to the community. This hurts the overall aesthetic look of a neighbourhood and drives down property values.

could potentially have environmental contamination lurking on or under the property. As a result, many older areas of communities have abandoned or vacant properties that hurt the character of the community by driving down property values and limiting the ability to achieve density to promote livable communities. Think about vacant or abandoned properties in your own communities. Many of them are likely brownfield sites that are difficult to re-develop. Note: Brownfields were discussed in the earlier chapter on green building. Developing on them gives credits on some green building rating schemes.

In the United States, the term came into widespread use in the 1990s as the impact of brownfields on communities was starting to become obvious. Many of the brownfield sites were found to be located in poor, minority communities where the abandoned properties limited the ability of investors to work on broad re-development community improvement projects. The costs were just too high for individual investors to take on the risk of incurring the heavy clean-up costs needed in order to re-develop the property.

As a result, new programs that brought together federal, state, local and private funding for brownfield re-development started that helped private companies direct re-development efforts. Public partners involved in the brownfield re-development projects helped defray the expensive costs associated with the clean-up of the sites and also helped with technical issues associated with some of the more difficult clean-up efforts.

Some may question why public tax dollars are involved with brownfield clean-up. However, brown-field sites lower property values and thus tax revenue. If a brownfield site can be re-developed, it provides a new taxable property, while also increasing the tax revenue that can be collected from the impacted neighborhood. At the same time, brownfield sites can be ticking environmental time bombs. The contamination can spread and cause larger problems. Tackling environmental clean-up at these sites during a brownfield re-development program provides the opportunity to eliminate long-term risk and more expensive clean-up efforts in the future.

One of the key elements of brownfield re-development is community involvement in land use decisions. Because public dollars are associated with the re-development of the property, it is appropriate to seek local stakeholder input. In the last 25 years, brownfield re-development has been a focus of community organization, re-development, and environmental justice. A number of new urban community groups have formed that focus on re-developing brownfields in their community. Citizens have come to understand that re-developing brownfields is good for the environment and good for their community. Vacant properties, which were once seen as neighborhood blight, are now seen as sites for re-development. Tremendous improvements have been made in communities all over the United States as a result of these re-development efforts. The re-development efforts provide opportunities to assess issues of environmental justice.

In Tampa, Florida, residents of one neighborhood, East Tampa, used brownfield development and special taxation rules as a way to revitalize a long-neglected community. The City of Tampa is a relatively new city. It grew rapidly from the late nineteenth century when it was a sleepy port and military base to a major US metropolitan region. The 7.5 square mile neighborhood of East Tampa (one of the largest neighborhoods in the city) has been home to a vibrant African American community since the early twentieth century. For generations, this middle-class community has been built around strong community assets like churches, schools, and workplaces. Many in the community worked in the port or in construction activities. However, discriminatory practices, such as the Jim Crowe rules and institutionalized racism led to community decline in some areas.

For generations, the city neglected basic amenities such as sidewalks, storm water drainage, and street repairs, while focusing on white or newly developed communities as the city grew. Abandoned properties were used as illegal dumps where outsiders dropped debris to avoid paying tipping fees at community waste collection sites. Dozens of brownfield sites existed that further plagued the neighborhood. The long-term neglect, dumping, and inability to develop brownfield properties created a crisis in the community.

By that time, the City of Tampa was well aware of the neglect of the past and was working to remedy problems. The city created a special taxation zone in East Tampa that ensured that a portion of the tax revenue generated in the neighborhood would be used for neighborhood improvements. Since the district was set up, many improvements have been made to reduce storm water flooding, improve road safety, and improve community policing. The City of Tampa even hired special police officers that were responsible for catching those responsible for illegal dumping in the community.

However, one of the key aspects of the re-development of East Tampa was the creation of a special brownfield district. Because there were so many properties in East Tampa that were abandoned, had well-known or potential pollution problems, and were impacted by illegal dumping, a large brownfield district was established that allowed the community to apply for brownfield development funds over wide swaths of property.

This effectively allowed East Tampa to create important re-development districts that encompassed housing along with commercial properties with mixed-use development. In this way, they were able to turn their brownfield problems into a community asset whereby developers were anxious to come into East Tampa to work on re-development projects. While not all of East Tampa is devoid of problems or brownfields, considerable improvements have been made in the community. The East Tampa work was done in partnership with stakeholders in the community who provided guidance as to the types of development that were needed. Indeed, the brownfield efforts prompted broader initiatives on public health, education, and environmental justice that improved the lives of many in East Tampa.

US EPA and environmental justice

Many of the current US brownfield and environmental justice initiatives are supported through the efforts of the US EPA's Environmental Justice division (Figure 11.5). The EPA has been involved with environmental justice issues since 1994, when President Clinton signed an executive order that stated in part, "... each Federal agency shall make achieving environmental justice part of its mission by identifying and addressing, as appropriate, disproportionately high and adverse human health or environmental effects of its programs, policies, and activities on minority populations and low-income populations... " Via this

Figure 11.5 Many countries have national organizations like the EPA that regulate air, water, and other pollution. In Yalong Bay in China, there is regular monitoring that takes place to ensure the protection of the environment.

action, environmental justice was integrated into the actions of the US government. The EPA's Environmental Justice office focuses on several initiatives:

1 It provides basic information on environmental justice topically and spatially by EPA region.
2 It manages the affairs of the National Environmental Justice Advisory Council, which includes academia, community groups, business and industry representatives, non-government organizations, state and local governments, and tribal governments and indigenous groups.
3 National environmental justice planning. The most recent plan, Plan EJ 2020, focuses on improvements from 2016–2020.
4 Grants and Programs. The office manages a number of different types of federal grants for communities and states including the Environmental Justice Small Grants Program, which "... provides financial assistance to eligible organizations to build cooperative partnerships, to identify the local environmental and/or public health issues, and to envision solutions and empower the community through education, training, and outreach." One of the programs that the office oversees is the Environmental Justice Showcase Communities Project. This effort focuses on pooling resources and expertise on solving environmental justice problems. The projects serve as demonstration projects for other communities.
5 Interagency Working Group. This group, managed by the EPA's Environmental Justice office, brings together members of other federal agencies to work on interagency

environmental justice issues. Participating offices include (besides the EPA), the Departments of Agriculture, Commerce, Defense, Education, Energy, Health and Human Services, Homeland Security, Housing and Urban Development, Interior, Justice, Labor, Transportation, and Veteran's Affairs. The General Service Administration, the Small Business Administration, and the White House Office are also involved.

What is innovative about this effort is that it not only includes the EPA, but it also crosses agencies of the Federal government to address environmental justice problems. In their handful of years of operation, they have funded dozens of projects and have provided several examples of successful resolutions of environmental justice problems. While there is no doubt that the United States has some significant environmental justice issues remaining, it is clear that the topic is being addressed in some way by the US government.

Many states and local governments also have environmental justice offices that work on more localized issues or themes. For example, Missouri's Department of Transportation has woven environmental justice into their decision-making framework. According to their website (quoted directly in the bullets https://www.modot.org/environmental-justice-0), environmental justice is important for transportations because it will:

- Make better transportation decisions that meet the needs of all people
- Design transportation facilities that fit more harmoniously into communities
- Enhance the public-involvement process, strengthen community-based partnerships, and provide minority and low-income populations opportunities to learn about and improve the quality and usefulness of transportation in their populations
- Improve data collection, monitoring, and analysis tools that assess the needs of, and analyze the potential impact on minority and low-income populations
- Partner with other public and private programs to leverage transportation-agency resources to achieve a common vision for communities
- Avoid disproportionately high and adverse impacts on minority and low-income populations
- Minimize and/or mitigate unavoidable impacts by identifying concerns early in the planning phase and providing offsetting initiatives and enhancement measures to benefit affected communities and neighborhoods.

This is but one example of environmental justice infused within state governmental management. Many governments are involved with trying to integrate environmental justice within their organizations.

Of course what continue to drive environmental justice in the United States are community activists like Hazel Johnson and researchers like Robert Bullard, who are shedding light on problems and striving to find solutions like those developed by Majora Carter and activists and government innovators in the City of Tampa.

Indigenous people and environmental justice

Of course, power and influence drive a conservable amount of decision-making about locally unwanted land uses. Powerful people and organizations can insert themselves into decision-making processes easily since they often have access to decision-makers and are informed about projects that may influence a region.

Native Americans, and native peoples throughout the world who have been in some way impacted by outsiders, have suffered many injustices over the centuries. While one could recount many issues of social and environmental justice problems of the past, the modern issues of environmental justice and Native Americans are complex. This is, in part, because Native Americans have a distinct relationship with the environment in ways that are not easily measured outside of the tribal environment. Thus, while many non-tribal members may be concerned about pollution exposure and brownfield re-development, some Native Americans are concerned about overall environmental degradation, impacts to sacred spaces, and hunting and gathering food from the land. Thus, when considering environmental justice issues in some areas of the world, it is important to recognize the unique relationships that some groups have with the land.

It is also important to note that many tribal lands are in some of the more environmentally difficult areas of the United States. Many are located in very dry or cold regions. Issues like global climate change, water withdrawal, and neighboring environmental disturbances can have disproportionate impacts than more temperate areas.

A flash point in indigenous rights in recent years has been the Amazon rainforest in Brazil. While there is a great deal of concern about the environmental damage caused by the deforestation, particularly in the area of climate change and biodiversity, another tragedy that is unfolding is the loss of indigenous life ways.

Many indigenous people have lost their lives over the last decade as illegal loggers and miners have encroached into their forest homes. In the fall of 2019, a 26 year old indigenous rights activist named Paolo Paulino Guajajara was shot in the head and murdered by illegal loggers. Prior to this, the leader of the Guajajara indigenous people complained to the Federal government in Brasilia about threats that they were receiving but it appears that little was done to protect them.

Indeed, the President of Brazil has said that the areas set aside for the indigenous peoples of the Amazon should be opened up for economic exploitation. Thus, it is unlikely that there will be much protection provided by the national government.

Unfortunately, according to Brazil's Missionary Council, violence against indigenous people in the country has spiked. A total of 135 were murdered over their reporting year in 2019 which is an increase over the previous year of 23%. Plus, just in the first nine months of 2019 there were 140 cases of land invasion, illegal exploitation of natural resources, or destruction of property which is twice the number over the previous year.

Clearly things in the Amazon are not going well under the current national leadership. Many in the government deny the reality of climate change and others seem to value short-term economic growth over the rights of native Brazilians. There is a growing call for a Brazilian divestment movement given the country's poor policy to native Brazilians and the environment.

Exporting environmental problems

One of the major themes of environmental justice that has emerged in recent years is the notion that developed countries are exporting environmental problems around the world,

Figure 11.6 Food is often labelled with an external benchmarking organization's logo. There are organizations that verify organic, fair trade, and GMO-free, among others.

leading to issues of environmental justice in the receiving country. This is done in the following ways:

1 Exporting dirty industries
2 Selling products banned in a developed country to a developing country
3 Taking advantage of overseas cheap labor that work in poor conditions to avoid paying workers a living wage in the home country
4 Exploiting resources of a country with poor environmental rules
5 Exporting waste, particularly hazardous waste, to other countries with lax regulation.

Many of us are powerless to address these issues in our daily lives. However, new organizations that evaluate environmental justice issues have emerged. For example, the Fair Trade label has emerged in recent years to note that the labor practices for the product are just. Plus, many companies have infused environmental justice within their international corporate environment and recognize that failure to act fairly in the world impacts not only the company's bottom line profit, but also the overall ethical reputation of the corporation (Figure 11.6).

Environmental justice around the world

While environmental justice issues have been fully integrated into key areas of national, state and local governments in the United States, the concept is relatively new in many other parts of the world. The following sections summarize some of the issues by region.

Environmental justice in Europe

Given the national fragmentation of Europe it may seem like environmental justice issues would be minimal since there is a certain degree of cultural homogeneity across the continent. However, immigration to Europe, as well as concerns regarding minority populations, have raised a number of questions about environmental justice. Two issues that have been raised of late have been minority populations that have moved into the region for work and the situation of the Roma, a migratory minority throughout many regions of Europe that has long been discriminated against.

Since the formation of the European Union, work rules have been liberalized across the continent to allow the migration of peoples from one region to another. What this means is that many have moved from poorer nations to wealthier nations to take advantage of job opportunities. Plus, immigration to many countries has been encouraged because Europe has had very low birth rates recently and the population of many countries has been flat or declining. The new immigrants come from all over the world. Many of these workers are often low-income individuals who end up living in areas with a number of social and environmental problems. As a result of this, new concerns have been raised about environmental justice and ethnicity throughout the continent.

At the same time, concerns have been raised about environmental justice issues associated with the Roma, who often live migratory lives or in permanent or semi-permanent settlements outside of cities and villages under the radar of nationality or local governance. Their situation often puts them situationally into difficult circumstance. Some do not live in homes or cities. Some live in substandard housing and shantytowns and thus they do not have access to infrastructure like potable water, sewer systems, and energy. Nor do they have regular educational opportunities or social services. The Roma can also be disproportionately exposed to floods and waste dumps in their settlements. They are not often welcome neighbors to those living in established non-Roma communities.

Environmental justice in Asia and the Pacific

Asia and the Pacific nations include vast areas of land and water that are home to a diverse array of culture. The environmental and social issues are complex and much has been written about problems or challenges in this important region. In this section, three case studies are highlighted to demonstrate some of the internal and external challenges faced here.

The Three Gorges Dam

In 1994, construction began on the largest hydroelectric project in the world, The Three Gorges Dam on the Yangtze River. It was completed in 2012. The dam can provide 22 500 megawatts of electricity, while also improving navigation and reducing flood problems. However, like any large engineering project such as this, there are real-world effects on people and communities living near the construction site and the area where the reservoir is located.

The dam has received considerable attention, not only for its engineering marvel, but also because the dam displaced over 1 million people. Over 300 villages and 140 towns were flooded by the time the dam was complete. Relocating this many people was not easy. The government did provide some compensation packages, but many were not pleased with what they received. Plus, many were evacuated far from their social network, outside of their home provinces. In addition, the receiving communities did not always appreciate the influx of outsiders. Evacuees did not usually have the right to decide where they wanted to move. Because of China's top-down approach to governing, there was little public debate on the problems. Many of the evacuees lost everything they had, as well as their jobs. Because of the problems, there were revisions to the compensation package for evacuees.

This case study demonstrates the problems inherent in any large engineering project anywhere in the world. Many people do not want undesirable activities in their neighborhoods because it reduces property values or limits their ability to earn a living. The Three Gorges Dam situation is obviously an extreme case, but at the same time, this issue crops up over and over again all over the world. Those with access to power make decisions that impact those without power. Promised compensation or benefits may or may not be forthcoming at the conclusion of the project.

In the last decade, China has been developing strategies to get away from its dirty energy sources, particularly coal. The development of hydropower is a key strategy in its green energy initiatives. Yet, the places where hydropower can be developed are in the more rural mountainous western and northern portions of the country, far from the energy-thirsty southern and eastern industrial cities. Some have questioned the fairness of relocating the mountain populations and flooding agricultural lands to provide energy to the more urban and industrialized cities. Similar questions have been raised in other parts of the world, most notably in Canada, where rural dams in Quebec provide tremendous amounts of power to eastern Canada and the eastern United States.

Bhopal and environmental justice in India

There are a number of environmental justice issues associated with India, a country that has struggled with enforcing environmental laws and regulations. There are many pollution problems, land ownership issues, agricultural policy challenges, and other issues that are challenges to their society. However, one of the most horrific environmental tragedies in the history of the world occurred in Bhopal, India. No discussion of environmental justice in Asia, or anywhere else in the world would be complete without addressing this disaster. Although it occurred over three decades ago in 1984, it still provides a cautionary tale of industrial safety, public health, and environmental justice.

Overnight in early December of 1984, a gas leak occurred at the Bhopal Union Carbide India Limited pesticide plant. The chemical that was released was methyl isocyanate, a chemical used to produce pesticides and other products. Exposure to even low levels of the chemical can lead to a number of respiratory problems and skin damage. It is toxic at just 0.4 parts per million in the atmosphere—a very low amount.

Within two weeks of the leak, up to 8000 people died with another 8000 dying at some point after exposure. Over 500 000 were injured in the leak with thousand incurring partial or permanent severe injuries. Nearly two-thirds of the 900 000 people living in Bhopal were in some way impacted by the gas leak.

While the exact cause of the gas leak is not known, many believe the problem was caused, in part, by the drive to bring production of new products to a country with weak environmental rules and regulations. It can be profitable for multinational companies to move to areas where labor costs are low, environmental regulations few and far between, and new markets plentiful. At the same time, India was pursuing an aggressive "green revolution" built around improving technical aspects of agriculture in order to gain greater yields. The chemical plant was a direct outgrowth of the Indian government's desire to expand production of chemicals that would improve agriculture.

So, regardless of the specific cause of the leak, in many ways, the leak was caused by both the Indian government's drive to modernize without appropriate regulations and by the lax maintenance and safety standards set in the plant, which would not be tolerated in chemical plants in other parts of the world.

The plant was built in a poor community of India with an ethnic minority of Muslims. After the disaster, a relatively modest compensation package was settled on by the courts in India and some of those who were believed responsible for the disaster were fined and jailed. However, many believe that the compensation was not enough to cover the long-term impacts of the disaster. Many people experienced lasting effects on their health from exposure.

There are obviously a number of environmental justice issues at play within this case study that include local decision-making as well as globalization challenges.

Tuvalu and global climate change

The island nation of Tuvalu is a small 10 square mile country made up of several small islands situated between Hawaii and Australia. It has a population of about 11 000 people who largely work in agriculture and fishing. Many Tuvaluans work in other countries and send funds back to their home. The country is largely isolated from major shifts in global economic conditions since it is so remote and most people earn their living from the island's abundant natural resources. There are few factories and limited port facilities, so Tuvalu has largely been immune to the sweep of globalization which was in part responsible for the disaster that took place in Bhopal, India.

Given Tuvalu's relatively idyllic setting, why are we concerned with environmental justice issues in a far-away place like Tuvalu?

Perhaps the biggest challenge for Tuvalu's long-term survival is global climate change. The highest point in the entire country is 15 feet (roughly 4.5 meters) above sea level. The average elevation is just 6.6 feet (2 meters), which means that half of the land in the country is below that elevation. Clearly any change in sea level is going to cause significant challenges to the long-term survival of the 11 000 residents living on the islands. Already, during extreme storm events, the island can be over-washed by storm waves and tides.

Based on what we learned in the chapter on energy and global climate change, most of the greenhouse gases associated with the problem are produced by major industrial or industrializing countries like the People's Republic of China and the United States. A country like Tuvalu produces a fraction of the greenhouse gases of these giant emitters of pollution, yet they are experiencing a disproportionate amount of damage.

Residents of Tuvalu are already experiencing the impacts of global climate change. There has been a rise of about 5 millimeters per year of water. Low lying areas that are normally dry during regular extreme spring tides are flooding. The islands are protected from wave erosion by living coral reefs which surround the islands. However, the temperature of the water surrounding the islands has increased concomitantly with CO_2-driven ocean acidification.

These two conditions have severely damaged the coral, thereby making the islands especially vulnerable to regular wave erosion and to severe storms, including hurricanes.

Many believe that if something isn't done to stop the steady rise in planetary temperature that Tuvalu will be uninhabitable soon.

Leaders of Tuvalu have pleaded with developing countries to try to solve their problem—a problem which they are not causing.

This example of environmental justice is clearly problematic. The existence of an entire country is threatened due to the widespread environmental pollution caused by other countries. While Tuvaluans do contribute to the overall global greenhouse gas budget of the planet, their impacts are disproportionate to their small emissions total. It is the disparity of cause and effect that makes Tuvalu such a troubling example of environmental justice in our globalized world.

Thinking about this case, how would you solve it? There are really only a few options for the future of the country:

1 The world addresses global climate change and solves the problem of greenhouse gas pollution
2 Tuvaluans evacuate the island and migrate to safer areas
3 Tuvaluans stay on the island and try to survive under difficult conditions.

Environmental justice in Africa

Africa is a highly diverse continent with a myriad of different ethnic and cultural identities. It also has extremely variable climates with hot, dry desserts in the north and south and tropical rainforests near the equator. There are numerous examples of environmental justice problems that fall within the themes identified in the previous case studies in Asia and the Pacific. However, one case is unique and highlights some of the long-term challenges of environmental justice in this region.

For decades, many countries in Africa have been impacted by hazardous waste dumps. In some cases, countries with manufacturing sites on the continent leave waste behind due to the lack of strong local and national environmental regulation. In other cases, waste is imported and dumped indiscriminately. However, one class of waste, electronic waste, is causing a variety of pollution problems.

We all love our electronic gadgets. I have two laptops, two desktops, a cell phone, two GPSs, an iPod, and an iPad. I also have two printers, several digital cameras, stereo equipment, and two televisions. Just a decade ago, I had one computer, one television, a stereo, and a cell phone. I more than doubled my electronic gadgets in just one decade. Of course, like most Americans, I update my electronics every few years to get models that do more—better memory, better screen, better sound, more aps, better service, etc. Technology is moving so quickly that we need to upgrade our equipment regularly just to stay current so we don't fall behind in trends or applications. However, my experience is a distinctly American experience. Many parts of the world are not so quick to upgrade electronics. Indeed, most in the world do not have the range of electronic gadgets that I have in my office.

The Western approach to consumer electronics, with its constant updating of products has tremendous implications for the waste stream. There is a huge amount of electronic waste that is produced each year on our planet—roughly 50 million tons. Some of this waste ends up on landfills, while some of it is recycled.

One of the challenges associated with recycling electronic waste is that it contains a great number of unusual chemicals and metals that can be toxic. Most individuals involved with professional electronics recycling firms are trained to ensure that they are not exposed to harmful materials during the recycling process.

The steps involved with electronics recycling include:

1 Breaking apart the machine to component parts
2 Collecting reusable circuitry or electronics
3 Separating out remaining materials for recycling into component parts
4 Determining which parts are inappropriate for recycling
5 Dumping unusable materials.

While there are rules that prevent nations from exporting or receiving electronic waste, some get around this issue by selling e-waste as repairable goods that could be refurbished. Over the years, tons of e-waste has ended up in many developing countries in Africa.

The problem, of course, is that developing countries often have lax environmental rules and regulations. Even if rules are in place, there is poor enforcement. What typically happens is that the waste is salvaged for usable parts or valuable scrap materials. The remaining unusable components are burned or left to pollute the land.

Recently general inspections of cargo ships leaving the European Union found that one in four contained illegal shipments of electronic waste. Many try to get around the rules banning the shipment of e-waste by claiming that the material was used goods that were bound for another market.

The challenge with e-waste is that it contains heavy metals, rare-earth elements, radioactive materials, and halogenated compounds that are known health risks (See Table 11.1). Since much of the e-waste is burned after usable materials are removed, many hazardous materials are released to the atmosphere, thereby exposing many unsuspecting people. The metals released in this process can linger in a region for generations, thereby causing long-term problems.

What makes the e-waste problem in Africa so frustrating is that rules are in place to prevent the shipping of e-waste to developing countries. Yet, the dumping continues. Certainly there are individuals in the developed world and in Africa who facilitate this trade. But the outcomes, pollution and public health problems, are the direct result of waste generated in the developed world.

Wealthy individuals or nations do not always feel the full impact of their consumerist lifestyles on others. The case of e-waste in Africa is a clear example of how our drive for new gadgets impacts the lives of others in ways we never imagined.

Environmental justice in Latin America and the Caribbean: oil pollution in Ecuador

Latin America and the Caribbean, like the rest of the world, has its share of environmental justice problems. One of the more noteworthy environmental justice cases in this part of the world is associated with the massive oil pollution problems that occurred in remote areas of Ecuador starting in the 1960s.

Table 11.1 Chemicals present in electronic waste and the potential health problems associated with them.

Metals	Health Risk
Arsenic	Poison, lung cancer, nerve conditions, skin diseases
Barium	Poisonous when oxidized. Muscle damage, heart, liver, and lung problems
Beryllium	Carcinogen, skin diseases, berylicosis
Cadmium	Poison, kidney problems, cancer
Chromium	Poison, DNA damage
Europium	Relatively new commercial product. Health impacts are not known.
Lead	Poison, variety of health problems ranging from learning disabilities to death
Lithium	Corrosive to eyes and skin
Mercury	Poison
Nickel	Carcinogen
Selenium	Selenosis
Yttrium	Carcinogen
Radioactive elements	
Americium	Carcinogen
Halogenated compounds	
Polychlorinated biphenyls (PCB)	Carcinogen, problems with reproductive and nervous systems
Tetrabromo-biphenyl-A (TBBA)	Toxic, hormonal disorders
Polybrominated diphenyl ethers (PBDE)	Potential hormonal disruption
Chlorofluorocarbon (CFC)	Harmful to upper layers of the atmosphere which protects us from skin cancer
Polyvinyl chloride (PVC)	Produces toxic chlorine gas when burned

Eastern Ecuador has tremendous oil reserves that have been exploited heavily over the last 50 years. The region where the oil is present in the subsurface is a remote part of the country characterized by the tropical rainforest common in this part of the Amazon basin. The residents present in the 1960s were indigenous tribes who were highly reliant on the resources provided by the rainforest ecosystem for survival.

When oil exploitation began in the 1960s, it was decided that Texaco, the company involved with oil extraction, would leave polluted drilling water in pits next to its 300 wells throughout the country. Millions of gallons of toxic water was dumped in the unlined pits. Texaco's practice in other areas was to pump the water back into the ground to where the oil was extracted. The dumping of the water in open air unlined pits was not practiced in other areas. Plus, the pipelines installed in the rainforest leaked and many gallons of oil were spilled. Texaco also burned significant amounts of oil, which produced pollution in some areas.

These practices were not the standard practices of Texaco in the United States where it had other oil operations. The poor practices were utilized in the Amazon because they were far from regulators and it provided an opportunity for the company to save money. It must be noted that the field was developed in partnership with local companies. Thus, not only were the outside companies complicit in the pollution, but Ecuadoran individuals involved with the project knew of the issues.

The local population has been greatly impacted by the pollution, which remains in many areas. Studies have documented the health problems of the individuals living in the area, although the oil industry disputes many of these. Plus, some of the pollution has moved downstream from the impacted areas to Peru. The development of the rainforest has made it difficult for indigenous tribes to maintain their traditional lifestyle. The problems in this region are still being litigated.

In order to prevent further damage to the region, the President of Ecuador, Rafael Correa developed an innovative plan whereby the country would forego developing a 4000 square mile of land in the Amazon rainforest if they could raise 3.6 billion dollars internationally for a trust fund. The plan had the benefit of not only protecting the fragile Amazon basin rainforest, but also trapping carbon underground, thereby limiting greenhouse gas emissions.

However, only 13 million dollars were collected toward the 3.6 billion dollar goal. In 2013, President Correa announced that development of the oil field would commence since they were unable to raise the funds. Taxes raised via oil revenue account for a significant amount of the budget of the country.

Environmental justice in a Globalized World

The case studies from around the world demonstrate that there are distinct challenges for environmental justice around the world. Disparities in wealth and power lead to environmental justice issues in almost every corner of the planet. While outsiders are sometimes responsible for the problems, often unknowingly causing the problem through consumption, local actors are also responsible for decision-making that leads to unfair activities.

Shedding light on these problems helps to solve them. As noted earlier in this chapter, environmental justice is a very new concept. Many of the problems that were reviewed in this chapter started prior to anyone conceptualizing the issue of environmental justice in a clear, organized way. Now that we understand the issue, many are working hard to solve existing problems and seeking to prevent new problems from happening.

Every generation these days seems to get some type of appellation that reflects the character of the time. Those born in the 1940s through the 1960s were the baby boomers. Those born from the mid-1960s through 1980 were Generation X, often characterized as a socially open MTV generation. Following this was Generation Y, widely called the Millennials, the group of people born from the 1980s through 2000. This is the group of people that are coming of age during a time of global economic deline. After them comes the current group. There is some disagreement about the name of this group. Some call it Generation Z in follow-up to Generations X and Y. Others have call it the Conflict Generation since they were born after the terrorist attacks of September 11th. Still others call it the iGeneration since it is the first time that a generation has fully been alive during the modern electronic digital age.

In the United Kingdom, some are calling the new generation, Generation Citizen. The reason for this appellation is because statisticians are finding that the current generation of young people are more involved with volunteerism than any generation since the 1930s.

Why is this?

Young people live in a very difficult time. Many do not see the same opportunities available to them as their parents. They have lower expectations for their future and do not necessarily have the same goals as those who came before them. They are finding greater value not in material things, but in relationships and simple living. They tend to care about the world around them in different ways from other generations (Figure 11.7).

Think about your own life and experiences. Do you think that individuals in your circle are making a difference? What about you? How do you get involved as a member of your community? Are you part of Generation Citizen?

Figure 11.7 All of my students are working to make the world a better place. What are you doing to try to improve the lives of others?

(Continued)

Faith-based organizations and environmental justice

Many environmental groups focus very clearly on environmental justice issues. However, faith-based organizations have been especially interested in issues of environmental justice in recent decades. This makes sense if you consider that many of these organizations are interested in broad issues of social justice and overall fairness. It is not a great departure from the teaching of most religions to incorporate environmental justice issues within their local, regional, national, or international mission. Indeed, in Chapter 3, we discussed the role of faith-based organizations in addressing climate change.

Yet religious organizations are doing more than just focusing on climate. Take for example, The World Council of Churches (WCC), which is an over-arching organizational body that focuses on unity within Christian churches around the world (Figure 11.8). They represent hundreds of church parishes and include organizations as diverse as the Church of England, the Methodist Church, the Lutheran Church in Liberia, and the Mennonite Church in the Netherlands. This group focuses on three broad environmental justice issues: water, climate, and poverty.

Figure 11.8 Father Mark Genszler, an Episcopal Priest, oversees the Garden at St. Marks which grows food for local food pantries.

The WCCs interests in water are implemented through the Ecumenical Water Network, which promotes access to water throughout the world. They seek to educate members about problems with access to water around the world and seek to make a difference in people's lives by promoting the idea that access to water is a human right. They have been very critical of water privatization issues around the world. They have also highlighted environmental justice issues associated with water distribution problems in the Middle East.

The WCC is also very active on climate issues. They, and other faith-based groups, have noted that climate change threatens creation and that Christians have a responsibility to take action to prevent damage to our planet. They are urging members to speak out and take direct action. Specifically, they are asking members to fast once a month in solidarity with those seeking to make a difference on the issue. They are also asking members to pray and take local action.

The third area of WCC's focus is poverty, which it addresses within a framework of poverty, wealth, and ecology. They encourage members to examine how disparities of wealth around the world impact the lives of others and to find ways that global wealth can improve the lives of others. The churches encourage AGAPE, which stands for Alternatives to Economic Globalization Addressing Peoples and Earth, by promoting conversations on issues including: "just trade, debt cancellation, financial markets, tax evasion, public goods and services, livelihoods and decent jobs, life-giving agriculture, power and empire, and ecological debt" (http://www.oikoumene.org/en/what-we-do/poverty-wealth-and-ecology).

Together, these three issues help to define the WCC as an organization deeply concerned about environmental justice around the world.

The Catholic Church and environmental justice

Most of the modern focus on environmental justice within the Catholic Church emerged after Pope John Paul IIs message for the World Day of Peace in 1990. In the introduction of his speech, he stated (http://www.vatican.va/holy_father/john_paul_ii/messages/peace/documents/hf_jp-ii_mes_19891208_xxiii-world-day-for-peace_en.html):

> In our day, there is growing awareness that world peace is threatened not only by the arms race, regional conflicts and continued injustices among peoples and nations, but also by a lack of due respect for nature, by plundering of natural resources and by a progressive decline in the quality of life. The sense of precariousness and insecurity that such a situation engenders is a seedbed for collective selfishness, disregard for others and dishonesty.
>
> Faced with widespread destruction of the environment, people everywhere are coming to understand that we cannot continue to use the goods of the

(Continued)

> Earth as we have in the past. The public in general as well as political leaders are concerned about this problem, and experts from a wide range of disciplines are studying its causes. Moreover, a new ecological awareness is beginning to emerge which, rather than being downplayed, ought to be encouraged to develop into concrete programmes and initiatives.

Since then, the message from the Catholic Church has been consistent on issues of environmental justice and the environment. For example, the Catholic Climate Covenant (CCC) is a coalition of Catholic organizations from around the world that seek to educate people about the issues of climate change, while also seeking individual and group action on the issue. They encourage people to take the St. Francis Pledge (http://catholicclimatecovenant.org/the-st-francis-pledge/). The pledge asks people or organizations to: Pray and reflect on the duty to care for God's Creation and protect the poor and vulnerable.

Learn about and educate others on the causes and moral dimensions of climate change.
Assess how we—as individuals in our families, parishes and other affiliations—contribute to climate change by our own energy use, consumption, waste, etc.
Act to change our choices and behaviors to reduce the ways we contribute to climate change.
Advocate for Catholic principles and priorities in climate change discussions and decisions, especially as they impact those who are poor and vulnerable.

What is especially worth noting in the efforts of the Catholic Church is that they have linked very clearly the impact of climate change and other environmental problems to poverty and the vulnerability of populations directly impacted by these issues.

The WCC and the Catholic Church are but two examples of faith-based organizations that focus attention on issues of environmental justice within their teaching. In fact, most religious organizations teach issues of environmental justice in some way. While some may not explicitly state that they are involved with environmental justice, they do promote the basic ideas of fairness, equality, charity, protecting creation, and caring for one's neighbor.

Think about your own community and experiences. How are religious organizations in your area committed to environmental justice? What can you find out about their initiatives? How involved are they with local, national, or international issues? If you are involved with a religious organization, how does your place of worship address issues of environmental justice? Do they have committees or community groups working on the topic? How can you get involved?

Over the last few decades, religious groups have been among the leaders trying to shed light on environmental justice issues. They have worked to solve many of the world's most troubling environmental justice problems.

12

Sustainability Planning and Governance

State and local governments are where the action is on sustainability. They are responsible for local decision-making that impacts the lives of residents. Since everyone lives within a local government of one form or another, it is worth examining the role of local governments to determine their overall role in advancing sustainability goals. They often are reacting to state, national, or international sustainability goals, while working at the grass-roots level to advocate for long-term sustainability of their region.

Local governments and their structure

Local governments take many different forms around the world. At the most basic level there is some access to local decision-making by all citizens. In prehistoric times, we lived with small tightly knit clan groups and decisions were made, if not cooperatively, certainly for the greater good of the clan. However, as we developed more complex societies, local governments became much more complex and we developed a range of approaches to managing local decision-making. Some forms of local government provide greater citizen input than others, but they all in some way have one form of local resident input. However, local governments also work with broader stakeholders in making decisions within a state or national context (Figure 12.1).

In most parts of the world, citizens have a right to select their local leaders who have the responsibility of implementing the community's will within the rule of law imposed by the local and higher-level governments. Communities may be served by an elected board or council along with a mayor. In some cases councils are appointed by a higher-order government authority. For example, in China, local leaders are often appointed by the central government. In some communities, the local elected leaders hire an administrator, such as a town or county executive, to manage the day-to-day operations of the community. Of course, there are many parts of the world today where local governance is very difficult due to political or social turmoil. These places are managed largely by external forces or warlords and there is very little opportunity to make improvements on sustainability-related issues in such places.

Introduction to Sustainability, Second Edition. Robert Brinkmann.

Companion website: www.wiley.com/go/Brinkmann/IntroductiontoSustainability

Figure 12.1 Political leaders work with a variety of stakeholders to make decisions that impact local communities, states, and nations.

In some places, there are special-purpose local governments that are organized for particular reasons. These may include school boards, sewer system boards, water boards, library boards, or public transit governance. All of the local government forms have some responsibility of ensuring the long-term sustainability of their community. They all work more or less with issues that fall within the three "E"s of sustainability: environment, economics, and equity.

The role of citizens and stakeholders in local government

Citizen participation is crucial to the success of local governments. Their input is important to the day to day workings of local governments. Without participation, democratic traditions in local governments can break down. Plus, citizens have a responsibility to oversee government processes via participation in public meetings to ensure that leaders do not misuse public funds or lands.

One of the most important aspects about citizen participation in local government is that the citizenry should be informed so that they make wise decisions on local issues and in voting for leaders. Thus, an educated public is an important aspect of successful local government. That is one of the reasons why so much attention is given to education in local communities. Plus, local governments have an obligation to ensure that they educate their citizens about key issues facing their community.

Community stakeholders

It is important to recognize that citizens of a community are not the only ones concerned with the actions of local government. Other stakeholders matter as well. A stakeholder is a person or an organization that may be impacted by the actions of a local government. At the most basic level, stakeholders are citizens within the community. But they also include organizations such as businesses, non-profit organizations, other government organizations, such as county or state governments, and any other person or organization that could be impacted by decision-making in the community. Thus, while local governments react strongly to local opinion, they are also impacted by external stakeholders who may or may not have the best interests of the community in mind.

For example, a county government may wish to build a landfill in a community and urge its approval in the community in which they want to build it. It might be the best thing for the county, but a bad thing for the local town government. However, a non-profit might wish to urge a community to impose green building standards in the community, which might be a good thing for the local residents. For this reason, it is important to understand that there are multiple voices of stakeholders in local decision-making that must be understood. Local governments react to the voices of citizens, but they also are strongly impacted by external pressures that can have positive and negative impacts.

In some instances, stakeholders may not have the best interests of the citizens in mind when engaging with a local government. For example, a local industry owned by outside parties may seek to expand operations. They may attempt to influence the local community to advance their operations that may lead to harmful impacts in a community.

When this occurs, local leaders can be put in conflicting positions of advocating for the needs of the voters, while trying to advance economic development opportunities within their community. External forces, particularly those with deep pockets, can seek to influence voters and local officials to allow activities that may not be in the best interest of the community.

Stakeholder work

Local governments often seek a great deal of input before making decisions on important things ranging from changing building codes to building a solar park. They often hold public meetings to get input from stakeholders, or residents and other individuals or organizations from inside or outside of the community that in some way may be impacted by the decision.

But how does a local government get input that is useful in local decision-making? Sophisticated governments have developed a variety of means to gain input from stakeholders in processes broadly called stakeholder analysis.

Stakeholder analysis can be conducted in many ways. They can include opinion surveys, interviews with key individuals, and public meetings with input. By far, the most common type of stakeholder analysis is done via public meetings whereby the public

(Continued)

is invited to discuss a particular issue. This can be done during regularly scheduled government meetings or a special meeting could be held to gain input on a particular issue.

Have you ever been to a public meeting that focused on an issue related to sustainability?

I have been to many over the years. One of the most interesting was one that was called by a local department of transportation. They were seeking input on bike lanes on roadways. The meeting was widely advertised and it took place during a time that was convenient for the public. However, I was the only person who showed up.

During that meeting, I was able to provide a great deal of input as to where I thought the bike lanes should be located. Of course, I tended to select areas that would be convenient for me, but I also suggested areas where I knew they would be welcome. Within a few months of that meeting, I saw bike lanes being built on the very routes that I suggested. That meeting taught me that it is very important to attend public meetings if you want to make a difference. Try it!

Of course, we have all seen videos of public meetings that turn into shout fests with bad public behavior. Individuals, for whatever reason, try to hijack a public meeting to make a broader political point. However, these meetings are rare in most communities. Usually, local government leaders seek to hear polite comment from the community in order to make wise decisions. When you hear about a public meeting on something that interests you, attend! You can make a difference just by showing up and speaking in support or against something. Your voice matters.

Boundaries and types of local governments

Local governments have a great deal of control over decision-making about things like land use, zoning, development, and infrastructure within their boundaries. They often have much more influence on local issues than state or national authorities. As such, they have great impact on a wide variety of key sustainability issues.

Local governments have distinct different boundaries and can be classified in different ways, depending upon population or areal size.

Cities, villages, and towns, all have clear local governmental structures with a high degree of command and control organization. They often have a mayor and an elected council responsible for decision-making on key issues in the community.

But urban areas are not just the simple boundaries where the main city is located. The impacts of the population of the broader urban area must be recognized. Outside of cities, there are also suburbs, which often have limited local government access. Instead, these areas are often managed by county governments which usually have limited local impact. Or, they are governed by community organizations (like condo boards) that require particular standards that may or may not be in line with sustainability goals. Often the purpose of these organizations is to maintain property values, and not for more altruistic sustainability goals.

These areas outside of city governance often have populations that are as great or greater than the primary city. For example, the City of Houston, Texas has a population of about 2.3 million, but the population of the greater Houston region has a population of 10 million. The 8 million people living outside of the City of Houston have limited access to the types of sustainability efforts that are underway within the city. Instead, these people live in a number of local government jurisdictions, often not coordinated with each other. Thus, while some fantastic work is being done in cities on sustainability, the broader suburban areas are not always brought into the mix of sustainability efforts. That is why the work of counties, metropolitan planning organizations, and states are so important.

County or other regional spatial public organizations have government organizations that manage many different types of government operations not often managed by local governments, such as pollution management, or high-traffic roadways. In addition, they play an important role in managing coordination of local governments within the counties. In many places, the counties take on the responsibility of managing or coordinating the sustainability efforts underway in local governments.

However, in many parts of the world, cities and their suburbs have expanded outside of a single county into multiple jurisdictional areas. Expansive cities like London, Los Angeles, and Mexico City require coordination beyond what can be done by a single city or county. To address these issues, regional planning agencies have evolved that coordinate efforts of multiple governments within such settings. These regional planning agencies work within metros, or complex urban areas, to coordinate complicated sustainability issues such as air pollution, food sustainability, and mass transit. In the United States, these places are called metros and are often managed by a metropolitan planning agency of some type.

In some places, state or national governments make efforts to coordinate local sustainability initiatives. For example, in China, great efforts are being made to reduce local air pollution via initiatives by the national government. These state or national initiatives tend to address large problems and often do not deal with traditionally more local issues such as zoning.

Thus, governments each have a role in sustainability management. Local, county, metro, state, and national governments each contribute in some way to broader sustainability initiatives.

In the United States, there is growing interest in managing urban areas within regions called metropolitan statistical areas (MSA). These are places that have a strong urban core, with economic and transit connections in the surrounding regions. The statistical areas use the county as a form of organization. Thus, the MSA includes the core urban county along with the surrounding counties to which it is connected. The New York MSA, for example, includes 25 counties, including ones in New York, New Jersey, Connecticut, and Pennsylvania. It also includes the well-known cities of Newark, Jersey City, Yonkers, Stamford, New Haven, Trenton, and While Plains.

Overall, there are 384 MSAs in the United States (Figure 12.2). Each has a core urban area of at least 50 000 and includes adjacent lands where there are economic and transportation connections. The five largest MSAs are the broad New York, Los Angeles, Chicago, Washington, and San Francisco regions. These cities expand far beyond their central urban core into the surrounding region. Thus, the place that is known as "Dallas" includes the city of Dallas as well as the surrounding communities where there are connections.

Figure 12.2 Downtown New York may center around the new World Trade Center complex, but in reality, the New York region extends for miles in all directions.

While local governments have full autonomous rights, there are many projects within metros that assist with regional sustainability. Take mass transit, for instance. It does not make sense for every local community to manage their own mass transit within a metro. Coordination must be undertaken to build connections and facilitate a solid regional transit system that allows the movement of people across the region. This cannot be undertaken by one local government, but must be coordinated within a region. Also consider basic services like water, sewage, electricity, and gas. While many communities within metros maintain their own infrastructure for these systems, regional coordination often saves money and promotes better use of resources over time. Pooling of resources also allows regions to work on large regional improvement projects that benefit all citizens. For example, efforts to improve water resources or expand sewer operations to create a state of the art green system may be expensive to undertake as a local government. Yet if multiple governments pool resources and create projects that can be shared, the region can benefit.

Leadership

Leaders of local governments, often called mayors, are usually supported by an elected board. These individuals may or may not be in support of sustainability initiatives. That is why leadership and public input on these issues is so critical. Without strong local government support, sustainability initiatives will not move forward.

However, local governments can be advised by stakeholders to move forward on sustainability projects. They may or may not do this, but the input of citizens and stakeholders is an important determining factor in advancing sustainability in some communities. Sometimes, important national leaders have major local impacts. Margaret Kenyatta, for example, the First Lady of Kenya, has been extremely active on a variety of local development initiatives including conservation, women's rights, and social justice.

Efforts to aid local governments on sustainability issues

Several different types of organization work to help local governments on sustainability issues. They range from international efforts like ICLEI for local governments, to the efforts of local or state non-profits that work on thematic issues like green building or smart growth. States also try to coordinate the efforts of local governments within sustainability networks on issues like improved building codes or greenhouse gas reduction.

The most widely recognized organization working on sustainable development with local governments is a group called ICLEI for local governments (Figure 12.3). This group

Figure 12.3 Many large and small communities are members of ICLEI including Vancouver, Canada, shown here.

emerged as an outgrowth of the UN's sustainability initiatives that developed after the Rio Earth Summit in 1992. Originally an arm of the United Nations and often called Agenda 21, it is now a separate non-government organization (NGO) that strives to coordinate sustainability initiatives around the world.

There are over 12 000 local governments that are members of ICLEI. While ICLEI provides a great deal of free information on their website, a local government must pay membership to gain access to a number of members-only benefits that include benchmarking tools, case study reports, and expert assistance.

Large and small communities, and communities in developed and developing nations are members of ICLEI. Thus, it is one of the few sustainability organizations that transcend development status. Indeed, there is a strong effort to match case studies and peer mentors within appropriate communities while trying to forge links between the developing and developed world. ICLEI also does a considerable amount of regional reporting and analysis to provide assessments and recommendations for sustainability initiatives across places like the entire continent of Africa.

While ICLEI is perhaps the best example of local government sustainability initiatives, the Florida Green Building Coalition's Green Local Government certification program is perhaps one of the best benchmarking tools available for local governments.

As was noted in a previous chapter, the Florida Green Building Coalition is a non-profit organization that developed in the state of Florida to manage the green building certification using the LEED building certification process. The issues in Florida are unique compared to the rest of the United States due to the subtropical setting of the region. This group not only puts a twist on how green buildings are certified in the state by taking into account things like the subtropical heat and threat to buildings from hurricanes, it also developed a system for assessing green local governments.

Like the LEED green building rating system, the local government rating is based on a point system within a number of different categories. Each component of local governments, such as purchasing and fleet maintenance has a series of points that can be earned. The points vary depending in importance, but it must be stressed that all aspects of local governments are covered in the scheme. Thus, all employees and all departments are engaged in the sustainability initiative.

The system was set up so that different sizes of governments can participate. Thus, scale issues are not significant in the process. Participation in the plan is voluntary, with smaller governments paying less for certification evaluation than larger governments. Across Florida, dozens of cities and counties of all sizes have been rated by the system. Because the Florida Green Building Coalition's local government rating system is quantitative, one may measure the broad impact of the system across the state. While not all local governments in the state participate in it, the scheme provides a means by which statewide sustainability efforts by local governments can be assessed.

There are also a number of less comprehensive initiatives that assist local governments with their benchmarking efforts. These initiatives tend to focus on one particular issue, such as storm water pollution reduction, food equity, or transit. While these initiatives may not be as comprehensive as some of the more broad-based initiatives, they do have tremendous impacts in some communities. For example, an effort geared toward the reduction of automobile use by the expansion of mass transit in local communities can be very significant to residents and regional economic development.

Scale and local governments

While the Florida Green Building Coalition's green local government rating system can be used across all the different sizes of government, there is no doubt that scale is an important factor in making a real difference at the local level.

Big cities, by some measures, are the most sustainable places on the planet. People tend to use fewer resources in cities than in other places. For example, take people who live in large apartment buildings compared to those of us who live in houses. They use less water and produce less pollution. They tend to drive less, if they drive at all. There is an efficiency of resource use that develops when people live in close proximity to each other. As one moves away from cities, basic resource use increases because it is much more inefficient to live apart from each other. We have to become much more self-sufficient because there is not a strong infrastructure or network of assistance that is available.

This same efficiency is true of governments. The larger the government, the more resources they have to manage sustainability issues. They can employ sustainability officers, afford to start new initiatives, and reorganize priorities with little impact to the basic services local governments provide. They can do this because they have a large talent pool among employees and volunteers from which to draw within their large governments and they have large budgets.

Governments of small towns, suburbs, or rural areas do not often have this flexibility. They typically have limited budgets and a limited staff. Thus, they are unable to provide a range of sustainability initiatives for their community. However, they often are able to specialize by focusing on some key areas. For example, some small towns have worked very hard to try to develop alternative energy and others have worked hard to develop community gardens and other local food options. But it must be stressed, larger cities have the resources to conduct a range of initiatives, while smaller communities are able to focus on just a few. Thus, smaller local governments, since they are unable to coordinate all-encompassing initiatives, no matter how green they try to be, are not as efficient as the larger governments.

As noted earlier, many, but not all, metros and counties work to coordinate sustainability initiatives in their regions. However, they are unable to compel all governments to participate. They tend to focus on very broad initiatives such as carbon reduction or transit. Thus, while they can play an important coordinating role on large important issues, they are often unable to coordinate detailed comprehensive initiatives better suited to local staff and leadership.

When considering these scalar challenges—local large governments, local small governments, and county and metro governments, it is clear that each has challenges in addressing key sustainability initiatives. However, there are good examples of case studies that demonstrate how each type of government confronts them.

Green regional development

Given the challenges with local implementation and management of sustainability initiatives, many have advocated for a more regional approach that takes into consideration the challenges of a broader area, but not always an urban and suburban landscape. In such settings, broad initiatives can be addressed with goals for the entire region. The benefit of these types of assessments is that they can allow for a degree of specialization within the region. For example, places with high density can focus on mass transit, while places with low density can focus on improvements in food access and quality. The regional approach means that not every single place has to address all areas of sustainability. Each area can specialize to do their part to make improvements and reach goals for the region. The regional approach also focuses financial and human resources in ways that allow the development of large projects. For example, it would be difficult for a small village to invest in a large solar farm or wind project. However, together, many villages within a region can share resources to invest in such large projects.

The regional approach requires considerable cooperation. Thus, many regional planning agencies have evolved in recent decades to address planning issues associated with sustainability. Many of these agencies started as part of regional planning efforts that were implemented in the 1960s as cities expanded outside of their traditional boundaries into suburban settings. However, in more recent years, these organizations have taken a leadership role in advancing sustainability initiatives at the regional level.

For example, some have conducted regional greenhouse gas inventories and others have developed mass transit initiatives. Others have evolved in recent years to focus on environmental problems such as water scarcity or pollution.

Pepper sauce, government approval, and citizen concern

Almost everyone loves Vietnamese hot Sriracha sauce, sometimes called rooster sauce (Figure 12.4). I use it in lots of different types of cooking and I also like to use it as a dipping sauce for spring rolls or as a table condiment for everything from eggs to burgers. A little bit in mayonnaise makes a delicious spread for veggie burgers. However, if you've ever worked with hot peppers, you know that they can be problematic. When I cook with fresh peppers, I always have to take extra care to avoid getting the seeds or pepper veins on my fingers. The hot oils produced in these parts of the peppers can easily find their way to eyes or other sensitive portions of the body. Plus, many of us find our respiratory system irritated from the fumes produced when peppers are cut or cooked. Some find themselves with burning eyes and some discover burning feelings in their lungs and sinuses. This can lead to difficulty breathing.

So, while I love hot Sriracha sauce, I am glad that I don't have to make it. And, I am also glad that I don't live near a Sriracha sauce factory!

When Hoy Fong Foods opened up a new 626 000 square foot factory in Irwindale, California, the food community rejoiced. The popularity of the company's signature Sriracha sauce led to increased demand and concerns over whether or not the company

Figure 12.4 What would your life be like if you lived or worked near a factory that made hot sauces?

could produce enough of the sauce for the market. The new factory produced jobs and greatly expanded Hoy Fong Foods' ability to meet the demand by tripling the output of the hot sauce.

However, when the factory opened, residential neighbors were not universally happy. They found themselves confronted with the very strong smell of the hot pepper. Some with existing respiratory difficulties said that the strong pepper smell made things worse. Residents complained that they could not venture outside without having problems like burning eyes, throat irritation, and headaches. Residents claimed that their children could not play outside and that they could not leave their windows open to get fresh air.

As a result, the City of Irwindale sued Hoy Fong Foods to stop production to protect the public's health. Eventually the suit was dropped when the company agreed to work on a filtration system to reduce the impact of the pepper fumes in the surrounding area.

The nature of the case made the press in the United States where Sriracha sauce is very popular. Many commenters on the situation noted how they felt that the city was out of bounds in trying to regulate a local business and that the owner had the right to run his business as he saw fit. They saw the city as too concerned with local complaints and felt that it was over-reaching its regulatory authority. However, others felt that the city should have been more aggressive in protecting the public health of local residents that were in place prior to the development of the new facility.

(Continued)

What do you think? Should officials have closed the facility to protect the local impacted residents who complained about the health issues or should the factory be allowed to emit the pepper smell? What would you have done if you were a local resident that was impacted by the fumes? What would you have done if you were the factory owner who moved the facility to the town? What would you have done if you were a local elected official? When the story broke, many elected officials around the country invited Hoy Fong Foods to relocate to their town. Would you want the company to relocate to your neighborhood? How could local governments work with companies like Hoy Fong Foods to reduce the impact within their communities?

NIMBY syndrome: not in my backyard

One of the challenges for any community is where to locate undesirable activities. Imagine this situation. Your community needs a new sewage treatment plant and there is no room in the present space to expand. A new site must be found. However, where to put it? No-one would choose to locate a sewage treatment plant in their neighborhood. The plant could lower property values and produce visual and sensory blight. At the same time, everyone in the community would benefit from the plant and it would be an opportunity for the community to grow and expand population and industrial and commercial development.

In these situations, local governments are faced with a very difficult choice. They either have to live with the existing sewage plant and limit growth, or they have to select a site that will certainly be unpopular within a segment of the population.

Anyone who works in local government is very familiar with these problems. The NIMBY (not in my backyard) phenomenon is found everywhere. It doesn't matter if it is a sewage treatment plant or a new power line. People do not like any change that will impact their lives. People who actively argue against projects in their neighborhoods are often derisively called NIMBYs.

For this reason, some have defined NIMBYs as people who try to exert an unreasonable demand on the community for collectively popular projects that seek to advance the greater good of the community. However, others have defined NIMBYs as individuals who are striving to protect the greater community from bad decisions by local officials.

Many times, such as in the case of the sewage treatment plant, the concerns are real. While the project is needed, the local impacts are very high. Concerns of the residents are real and must be considered.

Many citizens have a deep distrust of local governments and feel, whether based on reality or not, that local officials in some way benefit from new projects. There are plenty of examples of local communities undertaking or approving projects that are not necessarily needed for the long-term success of the community or approving projects that benefit a few. There are also examples of corruption where local officials are bribed or in some other way benefit from the approval of the project. Thus, the NIMBY motivations can arise in different ways.

However, when projects are needed, how can local communities deal with the NIMBY phenomenon?

1. Make sure that the government follows good practices in all matters to ensure that local officials and staff have a strong public record without corruption.
2. Ensure that the project is part of an overall community strategy that is part of a long-term community plan.
3. Develop long-term planning in the community to guide decisions on large projects that will impact the community.
4. Ensure that all information about the project is available to the public to create an environment of openness.
5. Provide opportunities for the public to comment on projects at public meetings and in writing. Ensure that responses are made to concerns.
6. Listen to suggestions from the public as to how the project could be modified or improved to limit impacts.
7. For large project, a citizen's advisory committee can work to advice elected officials on the project. The advisory committee should include representatives from a number of different stakeholders including elected officials, community staff, project experts, and citizens.
8. Create educational materials about the project that are available online and in print.
9. Bring in technical experts to outline the need for the project and how it will benefit the community as a whole.

No matter what the project, and no matter how good a job the local government does to ensure that the decisions are appropriate, some citizens will be unhappy with the outcome. Such is the nature of change.

Think about this in the context of wind energy. Everyone loves the idea of getting off the grid and limiting our reliance on dirty fossil fuels. However, would you want a giant windmill in your backyard? What about a sewage treatment plant or a Sriracha factory? We all want the benefits of these things (green energy, sewage treatment, and hot pepper sauce) but we don't want to be confronted with these land uses in our day to day lives. The reality is, however, that someone lives near these places. Every day local governments around the world must make difficult decisions on projects that may harm individuals, but help the greater community.

So, to some, NIMBYs are complainers who cause problems. They are seen as working for their own personal or monetary good and against the broader public interest of the community. Yet to others, NIMBYs are heroes who are taking on local governments and larger powers to fight projects that are locally unpopular. Yet, the best way for local governments to look at NIMBYs is to see them as stakeholders with an important voice on the future of the community. While agreement and understanding may never be reached between local government officials and the NIMBYs, it is important that their vision of the future of the community be considered in local decision-making.

Sustainable development

The idea of sustainable development, or development that meets the needs of the present without interfering with the needs of future generations, has been around for a number of decades. Yet how effectively have we advanced sustainable development around the world? There are no doubts that advances have been made in some parts of the world in areas like education, healthcare, and environmental protection. However, these advances are uneven. Plus, in our era of rampant globalization, it is important to examine how developed nations are impacting less-developed nations in their activities. Traditional development models follow a flow of goods and knowledge from the developed world to developing countries. However, this approach is neocolonial in fact, because it does not address the impacts of the behavior of the developed countries on their ability to be sustainable into the future. While there may be good intentions to focus on improvements, other processes of globalization can severely limit the ability of places to rise above their circumstances.

In many ways, there are many different types of sustainability based on one's perspective. In developed countries, we are worried about things like reducing carbon footprints, access to mass transit, and promotion of clean and healthy food. However, in developing countries, sustainability issues are much more focused on survival. Concerns over fresh and healthy water, dietary needs, and lifespan improvements are often central. The perspective one takes from one's life experience and values greatly impacts how we approach the way we view and implement our sustainability initiatives.

Some believe that sustainability is framed by our ideological thinking. How a capitalist or a socialist approaches sustainability varies considerably. We do not have any clear endpoint that defines sustainability and the way we move forward with initiatives is often a negotiated space in our cultures. Each region has their own way of finding a balance between environment, economy, and social equity. There are few places that are truly able to achieve this balance and there is always a continuous shifting of efforts to try to advance long-term sustainability based on local, national, regional, or international conditions.

The developed world, with its responsibility for global climate change and other harmful problems is perceived by many in the developing world as largely responsible for many sustainability problems, both global and local. As a result of this sentiment, there is distrust by many in the developing world about the overall notion of sustainable development since they see that the developed world is largely responsible for creating an unsustainable future for the planet.

Plus, some see sustainability initiatives advocated by the developed world as not particularly helpful in the long run. For example, many are concerned about how to feed the world during this time of rapid population growth. Some believe that the way to do this is to use technological advances like genetically modified crops (GMO) to enhance agricultural yields. They feel that significantly increasing the output of agricultural fields will allow us to feed the growing planet into the future. However, critics of this approach note that moving farmers into GMO food production makes them reliant on technological fixes that may not be sustainable or affordable into the future.

This conflict points to the issue that there are multiple ways of viewing sustainability that are not always compatible with each other. The idea of sustainability assumes that there is some appropriate convergence of economy, environment, and equity that we should try to achieve. But whose convergence? If the convergence were based on the very high-consuming developed world model, the world would be depleted of resources very quickly. But if the world were to strive toward a model of subsistence sustainability, there would be an outcry from the more developed nations. Thus, the notion of sustainability has considerable discord when applied to the idea of international development.

There are also two approaches to sustainable development that I call surfing sustainability and suffering sustainability. Surfing sustainability is the sustainability of the west. It includes things like organic food, green economic development, and plastic bottle bans. These things are great, but take them in contrast to those who are suffering sustainability. This is the sustainability of the developing world where people have limited access to resources and are trying to improve their lives. They sometimes have problems like war or famine imposed on them. Within this context, it is important to consider the individual location when thinking about what sustainable development means to a community.

As we will see, issues of globalization make the challenge of sustainability complex, as these different ways of viewing the world and its future meet.

Globalization

The world is an increasingly interconnected place. Globalization is the process by which ideas, products, media, and even pollution become around the world (Figure 12.5). This section will review the concept of globalization and look at different ways that it impacts

Figure 12.5 Globalization helps bring us together. Here my Chinese colleague is teaching me how to make dumplings with her American daughter in China. How is your world impacted by globalization?

the world. Following this, we will look at organizations that seek to promote sustainability across the planet in different ways. We will particularly examine the efforts of the United Nations.

Development of globalization

Globalization has been around as long as humans have travelled the Earth. Archaeological evidence shows extensive prehistoric trade routes in almost all corners of the world. For example, artifacts of worked shell pieces that originated along the Gulf of Mexico coastline have been found in the Midwest and throughout many corners of North and South America (Figure 12.6). In the Classical era of Europe, the cultures of Greece and Rome expanded their influence in many parts of Africa and Asia. In Asia, extensive travel between far-flung locations like India, China, Korea, and Japan created some unifying characteristics among the cultures. However, the slow pace of travel throughout much of history slowed the impacts of globalization and many regions maintained unique identities, while taking or rejecting elements of others.

The enlightenment and subsequent industrial revolution significantly increased the pace of globalization and its impacts. Empires formed around the world buoyed by the ability to travel great distances over ever shorter periods of time. The pace of cultural change increased as distances became less significant and as trade brought consumer goods from around the world to all areas of the planet.

There have been many concerns about the impact of globalization on local cultures throughout the last century. For example, the Opium Wars in nineteenth century China were largely caused by the British desire to sell opium, a highly addictive drug, to the Chinese population. While opium was used in China prior to the British involvement in the area, the extensive marketing of the drug pushed its use sky high. In our modern era, the clashes of western ideals with those of the Islamic world create significant cultural conflict.

Figure 12.6 These are projectile points that are about 3500–5000 years old! They were found on the campus of Michigan State University and were collected as part of a campus archaeology program. What artifacts might you find on your campus? (Photo courtesy of Lynne Goldstein).

Globalization has also caused significant environmental degradation. In the early twentieth century, feathered hats came into fashion in many areas of the world. International magazines such as Vogue, and new celebrity magazines showed models and starlets wearing exotic-looking hats that featured striking designs that included feathers. Yet, the impact of this international look was devastating for the bird populations in many parts of the world, most notably the Everglades of Florida.

Bands of bird hunters would extend across the vast wetland looking for birds such as the snowy egret. These animals flocked together in large colonies that lived in trees. It was easy for hunters to kill entire groups of birds. Many animals went extinct and populations declined as the shooters filled orders for hat makers around the world.

Feather demand is only one example of the impact of globalization on the environment. In our present era, we see that there is a great desire for shark fins for Chinese gourmets around the world and for rare minerals for the manufacture of modern electronics. We are global consumers and we gather resources everywhere. Unless a region has significant environmental rules and regulations, the impact of the demand for consumer goods can be significantly harmful to the planet.

Drivers of globalization

There are several drivers of globalization that impact its reach, several of which will be discussed below: Internet and communication, transportation, economic development, and transnational organizations.

Internet and communications

Perhaps the most significant driver of globalization today is the Internet and modern communication. When the Internet started a few decades ago, no one could have predicted its global reach or significance. Many people have social media accounts, even if they do not have access to their own computer. International websites provide quick access to nearly all knowledge available to mankind. Many of these sites offer quick translations thereby removing language as a barrier.

We also have global cable channels. News outlets such as BBC, CNN, Univision, and Al Jazeera offer information available throughout the world. Entertainment companies like Disney and Universal Studios offer a range of global diversions that range from theme parks to movies and television productions.

It's interesting to consider the range of impacts of this particular driver. There is no doubt that the global entertainment sector has a distinctly Western, indeed American, flavor. In contrast, the Internet is more open in most instances. Of course, there are cases around the world where the Internet is censored or limited—often to try to limit the influence of other ideas of culture on local population—but for the most part, the web is a relatively democratic place that captures a huge range of information. And that information is available for anyone to see. Ideas can be spread back and forth from one culture to another.

Perhaps the most interesting example of the use of the Internet and cell phones in demonstrating the role of globalization is the use of Twitter in the Green Revolutions that took place in some areas of the Islamic world in the early 2010s. Protesters from Tahir Square in Cairo posted photos and updates that could be read everywhere. People from Vermont to Durbin were following along and became part of the cultural change taking place in one particular location. In 2019, the Internet played a big role in organizing the protests in Hong Kong.

There is some concern over Internet Balkanization, or polarization, of ideas. We can easily enter an echo chamber where only ideas we approve of are heard and repeated. For example, if we are pro-oil or anti-drilling off the coasts of Alaska, we will only listen to or read information that agrees with our position. There is concern that the Internet can stifle true debate and reasoned decision-making. In addition, the Internet has been used to spread misinformation, most notoriously during the 2016 American presidential elections.

Transportation

Transportation is another important driver of globalization. While the Internet and media can instantly share ideas across computer platforms, the exchange of people and goods across the world is changing culture tremendously.

International travel is another major driver of globalization. People are moving for business, tourism, family, migration, and a number of other reasons. As they move, people bring their values and ideas, which can be in conflict with the host country.

I remember traveling to Latvia several years ago where I saw groups of men from other parts of Europe drinking heavily at all hours of the day. As it turns out there are tour companies that arrange such trips, to the dismay of some of the more conservative Latvians.

Migration of large numbers of Muslims to various corners of Europe, Central Americans to the United States, and Venezuelans to Peru have also created conflict. Similarly, the presence of US military bases in Japan and other places has caused stress and discord. All of this conflict and stress is due to distinct cultural differences.

Of course, not all movements of people cause conflict. Many travelers and immigrants are very successful in their efforts to experience and enter other cultures. But both the traveler and the host country are changed by their presence.

The movement of goods and services is, however, one of the key determinants of global cultural change. The growth of consumerism around the world is one of the key drivers of transportation.

Even our modern tourist industry is contributing to the impacts of globalization. Over the last several decades, improvements in modern communication, particularly the Internet, have greatly enhanced the ability of people to learn about other places and easily arrange travel.

Worldwide travel has thus increased. All over the world, efforts are underway to improve infrastructure and create amenities for the global leisure class. Some of these efforts, such as programs that promote ecotourism, serve as excellent models of sustainable development.

In addition, the protection of natural assets is seen as key to the success of tourism. Thus, tourism can promote the protection and preservation of important areas.

Many parts of the world have made interesting decisions to try to protect natural areas to promote tourism. In Florida, in the USA, for example, the state has opted not to allow commercial offshore drilling for oil (Figure 12.7). This is in contrast to other states that are along the Gulf of Mexico shoreline that have promoted the development of the offshore oil industry. Shallow and deep-sea oil platforms can be seen from some coastal areas and some of these platforms have caused significant oil spills.

Figure 12.7 Florida's beaches are so pristine because the state has banned the development of oil wells off shore.

In 2011, the Deep Horizon oil spill did untold damage to the Gulf of Mexico and its coastal areas. Some of the hardest hit areas were the fisheries and tourist regions of the northern Gulf of Mexico. Although Florida does not allow offshore drilling, its coastline was impacted when oil from the spill was carried there by currents and tides.

The development of oil resources off the coast of some of the northern Gulf Coast states limits their ability to develop tourism or sustain pristine natural habitat for other economic activities such as fishing or shrimping.

It is modern transportation that allows international travelers to visit these areas and bring with them their own culture, which can transform the local identity. Some have complained that the globalization of the world has created a sort of world culture, or a McCulture if you will, that is standardizing the world. While we are not anywhere near to having a uniform world culture, there is no doubt that some of the standardization that comes with global travel is making the world a smaller place.

Economic development

In today's economy, the world is a very small place. What happens in one market can influence the global economy. Companies such as Apple and Wal-Mart have global reach, influence, and significance, and even the smallest of business can be international in scope.

But the development of these businesses and industries can have unseen consequences. In 2012, a fire in a Bangladesh clothing factory caused the death of hundreds of workers. The conditions in the factory were rather poor and did not meet some international standards. When the fire occurred, it seemed like a local tragedy caused by lax factory standards.

However, upon closer inspection, it was revealed that several international luxury brand name clothing retailers produced clothing at the factory for consumption on the global market. Clearly the globalized labor market can deleteriously impact communities as world manufacturers try to find cheaper and cheaper ways of producing goods in a competitive global market.

The global market can also greatly influence over-exploitation of natural resources. The high price of gemstones, gold and other metals can start a flurry of over-mining in areas rich in such materials. The widespread and illegal gold mining in remote Guyana and Venezuela, for example, caused widespread environmental damage in the last decade. Significant pollution in these areas will have long-term consequences, not only for the environment, but also for human health.

As the world has become wealthier, the desire for high quality beef and fish has led to expansion of grazing herds and over-exploitation of the seas. Just a few decades ago, sushi was rarely seen in cities outside of Japan. Now, sushi restaurants can be found in all corners of the world. The taste for raw fish has led to the destruction of some fish stocks, most famously the blue fin tuna.

Of course, in some instances, globalization can greatly improve the lives of people through economic development. Take, for instance, the development of microcredit or loaning of small amounts of money for local and personal economic development.

Another example is the website, kiva.org, which matches individuals who are interested in loaning money to small business owners and individuals around the world. Many of the receivers of loans are opening or expanding businesses as diverse as clothing manufacturers to auto repair shops.

Transnational organizations

There are a number of large and small transnational organizations that promote or in some way influence sustainability. Perhaps the largest of these are the United Nations (UN) and the World Bank.

The UN was established in 1945 after the end of World War II to promote greater dialogue and understanding among nations. Currently, the UN has 193 member countries and offices all over the world. The headquarters in New York City is home to the General Assembly and most over-arching operations of the UN. Its iconic headquarters serve as a focal point for international debate on a range of issues associated with sustainability—from environmental protection to human health.

As noted in the introductory chapter of this text, the UN was responsible for the broad dissemination and acceptance of the term, "sustainability" through the publication of Our Common Future, sometimes known as the Brundtland Report. Published in 1987, the report clearly outlined problems with development trajectories and suggested that the world needed to make significant adjustments in order to ensure the long-term livability of our planet.

Since then, the UN has developed a number of key projects, including the Sustainable Development Goals, and various initiatives that emerged from the 1992 Earth Summit (formally called the United Nations Conference on Environment and Development (UNCED)).

The UN Millennium Development Goals (MDGs) focus on issues associated with poverty, human health, and broad social improvement (we discussed the MDGs in Chapter 2). While not explicitly sustainability goals, the MDGs focus on issues of great global concern that certainly impact the sustainability of society such as per capita income, eradication of particular diseases, and access to education and healthcare.

Great strides in overall development have been achieved through this effort and millions of people have gotten out of poverty throughout the world as a result. While much still needs to be done to improve the lives of others around the world, the MDG effort is proving to be a success.

The MDGs mainly focus on issues in developing countries. But they are not useful for assessing sustainability in many parts of the world. This is particularly true for some of the more developed countries, which have other issues associated with sustainability, such as consumption trends and income inequality

The 1992 Earth Summit efforts were fundamentally different from the MDG process. Instead of working on broad development goals outlined in the MDGs, the Summit turned its attention to a variety of environmental issues of concern. These included things like water supply, global climate change, pollution, and energy.

The summit was a significant event in the development of global environmental policy. Several important agreements were reached. These include the Convention on Biodiversity, the Kyoto Protocol, the Rio Declaration on Environment and Development, the Forest Principles, and Agenda 21.

The Rio Declaration was significant because it listed 27 principles that the nations agreed upon that framed the values of sustainability quite clearly. The principles are listed in the text box.

Since then, the UN hosted another Earth Summit in 2012, sometimes called Rio+20. This conference focused on creating broad international political support for sustainable development. The outcome document was called The Future We Want. It highlighted a number of key initiatives such as the development of sustainability indicators, and the expansion of sustainability management expertise in key areas.

The World Bank is another key multinational player on sustainability. It started after World War II to assist with the re-development of Europe after it was devastated by conflict. Their goal is to provide loans to developing countries in order to try to eradicate poverty. They tend to focus on large projects that require significant investment. However, they work on a number of projects that overlap with the UN's MDGs, such as reducing poverty, promoting gender equality, improving health and education, and environmental sustainability.

The organization has provided a number of loans to developing countries to advance development. While many of these loans are non-controversial and significantly improve the lives of people in receiving countries, there have been critiques of the World Bank. Some argue that the organization gives loans for projects that are not particularly sustainable, such as large dam or mining projects.

In addition, some argue that conditions given to borrowing nations advance unequal trade and favor trade and economic interests of developed or industrial nations. In addition, the Bank supports a number of private organizations in developing nations that do not necessarily advance the agenda of modern sustainability. Regardless, there is no doubt that the World Bank is concerned about sustainability and is striving to do a better job in ensuring that the projects they fund are environmentally sound. In addition, they have sought to fund projects that specifically address environmental needs in developing countries.

Nevertheless, the World Bank has been the focus of considerable protest and criticism at key international meetings such as the Davos World Economic Forum that is held every year in Switzerland. Many of the protesters rally against what they perceive to be the imposition of Western ideals and loans on non-Western developing nations. There is concern that the loans cause long-term economic dependency on the West.

There are also a number of other actors on the world stage. Some of these are well-regarded environmental groups such as the World Wide Fund for Nature. They often have targeted missions, such as The Rainforest Alliance, that allows them to have significant impact on particular issues.

Individuals can also have a tremendous impact on sustainability around the world. Perhaps the first true global environmentalists were Jane Goodall and Jacques Cousteau. Jane Goodall is a primate expert who has dedicated her life to the study and preservation of large primates in central Africa. Her work has brought attention to the pressing conservation needs in Africa. She has conducted a wide-ranging education effort and sought external funds from around the world in support of her work. While the situation for the primates is dire, her work certainly has globalized the local conservation issue.

The late Jacques Cousteau also brought environmental issues to a global audience. While Goodall worked on a geographically distinct issue, Cousteau focused attention on environments of concern to most—the oceans. His work emerged in the second half of the twentieth century at a time when very few understood the significance of the oceans.

Cousteau brought to light the interconnectedness of the world's oceans and demonstrated that the actions of one player can impact vast areas. He produced a number of highly acclaimed television documentaries that were widely seen around the world. In many ways, he created the notion that the oceans transcend boundaries and that they should be seen as a whole living ecosystem that should be protected and preserved for future generations.

Designing fashion in the remote mountains of upstate New York

John Schrader is a fashion designer educated in New York City. He has worked for a number of important fashion houses. However, he is now part of the overall global fashion supply chain. He works for a company called the Quacker Company that sells millions of dollars worth of clothing each year on QVC and on the Quacker Company Website.

The Quacker fashion line focuses on comfortable embellished clothing for women. Schrader designs many of the embellishments using colorful threads, sequins, and metallic or glass additions. The clothing line is very popular among middle class women and the Quacker Company seeks to develop a strong connection with its consumers. The company has sponsored meet-ups for fans of the clothing line and has also conducted cruises so that company employees can meet their customers.

While many fashion designers live or work in New York City, Schrader lives in a house in the country in the Catskill Mountains, several hours outside of New York City. There, he designs clothing in his home office. The designs are faxed to his boss in Philadelphia where they are approved or modified. From there, the designs are sent to factories overseas where samples are made. The samples are shipped back to the US where Schrader and his team approve them. After this, orders are placed.

On QVC, thousands of garments can be sold in a very short time period.

In this globalized world, Schrader's ideas can come directly to consumers without him ever leaving his Catskill home. He truly exemplifies the modern global economic system.

Of course many countries act as globalizing entities. Many of the developed nations provide support for sustainability and initiatives and export their ideas and values to other nations in the process. China, for example, has been providing significant aid to a number of developing nations for infrastructure projects.

Key UN Agreements that emerged from the 1992 Earth Summit

The following are key United Nations Agreements that emerged from the 1992 Earth Summit. Each one had a tremendous impact on global sustainability policy.

The Convention on Biodiversity. This is a legally binding agreement that requires signatory countries to develop national strategies for conservation and sustainability of biological diversity.

The Kyoto Protocol. The Kyoto Protocol is a binding agreement that emerged from an agreement called the Framework Convention on Climate Change. It sets targets for the emission of greenhouse gases of industrialized nations.

Forest Principles. This is a non-binding agreement that provides recommendations for the development of sustainable forestry.

The Rio Declaration on Environment and Development is a non-binding document that lists 27 principles that focus on the environment.

Agenda 21 is a non-binding implementation plan that focused on how to implement sustainable development by setting goals at the national, local, and global levels.

The principles adopted in the Rio Declaration in 1994

Principle 1

Human beings are at the centre of concerns for sustainable development. They are entitled to a healthy and productive life in harmony with nature.

Principle 2

States have, in accordance with the Charter of the United Nations and the principles of international law, the sovereign right to exploit their own resources pursuant to their

(Continued)

own environmental and developmental policies, and the responsibility to ensure that activities within their jurisdiction or control do not cause damage to the environment of other States or of areas beyond the limits of national jurisdiction.

Principle 3

The right to development must be fulfilled so as to equitably meet developmental and environmental needs of present and future generations.

Principle 4

In order to achieve sustainable development, environmental protection shall constitute an integral part of the development process and cannot be considered in isolation from it.

Principle 5

All States and all people shall cooperate in the essential task of eradicating poverty as an indispensable requirement for sustainable development, in order to decrease the disparities in standards of living and better meet the needs of the majority of the people of the world.

Principle 6

The special situation and needs of developing countries, particularly the least developed and those most environmentally vulnerable, shall be given special priority. International actions in the field of environment and development should also address the interests and needs of all countries.

Principle 7

States shall cooperate in a spirit of global partnership to conserve, protect and restore the health and integrity of the Earth's ecosystem. In view of the different contributions to global environmental degradation, States have common but differentiated responsibilities. The developed countries acknowledge the responsibility that they bear in the international pursuit to sustainable development in view of the pressures their societies place on the global environment and of the technologies and financial resources they command.

Principle 8

To achieve sustainable development and a higher quality of life for all people, States should reduce and eliminate unsustainable patterns of production and consumption and promote appropriate demographic policies.

Principle 9

States should cooperate to strengthen endogenous capacity-building for sustainable development by improving scientific understanding through exchanges of scientific and technological knowledge, and by enhancing the development, adaptation, diffusion, and transfer of technologies, including new and innovative technologies.

Principle 10

Environmental issues are best handled with participation of all concerned citizens, at the relevant level. At the national level, each individual shall have appropriate access to information concerning the environment that is held by public authorities, including information on hazardous materials and activities in their communities, and the opportunity to participate in decision-making processes. States shall facilitate and encourage public awareness and participation by making information widely available. Effective access to judicial and administrative proceedings, including redress and remedy, shall be provided.

Principle 11

States shall enact effective environmental legislation. Environmental standards, management objectives and priorities should reflect the environmental and development context to which they apply. Standards applied by some countries may be inappropriate and of unwarranted economic and social cost to other countries, in particular developing countries.

Principle 12

States should cooperate to promote a supportive and open international economic system that would lead to economic growth and sustainable development in all countries, to better address the problems of environmental degradation. Trade policy measures for environmental purposes should not constitute a means of arbitrary or unjustifiable discrimination or a disguised restriction on international trade. Unilateral actions to deal with environmental challenges outside the jurisdiction of the importing country should be avoided. Environmental measures addressing transboundary or global environmental problems should, as far as possible, be based on an international consensus.

Principle 13

States shall develop national law regarding liability and compensation for the victims of pollution and other environmental damage. States shall also cooperate in an expeditious and more determined manner to develop further international law regarding liability and compensation for adverse effects of environmental damage caused by activities within their jurisdiction or control to areas beyond their jurisdiction.

Principle 14

States should effectively cooperate to discourage or prevent the relocation and transfer to other States of any activities and substances that cause severe environmental degradation or are found to be harmful to human health.

Principle 15

In order to protect the environment, the precautionary approach shall be widely applied by States according to their capabilities. Where there are threats of serious or irreversible damage, lack of full scientific certainty shall not be used as a reason for postponing cost-effective measures to prevent environmental degradation.

Principle 16

National authorities should endeavour to promote the internalization of environmental costs and the use of economic instruments, taking into account the approach that the polluter should, in principle, bear the cost of pollution, with due regard to the public interest and without distorting international trade and investment.

(Continued)

Principle 17

Environmental impact assessment, as a national instrument, shall be undertaken for proposed activities that are likely to have a significant adverse impact on the environment and are subject to a decision of a competent national authority.

Principle 18

States shall immediately notify other States of any natural disasters or other emergencies that are likely to produce sudden harmful effects on the environment of those States. Every effort shall be made by the international community to help States so afflicted.

Principle 19

States shall provide prior and timely notification and relevant information to potentially affected States on activities that may have a significant adverse transboundary environmental effect and shall consult with those States at an early stage and in good faith.

Principle 20

Women have a vital role in environmental management and development. Their full participation is therefore essential to achieve sustainable development.

Principle 21

The creativity, ideals and courage of the youth of the world should be mobilized to forge a global partnership in order to achieve sustainable development and ensure a better future for all.

Principle 22

Indigenous people and their communities and other local communities have a vital role in environmental management and development because of their knowledge and traditional practices. States should recognize and duly support their identity, culture, and interests and enable their effective participation in the achievement of sustainable development.

Principle 23

The environment and natural resources of people under oppression, domination, and occupation shall be protected.

Principle 24

Warfare is inherently destructive of sustainable development. States shall therefore respect international law providing protection for the environment in times of armed conflict and cooperate in its further development, as necessary.

Principle 25

Peace, development, and environmental protection are interdependent and indivisible.

Principle 26

States shall resolve all their environmental disputes peacefully and by appropriate means in accordance with the Charter of the United Nations.

Principle 27

States and people shall cooperate in good faith and in a spirit of partnership in the fulfilment of the principles embodied in this Declaration and in the further development of international law in the field of sustainable development.

War and sustainability

As I am writing this, the world has more refugees as a result of conflict than at any time since World War II (Figure 12.8). The conflicts in the Middle East, the political difficulties in the Ukraine, and political repression in China and other parts of the world are all in the news this week. Each of these places has difficulty in trying to achieve sustainability initiatives when the governments and their citizens are working in a time of stress.

Afghanistan and the Islamic Republic of Yemen both have sustainability plans. Yet how can they achieve them under stressed circumstances? War and conflict are not conducive to the promotion of sustainability and there are many examples of how war can be disruptive to the environment and the greater good of populations.

Table 12.1 lists various impacts and effects of war on community sustainability. Conflict leads to challenges that range from the disruption of food supply to the breakdown of economies. The effects of these challenges can include famine, migration, family breakdown, and ecosystem disruptions. Based on these problems it is understandable why leaders in many parts of the world are unable to effectively address sustainability within their borders. Plus, some nations do not fully control all of the lands within their recognized borders. It is hard to implement national sustainability initiatives or strategies when civil strife divides a country.

Yet, some nations are trying to work on sustainability even during times of conflict. They recognize that they still need to move their nations forward on issues like water, energy, and food as a security issue. They may not be able to fully address all of the issues that they have identified in their sustainability planning, but they try.

Figure 12.8 Many in our families served in combat around the world. War has a negative impact on the environment and on society in many ways. This is my brother, Jim and my nephew John. Both are veterans, as is my nephew James who took this picture.

Table 12.1 Impacts of war on sustainability.

Impact	Effect
Disruption of food supply	Hoarding, lack of daily caloric needs, famine, migration, refugees
Disruption of water supply	Thirst, ill health, problems with hygiene, lack of irrigation for crops, migration, refugees
Disruption of sanitary sewer services	Pollution from human waste, public health problems, disease, migration, refugees
Disruption of government services and education	Breakdown of social fabric, lack of opportunity for youth, lack of basic government services such as garbage pick up, planning, building inspections, immunizations, transportation
Pollution	Pollution from ammunition (uranium tipped bullets for example), human waste from troops and refugees, corpses
Land clearing and encampments	Ecosystem disruptions, hydrologic changes, migration, refugees
Fuel use	Fuel shortages, pollution, greenhouse gas emissions, oil or gas spills
Loss of economic activity	Loss of homes, loss of jobs, family breakdown, migration, refugees
Loss of social fabric	Breakdown of government, breakdown of family structure, racism, conflict, refugees, migration

La Ruta del Cacao in Chuao, Venezuela: seeking awareness of over-development through action

Each year for the last several years, a group of intrepid runners have taken to the trails in a mountain on the coast of Venezuela to run a half marathon called La Ruta del Cacao. The trail takes runners on a 23 km route through some of the most unique forests in a cacao growing area in Venezuela known for producing some of the best cacao in the world. The trails extend from sea level to a peak of over 3000 feet, and then back to sea level. The region is remote and relies heavily on cacao production for survival. There has been growing concern over development and over-use of the land in the region.

The run is modeled after some of the famous European mountain trail runs. However, in this case, the organizers make the environment and the local population front and center in the race. They focus their attention in promoting environmental improvement, conduct a clean up of forests near the route, and facilitate tree planting along the coastal zone.

They look at the event as an opportunity to engage the local population and participants on the importance of environmental protection. They also produce a cultural presentation of dance to highlight the area's unique population.

This is an example of how an event totally unrelated to the environment can help sustainability and the planet. There are events that occur all over Venezuela that do not have such a mission. But, because the organizers decided to do a little extra work, improvements are made, people are educated, and differences are made in the lives of others.

Scallops and economic development

One of the problems facing the shellfishing industry is pollution and over-fishing. Scallops, a highly prized shellfish harvested in the Peconic Bay of Long Island were killed off in the 1980s due to a brown algae plume that some have attributed to polluted storm water runoff (Figure 12.9). This event destroyed the Peconic Bay shellfishing industry. Since the plume occurred, efforts have been made to improve the water quality in the waters of Peconic Bay and the water is now rather clean. But the scallops have not come back. As a result of this, the Cornell's Cooperative Extension of Suffolk County developed a variety of projects to help restore the scallop fisheries industry in Peconic Bay. Scientists developed a variety of techniques to breed scallops and sow the "seeds" of the scallops in Peconic Bay. Since they started this effort, the scallop industry has returned, bringing new jobs and providing a $3 million dollar economic impact.

Figure 12.9 Community leaders of Long Island showed up to help distribute Scallops in Peconic Bay, New York.

(Continued)

Scallops only have a 1–2 year lifespan with only one or maybe two reproductive cycles. They are different from longer-lived shellfish like lobster and clams. Thus, their long-term success requires maintaining a large population in Peconic Bay. While the scallops have returned to the point that they can be harvested, the population has not returned to pre-1980s levels. Thus, in 2011, Cornell Cooperative Extension of Suffolk County applied for a grant from the Long Island Regional Economic Development Council to expand scallop seeding operations.

This group was set up and funded by New York's Governor, Andrew Cuomo, to try to develop state funded projects that would improve the economy of the state after the 2008 economic downturn. The state was divided into several regions that competed for funding, and Long Island was one of the regions. Organizations competed for funding. The scallop project was selected by the Long Island group as one of the grant winners and they received $182 900 in funding to expand their operations.

The grant dollars allowed scallops to expand in the Cornell's scallop hatchery. Plus, they used the funds to expand their nursery operations in Peconic Bay, which allows the scallops to grow in a natural, but protected, environment. The scallop industry is only at about 10% of where it was prior to its collapse in the 1980s, so there is a great deal of opportunity for expansion (Figure 12.9).

Further reading

Biswas, A.K. and Tortajada, C. (Eds.) 2005 *Appraising Sustainable Development: Water Management and Environmental Challenges*. Oxford University Press. 223 p.

13

Sustainability, Economics, and the Global Commons

One of the main themes of sustainability is economy. How do we create a balanced future in which future generations can thrive, while sustaining a solid economy? Of course, this question has many traps. It depends on the future you envision, what is meant by thriving, and what one sees as a solid economy. This idea of economy within the framework of sustainability is perhaps the most difficult issue to address when considering sustainability.

The global commons

All the people on the planet share one Earth. We are unlikely, any time soon, to leave it to explore and colonize other planets. We have to find ways to survive on this planet without destroying it. Plus, we all share planetary resources, our global commons, and must find ways to protect them for the good of all.

In 1968, Garrett Hardin wrote an article in the journal *Science* called The Tragedy of the Commons. In his essay Hardin utilizes the village common as a metaphor for our modern age. In the past, village commons were used for grazing animals. They were called commons because the community shared the space. However, if one individual grazed too many sheep, it would destroy the common for the rest of the community, thereby creating a tragedy where the entire community is unable to use the space for grazing.

Hardin took that idea further and suggested that many in our society were depleting natural resources to the point that they were becoming unavailable or unaffordable to many in society. It is only through the management of the natural resources that we are able to maintain them. Without management, the resource can become exploited, which will lead to greater demand for the resource, which hastens the resource's destruction.

Water in Las Vegas is a good example of the tragedy of the commons issue (Figure 13.1). Las Vegas, Nevada is located in one of the driest areas of the United States. It is home to over 600 000 people and attracts millions for tourism. The city draws water from reservoirs in rivers created by dams. The water is considered a public resource and was used to support the growing population of the city.

As the casinos grew, so did the population. Grand hotels were built to encourage more growth. As the population and tourist numbers grew, so did the demand for water. Now, sadly, Las Vegas is running out of water. It is estimated that there are but a few decades of

Introduction to Sustainability, Second Edition. Robert Brinkmann.

Companion website: www.wiley.com/go/Brinkmann/IntroductiontoSustainability

Figure 13.1 The natural landscape of arid areas such as this one in southern New Mexico is not suitable for the development of large populations. Nevertheless, millions of people live in these areas and have to draw in resources, particularly water, from other areas.

Table 13.1 Common resources

Air	Water
Soil	Coastlines
Plants	Land
Animals	Oceans
Ecosystems	Lakes
Minerals	Rivers

water resources left to support the city. The very success of the city is driving its demise. Plus, the over-use of water resources in Las Vegas hurts surrounding communities.

In our modern era, we tend to look at constant economic growth as a sign of a strong economy. Yet, there are questions that many have raised as to the ability of our planet to fully absorb the resource needs of such high growth. This is especially true as world populations are increasing at the same time that many are rising out of poverty and expecting greater access to resources. Table 13.1 lists the major commons resources that have been impacted in recent decades.

The idea of a global commons is a good place to start to think about environmental economics since it addresses so many of the ideas that we will concern ourselves within this chapter.

Economic processes that put the Earth out of balance

While many people are involved with a "green economy" that seeks to promote sustainability within business, many are involved with more traditional economic activities that can put the Earth out of balance. Table 13.2 lists some of these activities (as quoted directly from Molly Scott Cato's book, Environment and Economy).

Thus, through the very act of living and working on our planet, we can, if exploitation proceeds out of balance with natural systems, do significant harm to the planet. Jared Diamond explores this idea extensively in his book, *Collapse: How Societies Choose to Fail or Succeed*. He noted how the residents of Easter Island caused widespread deforestation of their home because they opted to cut down trees to use them to transport the famous Easter

Table 13.2 Problems that emerge from particular economic causes. (From: Cato, M.C. 2011. *Environment and Economy*. Routledge, 263 p. Table from page 6).

Problem	Economic cause
Pollution	
Greenhouse effect/climate change (global)	Massive expansion of productive processes that emit greenhouse gases; increasing use of fossil fuels
Ozone depletion (global)	Emission of CFCs by manufacturers of aerosols and refrigerants
Acidification (continental)	Emissions from fossil fuel electricity generation plants; rapid expansion of personalized transport
Toxic contamination (continental)	A huge range of industrial productive processes
Renewable resource depletion	
Species extinction (global, regional)	Loss of habitat caused by displacement of subsistence farmers; pressure on land caused by population increase
Deforestation (global, regional)	Corporate pressure to use previously forested land for cattle grazing or biofuel production
Land degradation/loss of soil fertility ((bio)regional, national)	Loss of traditional systems of agriculture due to population displacement and movement away from subsistence agriculture
Fishery destruction (regional, national)	Excessively intensive fishing and industrial pollution
Water depletion ((bio)regional, national)	Expansion of demand due to changing lifestyle and expanded industrial production
Landscape loss	Population pressure and removal of subsistence farmers from land
Non-renewable resource depletion	
Depletion of various resources	Needed as inputs to productive processes
Other environmental problems	
Congestion (national)	Excessive material consumption

Island statues that rim part of the island's coastline. He points out that the society collapsed shortly after the deforestation of the island because the ecosystem that nurtured the society was destroyed.

The idea of collapse is a potent one that provides cautionary imagery for our future. Some have critiqued Diamond's work as overly simplistic. They believe that there are examples of societies finding ways out of difficult situations through adaptation and ingenuity. However, there is no doubt there are many problems on our planet that are brought about through economic activities that could lead to societal collapse in some areas. Indeed, the drying of the Aral Sea in Asia has caused a collapse of the coastal societies in that region. How long before Las Vegas is at risk?

Social and economic theories

In the last two centuries, much has been written about social and economic theories related to the environment. Most of these relate to capitalist and Marxist approaches to the economy. However, deep ecology and ecofeminism provide interesting alternatives to traditional theoretical approaches.

Neoclassical economics

The main economic theories in use in the world today come to us from neoclassical economics. This approach brings together the notion that a free market works best unregulated within the context of supply and demand.

Within this framework, consumers will make rational choices and producers will seek to maximize profits. Those who follow the neoclassical approach to environmental economics believe that market forces will eventually protect the environment and humans will make rational choices that will lead to improved situations. Thus, in the case of Las Vegas, the law of supply and demand would dictate that water costs would increase tremendously as resources diminish. When this happens, humans will make rational choices and leave the region, thereby reducing demand and costs.

Of course, there are many examples where humans do not make rational decisions. Consider the fate of the rhino and the African elephant (Figure 13.2). Even though there are very few of these creatures left that can produce ivory, they are being hunted down by poachers due to the high cost of the precious material. The value of the ivory to the poachers is so high that it is worth it to them to poach the animals. As a result of the activities of the African poachers, it is likely that the animals will be extinct in a generation or two.

The modern economic system in place throughout most of the world is called neoliberalism. Modern neoliberalism emerged in the last several decades throughout the world. It promotes open trade, economic freedom (limited government regulation), privatization, and open markets. Many in the environmental community are critical of neoliberalism.

Figure 13.2 Hunting without care for the future of the species is unethical. However, in many areas hunting is heavily regulated. My brother Charlie, shown in this photo from 1971, purchased a hunting license and followed strict hunting rules, including rules about the color of clothing one wore while hunting.

Environmental critiques of neoliberalism

Open trade and markets, combined with abundant access to credit, creates a potent catalyst for globalization and expanded economic activity around the world. Yet, there are many environmental challenges associated with neoliberalism.

1 *Consumerism.* The rise of neoliberalism has seen a concomitant rise in consumer culture around the world. People who were happy to live with a small environmental footprint now have access to the global market and as a whole we are using far more resources as a planet than ever before. Since the resources are finite, we are quickly depleting many important natural resources.
2 *Resource distribution.* The demand for trade of goods within a neoliberal open market system drives consumerism and thus resource exploitation (Figure 13.3). Thus, demands for resources are increasing, which is driving up the price of many resources and consumer goods. This is creating an unequal distribution of resources based on market forces.

Figure 13.3 Consumerism is one of the biggest challenges to sustainability. We are driven to want more by our greed and by reacting to advertising and trends.

3 *Inequality*. The inequalities associated with neoliberal policies can lead to global, national, and local inequalities based on access to resources and access to markets. Perhaps the best illustration of this situation comes from the world of food and agriculture. The drive for open markets in the food industry gives large multinational corporations access to markets around the world. This access can change traditional food production systems that value large farmers over small farmers. There have been many protests against the globalization of the food system in places like France, South Korea, and India. Indeed, in part as a result of neoliberal policies in India, tens of thousands of farmers have committed suicide over the last decade.
4 *Crony capitalism* and the rise of the oligarchy. One of the challenges of the neoliberalism is that it can lead to crony capitalism that favors a few over the masses. To reach and work within a global economy, special knowledge and skills are needed. As a result, not all are able to participate within a global economy. Plus, distinct political access is needed to make international business deals and gain access to some markets. The net result of this is that the wealth of the world is being managed by a limited number of individuals. The gap between the rich and the poor is increasing. After the fall of the Soviet Union, state run economies in many areas came into the hands of individuals who now run parts of Eurasia within an oligarchical system. An oligarchy is an organization or government that is managed by a handful of people, often in support of their own interests. Many have criticized neoliberalism because they feel that it can lead to oligarchical activities by governments. Indeed, many note that the presidency of Donald J. Trump highlights how powerful people can get away with serious ethical flaws. The combination of crony capitalism and inequality provide a unique crucible for corruption.

5 *Debt*. One of the outcomes of neoliberalism is national and personal debt. In a drive toward development, many countries must make significant improvements in infrastructure. Loans must be taken to advance energy production, build roads, and improve water and sewer systems. Taken together, these sound like great ideas. However, for a small, impoverished country to quickly develop Western-style infrastructure is difficult without burdensome loans. Many countries take on debt that they can ill afford from wealthy nations or organizations that set requirements on government policy. This debt, therefore, puts some national policies under the control of outside organizations in a strange form of colonialism. Individuals too can take on more debt than they can manage as they gain access to global consumer goods like televisions, cell phones, and computers. While access to these goods may seem like a good thing, there are distinct local and national challenges for emerging economies within the globalized neoliberal economic system.

Neoliberal economic policies are not in favor in all neoclassical economic systems. But, because of the challenges with traditional neoclassical economics approaches to environmental resources overall, a new subfield emerged in the last several decades called environmental economics.

Environmental economics

The field of environmental economics focuses largely on pollution and resource depletion. Specifically, environmental economists seek to understand how best to protect the environment from the challenges brought to society by economic activities that cause pollution or deplete resources.

One of the key elements of environmental economics *is valuing the environment*. What this means is that economists try to evaluate some of the direct, indirect, option, and existence value of the environment. The direct value is the value that can be derived from the resource. This would be something like iron ore from the Earth. Indirect valuing includes the ecosystem services that the environment provides. For example, salt marshes protect coastlines from coastal flooding and filter pollution. If they were gone, there would be very specific costs associated with flooding and pollution. Option costs are the costs for protecting the environment now for later consumption. Finally, existence value is the value that can be given to a place for its overall intrinsic value. Consider how much you would pay to protect the polar bear or fragile acres of the Amazon rainforest (Figure 13.4).

Cost-benefit analysis and its application in environmental economics

One of the ways that economists and businesses make decisions is to conduct a cost-benefit analysis. This helps decision-makers evaluate whether or not a project is justifiable within particular economic constraints. It also helps to compare competing projects to evaluate which one has the greater opportunity for financial gain.

To conduct a cost-benefit analysis, the benefits and disadvantages of a particular project are converted into real money values. Comparisons are made with alternative projects and with the costs of not completing the project. Costs and benefits are calculated for the effects of the projects on users or participants in the projects as well as non-users. External costs

Figure 13.4 What value do you place on the environment?

and benefits (those costs and benefits such as pollution or increased property values that impact those who did not choose to be impacted) must be calculated as well. Option values are also included in the analysis. An option value is the value of unused environmental resources such as forests or water. It can also include the value of transportation resources such as subways. The option value is the value of the resource whether it is used or not. In other words, we may value having access to subways, even though we may never use them.

Predictions are made of the projects' outcomes over a specific time. Once this is completed, decisions can be made as to the appropriateness of different options.

Cost–benefit analysis emerged in the middle of the twentieth century as a way to inform decisions around using public dollars for development projects. While environmental considerations were not immediately considered in the analysis, they are commonly considered today in many cost–benefit analyses utilized to make decisions that can impact resources or ecosystems.

However, inserting the environment into cost–benefit analysis can be difficult. How does one account for the costs of projects on the environment? This is especially true when one considers that there is very little expertise on valuing the environmental costs of projects in most areas of the world. This is largely because there is very little data available in many places on the environmental and social costs of projects. Plus, many places in the world do not have the same environmental ethos and value aspects of the environment very differently. Thus, the valuation of damage to a water supply in the dry Sahara would be very different to the valuation of the water supply in the wet, tropical Amazon.

Environmental impact assessment

An environmental impact assessment is another tool that can be used to evaluate the appropriateness of project. It is somewhat related to cost–benefit analysis in that the positive and negative impacts of a proposed project are evaluated prior to making any decision as to whether or not a project should move forward. However, unlike cost–benefit analyses, environmental impact analyses focus largely on environmental impacts. Social impacts are also assessed in some situations.

Many countries of the world require environmental impact assessments, to mixed effect. China, for example, requires environmental assessments for large projects, but there are only a few projects that have been halted or modified as a result of the reports. Many serious environmental problems, from pollution to mining accidents, have occurred, which has called into question the environmental assessment process and its effectiveness.

In the United States, environmental assessments are required for some projects that are likely to have some degree of environmental disruption. The report typically reviews the project, as well as how the project was presented to the public, along with opportunities for public involvement. It also summarizes key issues associated with developing the project and presents a range of alternatives to the project. It concludes with a review of the environmental consequences associated with project development. In some cases, a full environmental impact assessment evolves out of an environmental assessment. However, in most cases, project environmental assessments are used by decision-makers to evaluate a project.

The usefulness of environmental impact statements and environmental assessments in the United States is mixed. Those who seek to develop projects find them a frustrating, expensive, and lengthy hoop that they have to go through to get a project approved. Environmentalists find themselves frustrated by the regulatory process since most projects that complete environmental assessments are approved. However, it should be noted that the process of completing an environmental assessment educates developers about the appropriate development strategies for the project. For example, if a developer finds endangered species in the process of developing an environmental assessment, it will guide the project to try to limit the impact on the species. In other words, the environmental assessment process helps to create good behavior by project developers prior to a project moving forward to evaluation and approval.

Individual US states and local governments can also require addition assessments that are stricter than national requirements.

In the European Union (EU), environmental impact assessments are created for some large projects. EU Annex 1 projects are large projects like bridges; they always require an assessment. Smaller projects are classified as Annex 2. Individual countries can decide if Annex 2 projects require a full environmental impact assessment.

A strong focus of the EU environmental assessments is how to mitigate the environmental impacts of the projects to avoid negative outcomes. They also require that the project be described in non-technical terms to ensure that the public can make an educated decision as to whether or not to support the project. Another interesting aspect of the assessments is that they require projects to outline technical challenges or places where there is a lack of research or knowledge. This information helps to communicate regional and national research needs.

Environmental ethics

No matter which approach one takes toward environmental economics or the analysis one conducts in assessing projects or developing a new green economy, there are ethical dimensions that are in play in any decision that impacts the environment. Cost–benefit analysis and environmental impact assessments can only help make wise decisions for the environment within an economic context if society has an ethical value for the environment. Euros, pounds, and cents are certainly important considerations, but at the end of the day, our decisions reflect on our culture's ethical values.

In many ways, the study of environmental ethics emerged from the widespread destruction of the North American wilderness in the late nineteenth and early twentieth centuries. It was a time when we saw the widespread destruction of the buffalo and the extinction of the passenger pigeon. In Florida, when the federal government installed a game manager in the southern wild and wet marshes and swamps to try to prevent bird poaching for feathers to make women's hats, the manager was murdered.

Yet the public's concern about the impacts of our economic activity on the planet grew rapidly. Aldo Leopold, the author of *A Sand County Almanac* (published in 1949), wrote eloquently in the book about the challenges of trying to balance the desires of man with nature. He argued that we needed to develop an ethical framework that brings in an ecological conscience within our economic motivation.

In many ways, environmental ethics is codified within environmental rules and regulations. The laws help to define our collective environmental values at the national, state, and local levels.

The field of environmental ethics is often considered secular since it evolved largely out of the environmental conservation movement. However, it is important to note that many religious organizations are getting heavily involved with the environmental movement. For example, in 2014, the Pontifical Academy of Science, which consists of some of the leading scientists in the world who advise the church on scientific matters, hosted a conference called Sustainable Humanity, Sustainable Nature: Our Responsibility. No longer are churches, mosques, temples, and synagogues looking at the world as a human-centered space with limitless resources given to us by God. Now, religious organizations are bringing forward an ethical dimension that includes all of creation within a broader environmental and social ethic.

Green economics

Green economics differs from neoclassical and environmental economics in that it is a field that focuses specifically on reducing the impact of the economic systems on the planet. While the previous two focus on using basic economic theory to manage environmental systems, green economics is more of an applied field that promotes particular economic strategies to advance local, national, or international sustainability.

Green economics derives its name from the idea that there are particular economic activities that are better suited for long-term sustainability on the planet by providing goods and services that do not overly tax resources or damage ecological systems.

Green economics is sometimes discussed within the context of the green economy or green jobs. The green economy is loosely defined as the economic system that evolves from those involved with sustainability-related activities and green workers are those people who work within the green economy. The definition of what constitutes a green economic activity is often a matter of debate. For example, someone employed in the nuclear energy field is counted by some as a green job, while for others it is not. There are also problematic issues of greenwashing that can occur within this economic sector.

A number of assessment tools are in place to help evaluate goods and services associated with the green economy. Green buildings, organic food, and green hotelier associations are all examples of ways that organizations self-regulate their green economic activities.

In a political sense, the Green Party is somewhat aligned with green economics in that it does not reject capitalist economics. The four main tenants of the Green Party are stated within its four pillars, which include:

- Ecological wisdom
- Social justice
- Grassroots democracy
- Non-violence.

Some, particularly eco-socialists, criticize the Green Party for working within a capitalist framework.

Non-capitalistic economies

There are some that believe that capitalistic approaches are inconsistent with ideas of long-term sustainability of the planet. They believe that the demand for resources in a profit-driven world will destroy the planet and that the only way for the human species to survive is to find a new economic model.

One of the new models is called eco-socialism. This economic model is based on the teachings of Karl Marx and broad Marxist theory that suggests that the breakdown of environmental systems is caused by capitalist systems and associated imperial tendencies of capitalist governments. Eco-socialists urge the nationalization of industries and public ownership of land in order to protect resources. They are critical of traditional Marxist economies, with examples in the USSR, China, and Cuba, for treating the environment as a secondary issue not tied to social and economic inequalities.

Deep ecology

Deep ecology, while a philosophy and not an economic system exactly, does have an influence over economic ways of thinking about the environment.

Supply and demand

Supply and demand is a concept that dates back centuries. It is the idea that if demand increases and supply remains the same that costs will go up. The opposite is true: if demand decreases and supply remains the same, costs decrease. If supply increases, but demand doesn't change, the costs will also go down and if supply decreases and demand stays the same, prices increase.

(Continued)

Elephant ivory provides a good example of the problem of supply and demand in our modern world. In the early 1900s the population of the African elephant was approximately 3 million. Today, the population is about 700 000 and declining quickly. For example, in Chad, the population was once 400 000 and today there are less than 10 000 in that country.

The global demand for ivory is driving the decline in the elephant population. Most of the ivory sold on the world market originates from African elephants.

Ivory has long been prized as a fine decorative item. Its unique color and luster, and its ability to be carved into decorative items are evidenced through the many ivory artifacts that have been handed down to us through history.

In 1989, in an effort to protect the African elephant, the world came together to pass a global ban on the trade of ivory. Only ivory collected prior to 1989 could be traded on the open market. Yet, the demand for ivory remained high, particularly in China, where ivory is a very prestigious luxury item used for chopsticks and a number of decorative objects.

Once the ban was put in place, ivory became scarce, even though demand was high in the Chinese market. The costs for ivory on the black market went up due to the fall in supply. Today, the cost of ivory is over $1500 a pound on the black market.

The conservation of the elephant can be managed in part by managing the market and there are several efforts underway to do this. The Chinese government is trying to educate the public on the impacts of ivory purchases on the African elephant. They hope to reduce demand by making ivory a less desirable material for gifts and decorative objects.

There have also been attempts to flood the market with legally collected ivory by allowing the sale of ivory to some markets. However, this has had the unintended consequence of re-introducing a demand.

How else could we use the law of supply and demand to help conserve the African elephant? Can you think of how the law of supply and demand is used to conserve other resources like oil?

Ecosystem services

One way that economists have worked to protect the environment is to quantify the things that the environment does that we would have to pay for if the environment didn't do them.

In 2005, a group of some of the world's best biologists published a report on the state of the world's environment called the *Millennium Ecosystem Assessment*. In the report, the group reviewed the significance of the environment to human society and discussed the key ecosystem services the environment provides. They classified ecosystems services into four categories:

1 *Provisioning services*. These are services that are provided to us regularly for our use without any direct husbandry. Perhaps the most important provision service we get is the supply of clean drinking water. We also get a sizable portion of food from the environment, particularly fish. Game and a variety of foraged food like

mushrooms and berries are also provisioned in this way. Many people around the world also get fuel wood directly from the environment.

2 *Regulating services.* Without the regulating impacts of the environment, the Earth would be uninhabitable. Climate regulation, water purification, and pollination are all regulating services that occur without any support from human activity.
3 *Cultural services.* John Muir once wrote, "Everybody needs beauty as well as bread, places to play in and pray in, where nature may heal and give strength to body and soul." The environment provides us with great cultural services that include recreation, spiritual, and religious inspiration, aesthetic inspiration, education, sense of place, cultural heritage, and opportunities for tourism.
4 *Supporting services.* Of course, the environment provides us with many supporting services such as soil formation, primary production, nutrients, evolution, and the ground we live on.

The report notes that many of the ecosystems services provided by the environment are declining rapidly and there is concern over the future quality of life of our planet.

The importance of detailing and measuring ecosystems services in our modern globalized economy should be evident. Many ecosystems are destroyed every day in support of economic development and growth. Yet, if value could be put on the ecosystems, their importance to the global economy could become more significant to those involved with local decisions on land use.

In recent years, scientists have worked to quantify the value of ecosystems services by putting a dollar value on things like nutrient cycling or coastal protection. These values can be compared with values gained from the activities that would destroy the ecosystem.

Just think about a place like Chesapeake Bay. Here, development has destroyed many of the coastal wetlands that provide a nursery for fish, nutrient cycling, and protection from coastal erosion. Today, many would like to expand development on the bay to take advantage of the lucrative real estate market. However, more and more local decision-makers are taking into account the significance of the local wetlands to the economy of the region. They understand that if the wetlands are destroyed, it will have a direct economic consequence due to the loss of the valuable ecosystems services provided by the coastal marshes.

We can see the financial impacts caused by the destruction of ecosystems. In Chesapeake Bay, where coastal marshes were destroyed, coastal flooding is much more common. Tampa Bay and many other bays along the Gulf of Mexico have similar problems, especially nutrient pollution that cannot be mitigated by the extant wetlands. Some areas become so concentrated in nutrients during certain times of the year that it causes widespread fish kills, thereby hurting the recreational and commercial fishing industries.

Think about ecosystems services in your own region. What types of things are provided by the environment that most would normally take for granted? What would be the cost to your community if the ecosystems were destroyed? What types of activities have to be implemented in your community to mitigate the impacts of development?

Deep ecologists believe that the environment has an intrinsic value to the world as a whole. We therefore cannot quantify the values of pieces of the environment (resources, ecosystems services, etc.). The philosophy asks us to look at the environment in deeper ways that transcend normal methods of looking at it. If we do this, we will see that the environment is not just something that provides services for us. Instead, it is a whole that needs to be respected. Deep ecologists do not consider themselves capitalists, Marxists, or socialists. Instead, they believe that all of the world's economic systems are broken due to the overly complex and consumerist condition of the world. They advocate for simplifying our lifestyles, while attempting to live in harmony with our environment.

Ecofeminism

Ecofeminism is another movement that rejects traditional Marxist, capitalist, or socialist approaches to the environment. Ecofeminists assert that the modern industrialized world has been damaged via the masculine impulses of domination, exploitation, and collection. Men, who have been involved with most decisions in the last few centuries since the industrial revolution, are largely responsible for the problems that we now have with the long-term sustainability of the planet.

Ecofeminists argue that the current state of imbalance is a projection of masculine values. A return to more feminine values of nurturing and cooperation is seen as key to the long-term success of the planet. Some argue that the connection that women have to their families help them to appreciate and develop long-term visions for sustainability within their communities. Plus, the monthly menstrual cycle ties women to cycles of the planet.

Ecofeminism suggests that the breakdown of a structured binary way of looking at the Earth (good/bad, male/female, human/animal, etc.), will help us create a new, more holistic way of looking at the world that will promote sustainability.

Destruction regardless of theory

Humans cause environmental problems no matter what economic or theoretical system we live in or espouse. The very act of being human comes at a cost to the world. For tens of thousands of years, humans have caused problems, but the environment was able to respond by evolving and repairing itself. There is no doubt that some early humans caused extinction of animals and over-exploited resources. However, it is only since the industrial revolution that we have seen the widespread destruction accelerate. Today, some question whether or not the Earth can absorb the impacts of all of the activities taking place throughout it.

It is for this reason that some have called into question our modern way of doing things (Figure 13.5). What do you think? Do you have hope that we can maintain our current economic pace on the planet without long-term environmental damage? If not, what approach do you think we should use to try to ensure long-term sustainability?

Figure 13.5 There are many alternatives to traditional thinking about the environment as a resource. Which of the alternatives resonate within your values system?

Environmental economics: externalities

Within the context of environmental economics, many concern themselves with the issue of externalities within an economy. Externalities are things that are not accounted for in the true price of a good or service. If you are a mining company, for example, and you are polluting a stream, the cost of the clean-up or the cost to loss of tourism is an external cost. The external costs are often born by the public.

There are three ways to manage externality costs to try to limit their impact on the public:

1 *Regulations.* Instituting some form of regulation to limit things like pollution will lessen the impact on the environment and the costs the public has to pay for the production of the product. Managing a regulatory environment does have costs associated with it that must be borne by the public. Nevertheless, the external costs can be managed effectively if strong regulations are in place. The regulations may set strict limits on pollution or they may provide opportunities to trade pollution within a cap and trade framework. Cap and trade approaches to regulating greenhouse gases were discussed in Chapter 5. However, cap and trade can be used to try to reduce the impact of any pollutant.
2 *Taxes.* Taxing polluting products is an effective means to discourage sales, while generating revenue to bear the costs to clean up the pollution. Taxes have been used in Europe as a means to reduce gasoline consumption and fund environmental programs. This is the

same reason why bottled beverages are taxed via a bottle refund policy. In many cases, the revenue generated helps promote recycling or development of special environmental programs.

3 *Property rights.* Governments can put into place rules for the types of activities that can take place within any location. Thus, if a local community wants to limit the impact of a polluting industry, it can set particular property rights rules that grant land owners the right to clean air and water. In some cases the rights to this can be sold to allow organizations to pollute.

Measuring the economy

There are many ways that economists measure the economy to assess current economic situations as well as to predict future trends. These measurement tools are called economic indicators. They are measured at different time intervals and can be classified as leading or lagging indicators. Leading indicators are those that tell us how the economy is doing and where it is likely to go. Lagging indicators provide insight into how the economy performed in the past. Lists of leading and lagging economic indicators are presented in Table 13.3.

The indicators largely focus on traditional economic measures related to industrial output, employment, interest rates, and consumer behavior. They should be familiar to anyone involved in business or finance within the global economic market place. Yet, it is important to point out that none of the major economic indicators have anything to do with the environment. Instead, the list of indicators suggests that the major ways that we measure the economy rely on data that comes to us from the financial markets, not from environmental assessments.

Of course, from a business perspective, this makes sense. Most economists and business experts do not evaluate sustainability or the greenness of the economy. These are relatively new concepts and they have not informed the way the economy is evaluated. Instead, most economists rely on the pantheon of indicators that have been used successfully for generations. They offer ways to compare economies across space and time. Using the indicators, economists are able to evaluate long-term trends and assess economic conditions in different countries.

Yet many are suggesting that the indicators do not tell the whole story about what is happening in the economy. They believe that the traditional indicators do not evaluate things like environmental costs and impacts on the public. They feel that the new green economic indicators, sometimes called green growth indicators, are needed to steer our society in a direction that measures the economy in different ways.

The Organization for Economic Co-operation and Development (OECD) has been focused on green growth for a number of years. In 2014, they published a report called Green Growth Indicators 2014, which lists and measures a number of indicators listed in Table 13.4.

The list is strikingly different from the leading and lagging economic indicators presented in the previous table. The list is very comprehensive in developing a range of indicators that can be used to measure the health of the green economy from place to place and across time. Some of the indicators are traditional economic indicators, such as unemployment

Table 13.3 Common leading and lagging economic indicators used by economists to measure the economy. Note that the indicators measure traditional aspects of the economy that can be compared across space and time.

Leading economic indicators	
Money supply	This is a measure of the amount of money in play in the overall economy and includes everything from deposits to currency.
Weekly jobless claims for unemployment	The number of jobless claims increases when an economy is declining.
Average weekly hours	The number of hours worked by employees indicates the overall strength of the employment market and thus the overall economy.
Index of consumer expectations	This is a measurement of whether or not consumers plan to make major purchases in the coming years.
Building permits	This measures the robustness of the housing sector.
Vendor performance	This indicator measures the time it takes to deliver orders. Longer delivery times equates to higher demand and thus a stronger economy.
Manufacturers' new orders for consumer goods	Increases in orders for new goods indicates a stronger economy.
Standard and Poors Stock Index (S & P 500)	The price of major stocks collectively reflects on the strength of the economy.
Lagging economic indicators	
Length of unemployment	The length of time individuals are unemployed indicate the strength of the employment market in the past.
Changes in the consumer price index	This index measures the costs of consumer goods.
Ratio of consumer credit to personal income	This index measures the indebtedness of consumers.
Value of outstanding commercial and industrial loans	This index measures the indebtedness of commercial and industrial businesses.
The average interest rate charged by banks	The index measures the cost of consumer credit.

and gross domestic product (GDP). However, many of them are unusual within the context of traditional economic indicator assessment. For example, energy productivity is assessed, in part by the share of renewable energy within the overall energy sector in the economy. Environmental considerations are clearly measured within the context of wildlife, land use, pollution, and public health. Also, a number of demographic trends are measured as well, including life expectancy, population growth, and income equality.

Table 13.4 Green growth indicators. (Modified from http://www.oecd.org/greengrowth/greengrowthindicators.htm.) These indicators differ greatly from the indicators presented in Table 13.3 in that they measure a variety of social and environmental factors along with some of the traditional economic assessment variables that are listed in that table.

OECD green growth indicators	
Economic growth productivity and competitiveness	GDP growth and structure
	Net disposable income (or net national income)
	Labor productivity
	Multi-factor productivity
	Trade weighted unit labor costs
	Relative importance of trade
	Consumer price index
	Prices of food, crude oil, minerals, ores, and metals
Labor market, education, and income	Labor force participation
	Unemployment rate
	Population growth, structure, and density
	Life expectancy; years of healthy life at birth
	Income inequality
	Educational attainment; level of and access to education
Carbon and energy productivity	Production based CO_2 productivity
	Demand based CO_2 productivity
	Energy productivity
	Energy intensity by sector
	Share of renewable energy sources
Resource productivity	Multi-factor productivity reflecting environmental services
Natural resources stocks	Index of natural resources
Renewable stocks	Fresh water resources
	Forest resources
	Fish resources
Non-renewable stocks	Mineral resources; available stocks or reserves
Biodiversity and ecosystems	Land resources; land cover conversations and cover changes from natural state to artificial state
	Soil resources: degree of topsoil losses on agricultural land, on other land Wildlife resources
Environmental health and risk	Environmentally induced health problems and related costs
	Exposure to natural or industrial risks and related economic losses

(*continued*)

Table 13.4 (Continued)

OECD green growth indicators	
Environmental services and amenities	Access to sewage treatment and drinking water
Technology and innovation	Research and development expenditure of importance to green growth
	Patents of importance to green growth
	Environment-related innovation in all sectors
Environmental goods and services	Production of environmental goods and services
International financial flows	International financial flows of importance to green growth
Prices and transfers	Environmentally related taxation
	Energy pricing
Regulations and management approaches	Water pricing and cost recovery No indicators to date
Training and skill development	No indicators to date

The Organization for Economic Co-Operation and Development

The Organization for Economic Co-Operation and Development (OECD) evolved from its parent organization, The Organization for European Economic Cooperation (OEEC), which was founded in 1948 after the end of World War II to manage the Marshall Plan, which focused on the reconstruction of the European economy. The Marshall Plan is arguably one of the most successful economic development plans ever formed in that it led to the growth of robust economies throughout much of Europe. The original member nations were Austria, Belgium, Denmark, France, Greece, Iceland, Ireland, Italy, Luxembourg, Netherlands, Norway, Portugal, Sweden, Switzerland, Turkey, the United Kingdom, and Western Germany.

Due to its success, other nations joined the OEEC and transformed it into the OECD. The new members are Australia, Canada, Chile, the Czech Republic, Estonia, Finland, Germany (united East and West since 1990), Hungary, Israel, Japan, Korea, Mexico, New Zealand, Poland, the Slovak Republic, Slovenia, Spain, and the United States. The expansion of the OECD into North and South America, Asia and the Pacific, and Eastern Europe demonstrates the overall significance of the OECD in influencing economic strategies.

The modern mission of the organization is in part to "… promote policies that will improve the economic and social well-being of people around the world. The OECD provides a forum in which governments can work together to share experiences and seek solutions to common problems. We work with governments to understand what drives economic, social and environmental change..." The OECD is focused on helping its member governments in four main areas in the coming years.

(Continued)

- First and foremost, governments need to restore confidence in markets and the institutions and companies that make them function. That will require improved regulation and more effective governance at all levels of political and business life.
- Secondly, governments must re-establish healthy public finances as a basis for future sustainable economic growth.
- In parallel, we are looking for ways to foster and support new sources of growth through innovation, environmentally friendly "green growth" strategies and the development of emerging economies.
- Finally, to underpin innovation and growth, we need to ensure that people of all ages can develop the skills to work productively and satisfyingly in the jobs of tomorrow.

Clearly the infusion of green economic principles within one of the most important international economic development organizations in the world is a clear sign that sustainability is of growing concern among governmental leaders throughout the world.

Green jobs

One of the benefits of focusing on greening the economy is the development of green jobs. The United Nations defines a green job as "work in agriculture, manufacturing, research and development (R&D), administrative, and service activities that contribute(s) substantially to preserving or restoring environmental quality. Specifically, but not exclusively, this includes jobs that help to protect ecosystems and biodiversity; reduce energy, materials, and water consumption through high efficiency strategies; de-carbonize the economy; and minimize or altogether avoid generation of all forms of waste and pollution."

This definition encompasses many professional activities from farming to environmental consulting. However, the definition is a bit vague in that it does not lead logically to a clear classification of green jobs that can be easily quantified or assessed from one region to another.

In 2011, the Brookings Institute, a think tank and research center in Washington DC, published a document that addressed the classification of green jobs within a report called *Sizing the Green Clean Economy*. They used the term "clean" instead of "green" to avoid confusion with other green job classifications. However, the term clean or green are used interchangeably by many in the environmental field.

Brookings defines a clean job more simply than the United Nations. To them, a clean job is one in which workers are "engaged in producing goods and services that benefit the environment..." It is a broad definition, however, Brookings classifies green jobs within five categories: agriculture and natural resources conservation, education and compliance, energy and resource efficiency, greenhouse gas reduction, environmental management, and recycling, and renewable energy. Within each of the five categories, several segments are listed. A full review and description of the categories and segments is listed in Table 13.5.

As can be seen in the table, the range of jobs in the clean economy is rather striking and provides a far greater detailed classification system than any other green job assessment.

Table 13.5 Categories and segments of the green economy (from Brookings with slight modifications)

Category/segment name	Description
Agricultural and natural resources conservation	Establishments in this category work to conserve natural resources or natural food systems.
Conservation	Establishments in this segment manage public natural resources such as land, parks, forests, and wildlife.
Organic food and farming	Establishments in this segment process organic food, grow it on farms, and/or sell it.
Sustainable forestry products	Establishments in this segment make recycled paper or practice sustainable logging.
Education and compliance	Establishments in this category enforce or assist in the compliance of environmental laws or educate workers for jobs that benefit the environment.
Regulation and compliance	Establishments in this segment enforce or assist in the compliance of environmental laws.
Training	Establishments in this segment received a grant to train workers for the clean economy or are certified to do so.
Energy and resource efficiency	Establishments in this category make goods or provide services that increase energy efficiency.
Appliances	Establishments in this segment make energy-efficient appliances used for cooking, heating, cooling, and various consumer and industrial applications.
Battery technologies	Establishments in this segment make or develop batteries and other energy-storage technologies.
Electric vehicle technologies	Establishments in this segment make electric or hybrid vehicles—or supply them with specialized parts.
Energy-saving building materials	Establishments in this segment provide building insulation and weatherization services or make building materials that save energy (e.g. specialized windows, doors, insulation materials).
Energy-saving consumer products	Establishments in this segment make a wide variety of consumer products that meet energy-efficient standards (e.g. office products, computers, glass, shades) or provide energy-saving home repairs.
Fuel cells	Establishments in this segment make or develop technologies that convert hydrogen into fuel.
Green architecture and construction services	Establishments in this segment provide architectural or engineering services for building projects that meet stringent environmental standards.
HVAC and building control systems	Establishments in this segment make energy-efficient temperature control equipment or audit buildings for energy efficiency.
Lighting	Establishments in this segment make lighting equipment that meets federal Energy Star standards for energy-efficiency.

(continued)

Table 13.5 (Continued)

Category/segment name	Description
Professional energy services	Establishments in this segment provide certified energy-efficient professional services or services related to energy research or energy efficiency consulting and design.
Public mass transit	Establishments in this segment provide multi-passenger transportation to the public or school children, displacing less energy efficient single-passenger vehicle travel.
Smart grid	Establishments in this segment provide services related to electricity measurement and control.
Water efficient products	Establishments in this segment make products that conserve water/or prevent water leakage and waste.
Greenhouse gas reduction, environmental management, and recycling	Establishments in this category make goods or provide services that increase environmental sustainability.
Air and water purification technologies	Establishments in this segment make products that reduce or eliminate the pollution of air and water.
Carbon storage and management	Establishments in this segment develop technologies to eliminate carbon emissions from fossil fuel extraction and production.
Green building materials	Establishments in this segment make building products such as carpets, treatments, or wood products that are certified to be environmentally sustainable and non-polluting.
Green chemical products	Establishments in this segment receive certifications for refining chemical processes and ingredients (used in things like cosmetics, fertilizers, cleaning agents, and paints) to make end-products more environmentally sustainable.
Green consumer products	Establishments in this segment receive certifications or meet third-party standards for making consumer products that are on the cutting-edge of environmental sustainability (e.g. furniture, seafood, cosmetics, and surgical supplies).
Nuclear energy	Establishments in this segment generate nuclear power or provide technical services that monitor and/or reduce pollution.
Pollution reduction	Establishments in this segment make pollution control equipment or provide technical services that monitor and/or reduce pollution.
Professional environmental services	Establishments in this segment perform environmental consulting, research, or soil analysis.
Recycled-content products	Establishments in this segment specialize in making certified products out of recycled paper or metal.
Recycling and reuse	Establishments in this segment provide recycling services and wholesale distribution.
Remediation	Establishments in this segment provide environmental remediation and clean-up services.

(*continued*)

Table 13.5 (Continued)

Category/segment name	Description
Waste management and treatment	Establishments in this segment administer public sector air, water, and waste management and treatment services or provide those services directly.
Renewable energy	Establishments in this category make goods or provide services that facilitate the use of energy from renewable sources.
Biofuels/biomass	Establishments in this segment produce or develop energy from biological or agricultural materials.
Geothermal	Establishments in this segment generate or develop technologies that convert heat from the earth's core into energy or facilitate the use of such energy.
Hydropower	Establishments in this segment generate or develop power from dammed water.
Renewable energy services	Establishments in this segment provide professional or construction-related services to manage or implement renewable energy projects.
Solar photovoltaic	Establishments in this segment produce, develop, or install technologies that convert sunlight into electricity.
Solar thermal	Establishments in this segment produce, develop, or install technologies that capture and distribute heat from the sun for energy consumption.
Waste-to-energy	Establishments in this segment produce or develop technologies that convert waste into energy.
Wave/ocean power	Establishments in this segment produce or develop technologies that convert naturally flowing water into energy.
Wind	Establishments in this segment produce, develop, or install technologies or specialized components of those technologies that convert wind into energy.
Aggregate clean economy	The aggregate of all 39 clean economy segments.

Some may argue that some of the jobs may not be appropriately classified as a clean or green job (jobs in nuclear energy or school bus driver). Nevertheless, it provides a system by which to understand green job trends.

The Brooking's report provides some key summaries of the US green job environment between 2003 and 2010:

1 About 2.7 million people work in green jobs in a wide range of activities, most mature segments of the economy (manufacturing and public services like mass transit). Smaller numbers of workers are in emerging areas like renewable energy.
2 The growth in clean jobs is not as rapid as other areas of the economy. However, there is a rapid expansion in jobs in renewable energy.

3 About a quarter of the jobs in the clean economy are manufacturing jobs, many of which involve the export of manufactured goods. This is higher than other segments of the economy where 9% of the jobs are in manufacturing.
4 Workers in the clean economy are paid more than other workers.
5 Green jobs are everywhere in the United States, but there are some areas where there is a degree of specialization. Brookings classifies clean economies into the following: service-oriented, manufacturing, public sector, and balanced. For example, San Francisco has strengths in green professional services, whereas places like Louisville have distinct manufacturing strengths. State capitals are often centers of public sector jobs, while some places, especially some of the larger metros like Los Angeles and Atlanta have more balanced sectors.
6 Some metropolitan areas have distinct industry strengths that lead to clustering of particular types of green jobs, for example Los Angeles is a center of the photovoltaic industry and Chicago, the windy city, is predictably the center for wind energy jobs.

Profiles of green workers

Bhavani Jaroff is a chef, caterer, radio host, and food activist (Figure 13.6). She got into the food business while she was a college student at the College of Ceramics at Alfred University. She found upon arrival that there were limited offerings for vegetarians on campus, and designed and implemented a system that ended up feeding 125 students a day. Since that experience, Jaroff worked at a number of natural food restaurants throughout the northeast. She also opened up a catering business that focused on organic and healthy food.

Figure 13.6 Chef, caterer, radio host, and entrepreneur, Bhavani Jaroff.

When her youngest entered kindergarten, she got her masters in education and became a teacher at her daughter's school, where she focused on teaching holistically about food and transformed the school's cafeteria from serving traditional food into offering mainly organic food. She also merged gardening, composting, recycling, and food into the overall education of the students. The food program led her to develop a community service program where the students cooked for and fed the homeless in New York City. She now offers consulting, cooking classes, catering, workshops, and a variety of other services. Her radio show, *iEat Green with Bhavani*, is a staple on the Progressive Radio Network.

Jake Sacket works part-time for New York City's Edible Schoolyard organization while he is working on his undergraduate degree in Sustainability at Hofstra University (Figure 13.7). Edible Schoolyards started in 2010 to provide opportunities for kindergarten through fifth grade students in New York Public Schools to grow their own food and learn how to cook healthy meals. The idea for the program emerged out of concerns over the growing obesity epidemic in the United States and the lack of fresh fruits and vegetables in stores in many neighborhoods in the city.

Teachers work with students to plan and plant organic gardens that include fruits, vegetables, and grains. After harvest, they go to their kitchen classroom to prepare and

Figure 13.7 Jake Sackett working at New York City's Edible Schoolyard program.

(Continued)

eat the food. The experiences are integrated into lessons that address part of the New York State standards in the sciences and social sciences.

Jake tends and maintains one of the gardens at PS 6 and 7 School in Harlem where he weeds, harvests, and maintains container gardens while the students are away for the summer. In addition, he builds new above-ground garden beds to get ready for new students in the fall. He interacts with the teachers and gets to eat from the garden as one of his perks. "I like the job because I get to interact with young kids and help them understand that there are healthy food options in New York City," Sackett said. "Every day I incorporate the three 'E's of sustainability. I work on improving the New York City environment, I am making money in a green job, and I make food affordable so that children from all income groups can eat healthy food."

Rob Milyko has been working a green job for 37 years as a bus driver for Mountain Line bus service in Missoula, Montana (Figure 13.8). While it may not seem like it, driving a bus is a very green job. Rob states, "I feel good knowing every person on the bus is another car off the road." Employment in mass transit accounts for some of the greatest numbers of green jobs in the United States. While most of the mass transit jobs are in the large urban centers, smaller communities have mass transit services. Missoula, Montana, in the United States, with a population of approximately 70 000, has had mass transit services since 1977, offered by a publicly funded organization called Mountain Line.

Figure 13.8 Rob Milyko driving his bus in Montana. (Photo courtesy of Greg Siple.)

While Mountain Line's main mission is to provide a quality bus service to residents of Missoula, it is important to note that it has a strong environmental vision for the future. It hopes to:

- Significantly increase the use of transit
- Improve transportation options to reduce single occupancy vehicle use
- Create incentives for transit to reduce traffic congestion and improve community health
- Create partnerships of organizations to reduce vehicle miles traveled in the region. (Modified from: http://www mountainline.com/about-mountain-line/facts/.)

Mountain Line has made significant progress on their goals. Since the organization started, it has provided 22 million passenger trips. In 2012 alone, nearly 1 million passenger trips were generated. Milyko states, "We have some great environmental programs. We have a free ride program every summer for kids under 18. We encourage bicycling and have put in bike stations around town that are equipped with tools, parts, and air compressors. In addition, we are always looking for alternative fuel sources."

Milyko infuses sustainability into his life. He's a vegan, shops local, filters all purchases through a sustainability lens, and bikes throughout the year—even in Montana's tough winters. "Sustainability should be everyone's goal," he says. "Our planet and beings suffer immensely due to Western civilization's insatiable appetite for resources. It is the primary cause of many of our problems. It is easy to blame government and corporations, but it is up to every individual to cause change. I agree with Ghandi who said, 'Be the Change you want to see in the world.'"

Environmental impact statement for the New Hudson River Bridge

One of the major river transportation crossings in New York is the Tappen Zee Bridge. It was built in 1955 and had nearly 150 000 vehicles cross it daily. However, by the start of the century it was falling apart and was considered one of the most dangerous bridges in the country due to its poor condition. As a result, it was demolished in 2018 after the completion of a new bridge.

Prior to starting construction, however, an environmental impact statement had to be written. It is a good example of a quality environmental impact statement and you can view the statement here: http://www.newnybridge.com/documents/feis/.

The organization of the document is worth reviewing. Following an executive summary, there are chapters titled:

Chapter 1. Purpose and Need
Chapter 2. Project Alternatives
Chapter 3. Process, Agency Coordination, and Public Participation
Chapter 4. Transportation

(Continued)

Chapter 5. Community Character
Chapter 6. Land Acquisition, Displacement and Relocation
Chapter 7. Parklands and Recreational Resources
Chapter 8. Socioeconomic Conditions
Chapter 9. Visual and Aesthetic Resources
Chapter 10. Historic and Cultural Resources
Chapter 11. Air Quality
Chapter 12. Noise and Vibration
Chapter 13. Energy and Climate Change
Chapter 14. Topography, Geology, and Soils
Chapter 15. Water Resources
Chapter 16. Ecology
Chapter 17. Hazardous Waste and Contaminated Materials
Chapter 18. Construction Impacts
Chapter 19. Environmental Justice
Chapter 20. Coastal Area Management
Chapter 21. Indirect and Cumulative Effects
Chapter 22. Other NEPA and SEQRA Consideration
Chapter 23. Final Section 4(f) Evaluation
Chapter 24. Response to Comments on the Draft Environmental Impact Statement.

As can be seen in the list of chapters, the Tappen Zee Bridge environmental impact statement provides a very detailed and holistic assessment of the range of issues that may be of concern in the construction of the bridge. This is not especially surprising given the significance of the project. This is one of the largest public works projects to be undertaken in New York in some time. However, it is worth noting that the themes that emerge in the report align with the three "E"s of sustainability.

14

Corporate and Organizational Sustainability Management

There are many businesses that have made a significant contribution to sustainability. They have looked at their business processes and tried to find ways to deliver goods and services, while considering issues of sustainability. Some businesses infuse the three "E's" of sustainability, environment, economics, and equity, within their corporate decision-making.

Cognitive dissonance

It might seem strange to think about corporate sustainability. Many of us in the green movement are much more comfortable embracing organizations and businesses that are explicitly green, such as organic farms or the recycling industry. However, we all take advantage of goods and services that have the potential to do damage to the environment in some way—whether through their production or use.

The challenge with sustainability is not necessarily to get rid of the goods and services that we rely on in our day-to-day lives, but how to make those goods and services more sustainable. Think about an oil company (Figure 14.1). We all understand that oil companies produce products that are potentially damaging to the environment. However, can an oil company move toward sustainability while still being an oil company? What about a pesticide service, or a battery manufacturer, a shipping company, or a large international retailer?

While it might be uncomfortable to think about how to "green" an oil company or chemical plant, it is important to understand that many industries can make a great difference in trying to make our planet a much more sustainable place no matter what they do.

Indeed, many organizations, from large corporations to mom-and-pop businesses, have made great strides in trying to find ways to improve their businesses to advance sustainability.

Some of the most important jobs in the field of sustainability are with companies that are seeking to make their operations more sustainable. It might seem odd that an organization like a major chemical company, a car manufacturing firm, an oil company, or a shipping company hires sustainability experts. However, in many ways, these are the most important places to infuse sustainability ideas.

Introduction to Sustainability, Second Edition. Robert Brinkmann.

Companion website: www.wiley.com/go/Brinkmann/IntroductiontoSustainability

Figure 14.1 How can we make the procurement of gasoline greener in all phases of its production?

Why are businesses concerned with sustainability?

Profit

Many businesses are attracted to sustainability because sustainability practices save money. Energy and water conservation are often winning propositions to those concerned with profits and the bottom line. While some sustainability initiatives have a high initial investment, the payback is great over time.

Public relations

There is no doubt that "going green" is popular with the public. Some companies are interested in advancing sustainability within their industry to promote a particular company image. How many times have you seen organizations advertise their sustainability initiatives or things that they are doing to preserve the environment? While the initiatives may be initiated for altruistic reasons, public relations departments around the world promote environmental and sustainability projects to the public.

Altruism

Many companies develop sustainability purely for altruistic reasons. They have a personal or corporate will to "do the right thing." While improved profit and public relations may be an outcome from the decision to embrace sustainability the main reason to support sustainability goals comes from the will to do good. The source of the will to do good can come from many motivating sources such as religious convictions, social responsibility, or an environmental ethic.

Figure 14.2 What and where we buy as consumers matters. We can choose to become more deliberate in limiting our consumption and being judicious about our purchases.

Concern over the long-term sustainability of the industry

There are other organizations that are embracing sustainability because they are concerned with the long-term ability of their organization to offer goods and services (Figure 14.2). The very resources that they may need to deliver their goods or services may be at risk. They will promote sustainability solely to protect the long-term viability of their business. Just think about a producer of fine furniture made from tropical woods like mahogany. Is it in the greatest interest to buy mahogany on the open market without concern over the viability of the tropical forests where mahogany trees grow? Or it in the best interest to work to promote the conservation of the tropical forests to ensure a long-term steady supply of mahogany wood?

Professional standards and norms

Some in the business community embrace sustainability because it is the professional standard in their field. They may be in an industry where it is standard practice to operate a "green" business. For example, if one is in the organic food business, it would be expected that sustainability would infuse all operations of the business.

Total quality management and sustainability

One of the key developments in organizational management in the twentieth century is the idea of total quality management or TQM. To understand TQM, it is important to understand that the idea emerged when many Western countries realized that their products and services were not of the same quality as those exported from Asia. After World War II, Asian products were considered inferior to many in the West. However, in the late twentieth century, Western countries found themselves in a losing trade battle with Asia due to the high quality of goods coming from that region. Products were far superior and made much more inexpensively.

Western corporate leaders started to examine the broad management of Western industries and realized that greater assessment of all areas of management was needed to make improvements. They studied their eastern counterparts and realized that the most successful companies focused heavily on assessing all aspects of business operations. They found that quality assurance at every management level was key to delivering excellent products and services to consumers at reasonable prices.

W. Edwards Deming is often considered the father of TQM. In his 1986 book, *Out of the Crisis*, Deming outlined 14 major points to improve management. It is these 14 points that many companies relied on in re-thinking how to institute corporate change to improve the competitiveness of Western companies in the late twentieth century. In many ways, Deming's 14 points provide some of the earlier management approaches to corporate sustainability This is not all that surprising, since Deming wrote the book, *Out of Crisis*, just as the concept of sustainability was emerging out the United Nation's Brundtland Report. Indeed, in 2011, Joseph Jacobson made this point when he revised Deming's points through a sustainability lens to demonstrate how closely aligned TQM is with sustainability. A comparison between Deming and Jacobson's points is shown in Table 14.1.

When the two 14-point schemes are compared, it is easy to see why some argue that the TQM approach to business led to broad acceptance of sustainability principles. Each approach requires managers to change the focus from looking at rapid production for monetary gain. Instead, they suggest that only through a deeper examination of all aspects of business will an organization be successful. Products and services must be produced, while allowing creativity of workers to emerge, and focusing in on the quality of product and quality of experience for workers.

There is no doubt that Deming's 14-point business model had a tremendous influence on the management of Western companies. Organizations all over the world have applied TQM with great success by examining the management of all operations in order to ensure that each part of the organization is running successfully and that each is integrated into the whole. Sustainability and sustainability assessment fits nicely with this focus on holistic assessment of organizational operations. The complexities of themes that are involved with sustainability (economic, social, and environmental) require exactly this type of holistic assessment.

People, planet, and profits

There are many in the business community who understand that the future of the world will be shaped by the actions they take today. They can make decisions that leave the world a worse place than when they started their activities, or they can try to make improvements in the world while they are conducting their business.

Table 14.1 A comparison of W. Edwards Deming's 14 points for transforming business with Jacobson's 14 points for transforming businesses within a sustainability lens.

Point	Deming's TQM point	Jacobson's sustainability point
1	Create constancy of purpose toward improvement of product and service, with the aim to become competitive, to stay in business and to provide jobs.	Create constancy of social responsibility and sustainability
2	Adopt the new philosophy. We are in a new economic age. Western management must awaken to the challenge, must learn their responsibilities, and take on leadership for change.	Reject waste and defect
3	Cease dependence on inspection to achieve quality. Eliminate the need for massive inspection by building quality into the product in the first place.	Reject inspection—build quality into design.
4	End the practice of awarding business on the basis of a price tag. Instead, minimize total cost. Move towards a single supplier for any one item, on a long-term relationship of loyalty and trust.	Use quality criteria to award social responsibility and sustainability
5	Improve constantly and forever the system of production and service, to improve quality and productivity, thus constantly decrease costs.	Constantly improve sustainability and social responsibility performance
6	Institute training on the job.	Develop environmental and social responsibility training programs
7	Institute leadership. The aim of supervision should be to help people and machines and gadgets do a better job. Supervision of management is in need of overhaul, as well as supervision of production workers.	Develop environmental and social responsibility management
8	Drive out fear, so that everyone may work effectively for the company.	Drive out fear, punishment, and punitive actions
9	Break down barriers between departments. People in research, design, sales, and production must work as a team in order to foresee problems of production and usage that may be encountered with the product or service.	Breakdown functional barriers

(*continued*)

Table 14.1 (Continued)

Point	Deming's TQM point	Jacobson's sustainability point
10	Eliminate slogans, exhortations, and targets for the world force asking for zero defects and new levels of productivity. Such exhortations only create adversarial relationships, as the bulk of the causes of low quality and low productivity belong to the system and thus lie beyond the power of the work force. a) Eliminate work standards (quotas) on the factory floor. Substitute with leadership. b) Eliminate management by objective. Eliminate management by numbers and numerical goals. Instead substitute with leadership.	Eliminate targets and slogans—it's a lifestyle
11	Remove barriers that rob the hourly worker of this right to pride of workmanship. The responsibility of supervisors must be changed from sheer numbers to quality.	Eliminate numerical quotas—it's about everyday quality and integrity
12	Remove barriers that rob people in management and in engineering of their right to pride of workmanship. This means, inter alia, abolishment of the annual or merit rating and of management by objectives.	Remove barriers from hourly workers
13	Institute a vigorous program of education and self-improvement.	Train vigorously
14	Put everybody in the company to work to accomplish the transformation. The transformation is everybody's job.	Create a supportive management structure that embraces sustainability and social responsibility

The infusion of the three "E"s of sustainability (environment, equity, and economics) within this business ethic is often renamed the three "P"s of people, planet, and profits. This concept, sometimes called the triple bottom line, emerged in recent years as the business community became more concerned over the long-term impacts of environmental and social degradation. This ethic argues that profits can be generated only by ensuring the long-term health of the planet while maintaining strong, healthy societies.

The people component of the three "P"s focuses largely on social responsibility. While each part of the world has distinct policies and procedures that govern their society, including things like labor laws, environmental rules and regulations, and corruption, there is no doubt that there is a growing global business ethic of social responsibility. Companies are increasingly concerned with the conditions of workers in their factories. Issues like the age of workers, working hours, pay, and worker safety have become international concerns. At the same time, there is also concern over the impact of business practices within communities where factories are located or where resources are extracted for materials that are part of the supply chain. Businesses are increasingly looking at the role of their activities on communities that are impacted in some way by their operations.

For some time, the globalized business community focused largely on the profits that could be generated by offshoring their operations to cheaper labor markets in the developing world. While some may have looked negatively at the social outcome of those decisions, overall the practice was looked on as a favorable way to earn profits. Now, however, businesses are finding themselves confronted with the ethical reality of the outcome of such practices. Their brands are linked with things like child labor, poor working conditions, and rampant environmental pollution. Reputations of some businesses have been hurt when things like poor working conditions in offshore factories are revealed. More and more, companies are focusing on ensuring that social responsibility is part of their business practice in order to avoid the problems that emerged in the early days of modern globalization. It is not just the right ethical thing to do, it is good for business.

The planet component of the three "P"s focuses mainly on environmental responsibility. Within the globalized world, it is easy to not concern oneself with the environmental impacts of consumption. We might buy a beautiful piece of jewelry without thinking of the miners, the mine, or the pollution associated with processing ore. We do not feel the impact of consumption. The same is true at the corporate level. A business may need to order a significant amount of wood for a manufacturing process. They will do so on the global market without thinking of the impact of the purchase. The goal is to buy the best product for the cheapest price.

However, because of widespread global pollution and environmental deterioration, many in the business community are trying to move away from an unthinking attitude about the impacts of their decision-making. In contrast to many political leaders around the world, most of the world's corporate leaders have come to terms with many of the important environmental challenges we face, such as global climate change, rising sea levels, pollution, and loss of biodiversity.

These business leaders understand that they have a responsibility to not destroy the planet and the very things they need to ensure the long-term success of their businesses. They cannot succeed in a ruined, environmentally unstable planet. They believe they have a responsibility to generations to come to protect the Earth in the best way they can.

Figure 14.3 Many feel as if the business community is too tight with politicians. What do you think? How do business interests drive political decisions in your community? (Photo courtesy of J. Bret Bennington.)

Of course, the profit component of the three "P"s is the one that is easiest to comprehend. Businesses focus on profit generating. Old models of profit generation, particularly within a globalized world, focused heavily on extracting wealth without considering social and environmental consequences. Many argue that the three "P"s provide a context for enhancing profits while focusing on social and environmental responsibility (Figure 14.3).

Yet how can an international company succeed for generations in the face of global climate change, growing income inequality, and social disruptions due to conflict? Many corporate leaders believe that the business community has a responsibility to make significant changes in order to ensure the success of their businesses within a shifting environmental and social landscape by infusing sustainability and the people, planets, and profits ideal into their corporate agenda.

Indeed, many leaders argue that changing business as usual attitudes to a sustainability-oriented framework is the key to success in our era. They believe that sustainability concepts provide a way for corporations to become more innovative. New approaches using sustainability give businesses a competitive edge, which will lead to greater profits and corporate value.

Many around the world have seen the impacts of widespread economic development. In the developing world, societies have seen great gains, yet many question the standard capitalist model that created great inequality, pollution, corruption, and public health problems in their nations. Those in the business community who espouse sustainability as a driving ethical principal understand that they have a responsibility to right the wrongs of the past by making strong improvements in their businesses. They seek to make a fairer and cleaner world, while at the same time operating businesses that generate profits.

Ray Anderson, the father of the green corporation and the growth of green corporate environmentalism

One of the leading voices urging stronger support for infusing sustainability within corporate culture was Ray Anderson (1934–2011). We discussed him briefly in Chapter 3. His advocacy for greening businesses helped to transition some areas of the global business culture from profit-focused to Earth-focused. His work in greening his own company led to his notoriety and work with other causes. He worked on a number of important climate change initiatives in the United States and internationally, most notably chairing the President's Action Plan committee in 2008.

Mr. Anderson was the founder of one of the largest carpet companies in the world, Interface Carpeting, which is located in Atlanta, Georgia. It is the leading producer of carpet squares, or modular carpeting for commercial and home uses. If you've been to a major airport or hotel, you've probably walked on an Interface Carpeting product. Modular carpeting is very popular for high-volume areas because worn or damaged pieces can easily be replaced without replacing the entire carpet. Thus, the innovation of carpet squares, or modular carpeting, was a sustainability initiative on its own.

However, in 1994, Ray Anderson read, *The Ecology of Commerce*, by Paul Hawking. The book focuses on why we need to transform our thinking about the environment. Hawking believes that the age of the industrial economy is coming to an end just at the time that we are facing important ecological crises. Businesses, he argues, need to transform their purpose from profit generating, to increasing the "well-being of mankind." He noted that our modern business practices are out of step with the fundamental way that natural economics work. Our current economy focuses greatly on over-exploitation and destruction. Hawking believes that we need to have a more ecological model of commerce that looks at the systems of production to assess environmental issues.

He, and many others in the business community, began to question the overall ethics of the corporate world in light of the kind of world they were leaving for future generations. Many began to ask whether the drive for profit without fully looking at the impact on the environment was in the best interest of the planet.

In many ways, he argued, businesses are exploiting natural capital in ways that damage future generation's abilities to survive and thrive because natural capital is not valued appropriately in traditional economic models. For example, we do not account for nature's genetic diversity, its ability to provide clean water, and its ability to absorb pollution.

Hawking's influence on Anderson was tremendous. He instituted a program at Interface Carpeting called "mission zero" that sought to eliminate any negative impact on the planet by 2020. He did this by looking at all aspects of the carpet business from manufacturing and supplies to sales and travel. He focused heavily on investigating the materials used to manufacture the carpeting the company sold to their customers. In doing so, he converted the supply chain used by Interface Carpeting into one that was more sustainable by reducing the use of harmful glues and solvents and by purchasing renewable materials and fibers. He transformed Interface Carpeting from a company fully focused on producing carpeting and profits into one with a sustainability focus that ended up with net-negative greenhouse gas emissions.

Anderson wrote about his efforts in a book that is considered one of the most important sustainability and business books of the last few decades, *Confessions of a Radical*

Industrialist: Profits, People, Purpose: Doing Business by Respecting the Earth. He spoke to many corporate groups around the world and had a huge influence on many corporate leaders. He spoke widely about the importance of having a planet-centered vision for industries and how this can be done while still earning profits.

Anderson's legacy

While there are certainly many corporate groups that do not fully embrace the tenets of sustainability, many have moved forward with all or parts of Ray Anderson's message—you can make profits while being environmentally responsible and sustainable. Do an Internet search for any corporation you can think of along with the word, "sustainability." I would bet that you will find some website that addresses the organization's sustainability initiatives.

Since Anderson's death in 2011, corporate sustainability has gone relatively main-stream. It is normal for major corporations to publish sustainability reports, have websites dedicated to sustainability, and work with sustainability consultants or have a sustainability officer on staff. Many corporate leaders recognize that sustainability initiatives are not just good for saving money or for public relations. Infusing sustainability within a corporate culture is good for the long-term health of our planet and our society.

Greenwashing in the corporate world

Of course, one of the concerns with the infusion of sustainability within the corporate world is the risk of greenwashing, which is the use of sustainability ideas to promote a green image for the company, particularly when it is undeserved. Greenwashing is often used by marketing officials to encourage consumers to think that a product or service is considered green via advertising that uses images and music associated with nature and environmentalism. The amount of money spent on greenwashing campaigns may be more than the amount of money spent on actual green or sustainability initiatives.

We've all seen the advertisements for things like dishwashing soap or automobiles that give you a sense of the outdoors and nature, even though the products have little to do with or may even damage the environment.

The problem with greenwashing is that it confuses the message of the sustainability movement. If a polluting chemical company is using the same language and music as a major environmental organization within advertising campaigns, the message of both get mixed. The mixed messages do damage to the environmental organization while creating a greener image for the company. This leads to overall wider public distrust of all parties.

Can you think of any greenwashing issue that you have run across in your life? How did it make you feel about the product? Do you think greenwashing hurts the sustainability movement?

Green consumers

Many of us consider ourselves green consumers. We recognize that where we spend our money influences consumer trends. Thus, if we act as a group, good things can happen. However, who are green consumers? According to BSD Global (http://www.iisd.org/business/markets/green_who.aspx), green consumers have three major traits:

1 They have an interest in expanding their green lifestyle and are sincere about their green personal ethic
2 They see their green efforts as inadequate (don't we all!)
3 They recognize that companies are complex and must be looked at for their commitment to sustainability.

Not every company will be perfect, but they must be seen as doing something positive for green consumers to feel good about purchasing their product or service. Yet it must be noted that green consumers have been found to be not all that green in practice. They tend to talk the talk, but not walk the walk. They may espouse a green ethic, but live a life that is highly unsustainable. We've all seen the contradictions present with some in the sustainability community. They may buy local produce while driving a gas guzzling car. These consumers tend to focus on issues that they can manage within small decisions in their daily lives.

It is well-known that many green consumers don't necessarily want to make the sacrifices that one needs to make to be sustainable. They want a very easy fix. That is why convenience is an important issue with green consumers. At the same time, green consumers are interested in learning more and being educated about products and services. They tend to be better educated than the average consumer and do not mind reading about issues in product materials or websites.

Importantly, they do not trust companies based on what they are saying. They are highly sensitive to greenwashing and are distrustful of broad claims by industry. They want some type of third party verification of the efforts of the company.

Companies often seek to reach green consumers. They make this effort because green consumers tend to be wealthier than the average consumer. They are willing to pay more for a "green" product. They have more income to pay for higher-end organic or locally derived products. They often shop at more expensive stores and purchase luxury items.

Generally, green consumers tend to be young adults. What is important about this fact is that companies have the opportunity to build lifelong brand loyalty with these consumers. However, the consumers must feel that the product or service is truly green and continually trying to make improvements in sustainability initiatives.

At the same time, green consumers react like other consumers in many ways. While they are willing to pay more for a high-quality green product or service, they will not pay much more. They also have to feel like they are getting a quality product or service. If they think their purchase resulted in a product or service that had limited quality, they will not continue to purchase the product. Indeed, a low-quality product, green or not, will hurt the brand.

While some consumers will shop at specialty green stores (bricks and mortar or online) that offer goods and services, most want convenience. They will not go out of their way to buy green. That is one of the reasons why so many green products are finding their way into mainstream stores. It is not that unusual to find organic or green products in a standard grocery store. Some of these products are from manufacturers that are well known for their green products, while others are emerging from established manufacturers who recognize that consumers have a strong desire to make greener purchases. Many have developed green or organic products that sit side by side with their standard products. While consumers remain concerned about greenwashing, there is a growing understanding about the difference between truly green products and services, and greenwashing efforts to promote products using inappropriate environmental platitudes.

Global Reporting Initiative

In order to avoid problems of greenwashing and provide internationally comparable statistics on sustainability, an organization called the Global Reporting Initiative was formed in 1997 in the United States in partnership with two non-profit organizations concerned about measuring and assessing sustainability. With the support of the United Nations, the organization relocated to Amsterdam and focused on broadening sustainability reporting and education across many areas of the global economy. The organization is global in scope and seeks to promote sustainability reporting by corporations and other organizations using internationally agreed upon standards.

The reporting guidelines include elements of economic, environmental, and social aspects of sustainability (Table 14.2). There are distinct assessment indicators that can be used within the economic and environmental themes. Four subcategories within the social category also have assessment indicators that are measured. What is useful to know is that like the other assessment tools noted in this book like the Leadership in Energy and Environmental Design (LEED) green building assessment system, the Global Reporting Initiative regularly evaluates and updates its indicators to ensure that its program is current with appropriate sustainability practices.

To date, over 6000 organizations have utilized this reporting tool to disclose their sustainability practices and initiatives. Each has submitted a report to the Global Reporting Initiative, which makes it available on its website here: http://database.globalreporting.org/. The website is searchable by publication year, organization size, organizational sector (i.e. agriculture, aviation, healthcare services, etc.), and region (Africa, Asia, Europe, Latin America and the Caribbean, North America, and Oceania).

The searchable database of reports is very helpful to find particular reports that may be of interest to your own particular area of concern. For example, a search of small healthcare product companies in Europe generated several distinct reports. The reports detail quantitative indicators as noted within Table 14.2 and evaluate particular corporate sustainability goals. The reports also highlight interesting initiatives or special challenges for the organizations.

Sustainability reporting in the S & P 500

Standard and Poors, an investment firm, has been tracking some of the top US companies in the world in a US stock index. The index started in 1923 and is used as a means to track the overall state of the stock market and it is used and reported internationally to gain an understanding of the health of the economy in the United States and the world, since it does include non-US companies. It uses a weighted index to assess 500 of the top stocks in play in the US stock market. It is similar to other generalized indices like the Dow Jones Industrialized Average, although the index has distinct selection criteria for inclusion similar to lists of top firms like the Fortune 500 firms.

Table 14.2 Categories of assessment for sustainability reporting by the Global Reporting Initiative. (Modified from https://www.globalreporting.org/resourcelibrary/GRIG4-Part1-Reporting-Principles-and-Standard-Disclosures.pdf)

Category	Subcategory	Themes
Economic		Economic performance
		Market presence
		Indirect economic impacts
		Procurement practices
Environmental		Materials
		Energy
		Water
		Biodiversity
		Emissions
		Effluents and waste
		Products and services
		Compliance
		Transport
		Overall
		Supplier environmental assessment
		Environmental grievance mechanisms
Social	Labor practices and decent work	Employment Labor/management relations
		Occupational health and safety
		Training and education
		Diversity and equal opportunity
		Equal remuneration for women and men
		Supplier assessment for labor practices
		Labor practices grievance mechanisms
	Human rights	Investment
		Non-discrimination
		Freedom of association and collective bargaining
		Child labor
		Forced or compulsory labor
		Security practices
		Indigenous rights
		Assessment
		Supplier human rights assessment
		Human rights grievance mechanisms

(*continued*)

Table 14.2 (Continued)

Category	Subcategory	Themes
	Society	Local communities Anti-corruption Public policy Anti-competitive behavior Compliance Supplier assessment for impacts on society Grievance mechanisms for impacts on society
	Product responsibility	Customer health and safety Product and service labeling Marketing communications Customer privacy Compliance

Table 14.3 Questions (with some slight modifications) used to assess S & P 500 companies by the Governance and Accountability Institute

Question 1. Does it matter if companies report on sustainability—and does it make a difference if they report according to the Global Reporting Initiative guidelines?
Question 2. Does reporting on sustainability have an impact in the capital markets (and among investors)?
Question 3. What other tangible benefits do companies receive from reporting?
Question 4. Who really cares? And does it really matter?

The companies in the S & P 500 are often analyzed to understand business trends and evaluate company policies. In 2012, the Governance and Accountability Institute surveyed sustainability practices within the S & P 500 companies using the Global Reporting Initiative framework and the questions listed in Table 14.3.

First it must be noted that the Governance and Accountability Institute found that most of the companies in the S & P 500 do publish some type of corporate social responsibility report. Of those that do some type of reporting on social responsibility, most are using the Global Reporting Initiative process. This indicates that the corporate world is strongly embracing its role in promoting some degree of social responsibility and environmental sustainability. The doubling of reporting in such a very short, one year, time frame indicates that this is a very rapid trend that is likely to continue.

The Institute then compared the performance of companies that engage in some type of sustainability or social responsibility index with those that do not. What they found is that companies that are involved in reporting have a "considerable advantage when compared to non-reporting peers." The advantage came in the form of inclusion on corporate indices and lists such as inclusion in the Dow Jones Sustainability Index.

Dow Jones Sustainability Index

The Dow Jones Sustainability Index started in 1999 to benchmark corporate stock performance of companies that were focusing attention on sustainability within their corporate decision-making. The Index evaluates companies based on a number of sustainability criteria and selects companies for inclusion based on their overall sustainability commitment. It is evaluated globally as well as within regions (Europe, Eurozone, North America, United States, Asia Pacific, Emerging Markets, Korea, and Australia).

Each year, the Index selects a leader within the industry groups it evaluates. The 2019 leaders are listed in Table 14.4. Most of the leaders are from Europe with Asia and North America as close contenders. The remainder come from Australia and South America. What is interesting about the index is that it compares companies within industry groups. Thus, banks are compared with banks and car companies with car companies. In this way, companies that are seen as traditionally consuming and not particularly green can be compared with each other in order to try to make improvements within that particular industry. Providing competition among companies within industry groups to achieve high rankings pushes each industry group to make improvements. This approach has caused some criticism of the Index.

The leading companies are evaluated on a complex set of criteria within economic, social, and environmental dimensions. The Index uses a self-reporting evaluation tool that consists of a series of questions on everything from board activity to environmental indicators used for self-evaluation and reporting to stakeholders. According to the Dow Jones Sustainability Index website there are five ways corporations infuse sustainability into their practices:

1 *Strategy*: Integrating long-term economic environmental and social aspects in their business strategies while maintaining global competitiveness and brand reputation.
2 *Financial*: Meeting shareholders' demands for sound financial returns, long-term economic growth, open communication, and transparent financial accounting.
3 *Customer and Product*: Fostering loyalty by investing in customer relationship management and product and service innovation that focuses on technologies and systems, which use financial, natural, and social resources in an efficient, effective, and economic manner over the long term.
4 *Governance and Stakeholder*. Setting the highest standards of corporate governance and stakeholder engagement, including corporate codes of conduct and public reporting.
5 *Human*: Managing human resources to maintain workforce capabilities and employee satisfaction through best-in-class organization learning and knowledge management practices and remuneration and benefit programs. (From: http://www.sustainability-indices.com/sustainability-assessment/corporate-sustainability.jsp.)

The assessment tool that is used evaluates how effective the companies are at these five themes.

Table 14.4 The top industry group leaders listed by the Dow Jones Sustainability Index for 2019. Note that most of the leaders are in Europe and that comparisons are made within industry groups. (Source: http://www.sustainability-indices.com/review/industry-group-leaders-2013.jsp)

Company name	Industry group	Country
Pirelli & C SpA	Automobiles components	Italy
Peugeot	Automobiles	France
Banco Santandar	Banks	Spain
Leonardo	Capital goods (aerospace and defense)	Italy
Owens Corning	Capital goods (building products)	USA
Ferrovial	Capital goods (construction and engineering)	Spain
Signify	Capital goods (electrical components and equipment)	United States
SK Holdings Co	Capital goods (industrial conglomerates)	South Korea
CNH Industrial	Capital goods (machinery and electrical equipment)	United Kingdom
ITOCHU Corp	Capital goods (trading companies and distributers	Japan
Waste Management Inc	Commercial services and supplies	United States
SGS SA	Professional services	Switzerland
Sumitomo Forestry Co Ltd	Homebuilding	Japan
Arcelik AS	Household durables	Turkey
LG Electronics Inc	Leisure equipment and products and consumer electronics	South Korea
Moncler SpA	Textiles, apparel, and luxury goods	Italy
Star Entertainment Grp Ltd	Casinos and gaming	Australia
Hilton Worldwide Holdings Inc	Hotels, resorts, and cruise lines	USA
Sodexo SA	Restaurants and leisure facilities	France
UBS Group AG	Diversified financial services and capital markets	Switzerland
Banpu PCL	Coal and consumable fuels	Thailand
Saipem SpA	Energy equipment and services	Italy
Thai Oil PCL	Oil and gas refining and marketing	Thailand
Enagas SA	Oil and gas storage and Transportation	Spain
PTT Exploration and Production PCL	Oil and gas upstream and integrated	Thailand
CP ALL PCL	Food and staples retailing	Thailand
Thai Beverage PCL	Beverages	Thailand

(*continued*)

Table 14.4 (Continued)

Company name	Industry group	Country
Thai Union Group PCL	Food products	Thailand
British American Tobacco PLC	Tobacco	United Kingdom
Abbott Laboratories	Healthcare equipment and supplies	USA
Cigna Corp	Healthcare providers and services	USA
Colgate-Palmolive Co	Household products	USA
Unilever NV	Personal products	United Kingdom
Allianz	Insurance	Germany
Alcoa Corp	Aluminum	USA
PTT Global Chemical PCL	Chemicals	Thailand
Grupo Argos SA/Colombia	Construction materials	Colombia
Billerud Korsnas AB	Containers and packaging	Sweden
Teck Resources LTD	Metals and mining	Canada
UPM-Kymmene Oyj	Paper and forest products	Finland
Hyundai Steel Co	Steel	South Korea
Alphabet Inc	Interactive media, services, and home entertainment	USA
Telenet Group Holding NV	Media, movies, and entertainment	Belgium
Biogen Inc	Biotechnology	USA
Agilent Technologies Inc	Life sciences tools and services	USA
GlaxoSmithKline PLC	Pharmaceuticals	United Kingdom
Dexus	Real estate	Australia
Wesfarmers Ltd	Retailing	Australia
ASE Technology Holding Co Ltd	Semiconductors and semiconductor equipment	Taiwan
ATOS SE	IT services	France
SAP SE	Software	German
Cisco Systems	Communication equipment	USA
Hewlett Packard Enterprise	Computers and peripherals and office electronics	USA
Delta Electronics Inc	Electronic equipment, instruments, and components	Taiwan
True Corp PCL	Telecommunication services	Thailand
Air France—KLM	Airlines	France
Royal Mail PLC	Transportation and transportation infrastructure	United Kingdom
Terna Rete Elettrica Nazionale SpA	Electric utilities	Italy
Naturgy Energy Group SA	Gas utilities	Spain
Engie SA	Multi and water utilities	France

Sustainability reporting

Many companies regularly publish sustainability reports, often on an annual or semi-annual basis, on the operations of their businesses. They review the activities of the company, summarize the company's efforts toward sustainability, and highlight goals for the coming reporting period.

International Organization for Standardization (ISO): ISO 14000 and ISO 26000

One of the major international organizations focusing on sustainability is the International Organization for Standardization, or ISO. This group has been around since 1947, when it formed to promote international standards associated with industrial and commercial operations. Today, most countries of the world voluntarily conform to ISO standards in most operations.

ISO operates thousands of standards and third-party verification of standards. They range from basic standards like materials codes for ordering forms to quality management standards. In addition, they operate many standards for materials, parts, shipping, and manufacturing. One must purchase access to the standards. The purchase supports the work of the organization. Some have criticized the costs of getting access to the standards, particularly those involved with open source operations. Regardless, ISO remains the main international third-party organization focused on developing and maintaining standards in business.

Two of the key suite of standards that ISO developed that are of concern to those of us interested in sustainability are ISO 14000 and ISO 26000.

ISO 14000

This suite of standards focuses on environmental management and includes all business operations associated with the environment so that organizations limit their impact on the environment, while also complying with local, national, and international laws. The assessment may include everything from energy and waste to greenhouse gas emissions and environmental communications. Using ISO 14000 guidelines provides a framework for organizations to not only ensure legal compliance of the law while trying to limit environmental impacts, but also to continuously update and evaluate policies and procedures for all aspect of operations.

One of the key points of the ISO 14000 process is that goals are established for improvements. By setting goals, new policies and procedures and put into place to ensure compliance. Then, progress toward the goal is evaluated to see if the appropriate policies or procedures are in place. Modifications in operations can be instituted to make improvements.

ISO 26000

This suite of standards focuses on social responsibility. While most other ISO standards can be certified, the ISO 26000 standards are voluntary. Organizers are encouraged to follow ISO guidelines, but ISO does not certify organizations as compliant with ISO 26000 standards.

The standards focus largely on workers, the environment, and communities within seven main areas:

1 *Organizational governance*. This area focuses on how an organization is governed by its owners, board, shareholders, workers, and other stakeholders.
2 *Human rights*. ISO provides a variety of guidelines on human rights. They focus on areas like civil rights, access to the political process, economic freedom, education, and a variety of other human rights issues.
3 *Labor practices*. This area includes a variety of issues related to how an organization manages its workers.
4 *The environment*. The focus of this theme is the impact of the organization on the environment. This may include resource use, particularly energy, as well as impacts of organizational activity on the natural environment.
5 *Fair operating practices (FOP)*. FOP standards are associated with organizational ethics within and outside of the organization.
6 *Consumer issues*. The standards in this theme focus on how the organization interacts with consumers. It includes issues such as contracts, marketing, and product information.
7 *Community involvement and development*. This is perhaps the most complex theme of the group of seven. This area involves how the organization interacts with other organizations where they operate and/or do business. It includes interactions with governments, other businesses, and any other organization. The theme focuses on how the organization is making a positive contribution to broader society within their community.

ISO 26000 is relatively new; it was released in 2010. Yet it is hoped that it will have a strong influence on corporate responsibility. There have been many well-publicized instances of bad corporate behavior in the area of social responsibility that include issues of child labor, human trafficking, and poor working conditions. This scheme is an attempt to try to alleviate some of the international human rights problems that are emerging in this era of widespread globalization.

Case studies of sustainability at the corporate level

We could look at many companies to see what they are doing toward sustainability. In this section, two case studies demonstrate that companies are transforming how they are doing business and that they are changing the way they see their role on the planet. The two examples are Walmart and Unilever. One is a major retailer of consumer goods and the other is a major producer of a number of different types of products, including personal care products and food. Many other companies could be examined. As noted earlier, many major corporations regularly produce sustainability plans. Companies such as Ikea and Chevron state their commitments to sustainability in reports on their activities.

What is important to note is that most major corporations are attempting to make their operations more sustainable in some way. This, of course, can produce a tension for those of us concerned about sustainability. Can corporations that seek to derive profits on labor and natural resources ever be truly sustainable?

Certainly many believe that consumer culture should be critiqued as largely unsustainable and corporations should be criticized for not doing enough or for utilizing their sustainability initiatives for greenwashing purposes to promote their products under a sustainability banner. But it is of utmost importance to recognize that issues of sustainability are finding their way into corporate culture and thereby into our consumer culture. While it is crucial that we recognize that consumerism and consumer culture in itself is often considered unsustainable, we must also recognize the reality of our world and strive to make existing conditions better.

The infusion of sustainability within the upper echelons of corporate boardrooms is important for several reasons:

1. Corporate leaders have a greater sense of responsibility toward the environment and society and are concerned with the impacts of the decisions. There is a growing sense of the role of major corporations in the world. Corporate leaders are reacting to critics of global corporate identity by seeking to find ways to ensure that their business activities are ethical.
2. Companies compete for verifiable sustainability initiatives. One of the most interesting aspects of sustainability in the corporate world is that competition has emerged among corporate leaders over their sustainability initiatives. Some may compete to be the leader in the use of green energy, while others may compete on social equity measures. Regardless, the use of competition in infusing sustainability within corporate initiatives is a welcome addition to the sustainability discourse. Many students might be familiar with campus sustainability contests such as recyclemania that urge student groups or residence halls to compete with each other for improved recycling or lower energy consumption.
3. Companies influence consumers and other companies. If a corporation is a major manufacturer of consumer products like clothing and opts to market its clothing as made with sweatshop-free labor or organic materials, consumers will react positively or negatively. If the brand is a success, suddenly other organizations are striving to mimic the success of the product. Consumers react to this by purchasing the popular product thereby infusing greater sustainability within the consumer culture.
4. Corporations influence manufacturing processes at the global scale. One of the striking issues in our modern world is that the more affluent Western world has exported the manufacturing processes of consumer goods to poorer nations. Unfortunately, many of these countries do not have the strong environmental and social rules and regulations of the Western countries. This means that pollution and social problems occur in some areas that are manufacturing consumer goods bound for Western markets. However, many companies are setting standards for manufacturing to limit social and environmental problems wherever the materials come from for export. This influences the global supply chain by influencing other companies to meet or exceed standards set by their competitors.
5. Corporations influence packaging and shipping. Consumer goods are shipped all over the world. To do so, they have to be packaged and arrangements need to be made for transport. In recent years, there has been a considerable effort underway to cut energy use for

shipping products by using smarter transportation planning. In concert with this, many companies have looked at their packaging to try to reduce materials used in packaging and to reduce the weight of materials to cut shipping costs.
6 Corporate leaders influence national and global policy. If we look at the history of the world, we can see that the last several centuries have seen the rise of the corporate entity as a major political and social force. There is no doubt that major companies have significant influence on political discourse in many countries of the world. In the United States, for example, the Supreme Court has decided that corporations have some of the same rights as individuals. And in some small countries, one or two corporations are the driving force that influences the majority of the economic activity taking place within the borders. Thus, corporations that embrace and support sustainability can have a strong influence on national and global policy. Most of us are familiar with Bill Gates, the former head of the Microsoft Corporation. He is currently using his influence on a number of sustainability and public health issues in the developing world. His connections with other corporate and political leaders has helped to influence the development of sound health and sustainability policy in many countries.

As we examine the two case studies, keep these issues in mind. Also, please dig deeper into corporate sustainability by examining the websites of these organizations. This text provides brief summaries and much more is available on the Internet. In addition, take a look at the websites of other corporations that interest you. Find out if they are doing anything important to promote sustainability within their industry.

Walmart

It might seem odd to discuss Walmart within the context of sustainability, but the company is actually one of the first major corporations to embrace sustainability in a big way (Figure 14.4). Given that Walmart is the largest public corporation in the world, it has a tremendous influence on many aspects of our modern consumer culture from resource use and manufacture, to transportation and the retail environment. Walmart operates in 27 countries and has over 11 000 stores.

Walmart's main three sustainability goals are:

1 *Energy*: Reduce energy intensity and emissions
2 *Waste*: Eliminate waste
3 *Value Chains*: Improve sustainability in value chains

To do this, Walmart employs sustainability experts and consultants to help them examine all aspects of their operations throughout the world in order to assess where to make improvements. The sustainability initiatives fall within a broader corporate global sustainability initiative that includes issues such as fighting hunger, volunteerism, hiring veterans, emergency preparedness, and opportunities for employees.

In terms of energy, Walmart receives approximately 25% of its energy from renewable sources, which accounts to approximately 2 billion kilowatts of power. While this is short of its overall goal of 100%, progress is being made. Each year, Walmart is developing new projects that address specific issues related to green energy or energy efficiency.

Figure 14.4 Walmart is well known for its sustainability initiatives.

For example, it has built approximately 470 renewable energy systems on its facilities that provide 15–30% of a store's energy use. In addition, it uses fuel cells at 42 stores and has a wind turbine at its Red Bluff Distribution Center in California where it provides 15–20% of the energy needed for the site's operations. It must be noted that Walmart is both a green energy producer in that it produces green energy at its operations and a green energy buyer. Since 2012, it has doubled its purchases of green energy from green energy providers. Overall, it hopes to advance the development of 7 billion kilowatt hours of renewable energy around the globe by the end of 2020. This more than triples its current green energy use and gets Walmart to approximately 75% green energy consumption using current energy use statistics.

Of course, the company is also focusing heavily on energy efficiency. They are replacing traditional lighting with energy efficient LED lights and have committed to reducing energy intensity use of stores by 20% (when compared with 2010 baselines) by the close of 2020. In addition, they have reduced refrigerant emissions by 8% since 2005.

These energy initiatives and targets are excellent examples of how companies set long-term goals to make a difference in sustainability. While Walmart is often criticized for a number of its practices, it is clear that the company is making a substantial effort to reduce energy consumption and promote alternative energy development.

While Walmart has made a significant effort on all of its stated goals, perhaps its most significant contribution is the development of the Walmart Sustainability Index which evaluates its suppliers of products. The index was developed in partnership with The Sustainability Consortium, which is a group that brings experts from different areas together to encourage good sustainability practices.

What is important about the index is that it evaluates suppliers on four major themes:

1 Theme 1: Energy and climate
   (a) Has the organization conducted a greenhouse gas inventory and measure emissions?
   (b) Have the emissions been reported to the Carbon Disclosure Project (an organization that assists corporations in disclosing greenhouse gas emissions)?
   (c) What are the most recent greenhouse gas emissions?
   (d) Are there targets for greenhouse gas reduction and what are they?
2 Theme 2: Material efficiency
   (a) How much solid waste is generated by the organization each year?
   (b) What are the waste reduction targets?
   (c) What amount of water is used by the organization each year?
   (d) What are the water reduction targets?
3 Theme 3: Nature and resources
   (a) Has the supplier developed purchasing guidelines that include environmental compliance, employment practices, and product safety?
   (b) What third party certifications do your products have that insure sustainability or responsibility?
4 Theme 4. People and community
   (a) Does your organization know all of the facilities that are involved in the production of your product?
   (b) Does your organization evaluate the quality and capacity of production of all facilities involved in the production of your product?
   (c) Does your organization evaluate social compliance at the manufacturing level?
   (d) Do you work with suppliers to deal with social compliance issues and document problems and improvements?
   (e) Do you promote, through investment, community development in areas where you source materials or operate?

Clearly, by asking suppliers these questions and others specific about their products, Walmart is very broadly influencing their supply chain and at the same time the supply chain of all producers around the world. Since Walmart does not really make anything, this effort strongly influences not just its corporate culture, but the corporate culture of organizations all over the world.

To date, the index has been used to evaluate 200 of Walmart's largest merchandising categories and there are plans to expand the index effort into most of what the organization sells. Given that Walmart has over 100 000 suppliers, this is a significant initiative that has the potential to transform manufacturing and retail trade. Walmart buys 70% of its materials from vendors who use the index.

Unilever

Unilever is a producer of over 400 different brands that include personal care, food, and home goods. They frame sustainability very clearly within their corporate vision statement, which is, "Our purpose is to make sustainable living more commonplace."

If you compare this vision statement with that of other companies, I think you will agree that it provides a very clear commitment to sustainability that is unique. It clearly defines the company as interested not just in providing goods to consumers, but in making the world a better place while making a profit.

Unilever has an online sustainability plan that is divided into nine main areas within three themes. Each theme has subgoals that are quantified within the plan:

1 Improved health and well-being. Goal: By 2020 we will help more than a billion people improve their health and hygiene.
   (a) Health and hygiene.
      (i) Healthy handwashing habits for life
      (ii) Provide safe drinking water
      (iii) Toilets for a better tomorrow
      (iv) Improve oral health
      (v) Building body confidence and self esteem
      (vi) Ready to respond to disasters and emergencies
   (b) Improving nutrition.
      (i) Responsibly delicious
      (ii) Nutritious diets
2 Reducing environmental impact. Goal: By 2030 our goal is to halve the environmental footprint of the making and use of our products as we grow our business
   (a) Greenhouse gases.
      (i) Our greenhouse gas footprint
      (ii) Protecting our forests
      (iii) Global climate action
      (iv) How we're becoming carbon positive in our operations
      (v) Reducing transport emissions
      (vi) Climate friendly freezers
      (vii) Innovating to reduce greenhouse gases
   (b) Water use.
      (i) Our water footprint
      (ii) Working with suppliers and farmers to manage water use
      (iii) Sustainable water use in our manufacturing operations
      (iv) Water-smart products for water-stressed living
   (c) Waste and packaging.
      (i) Our waste footprint
      (ii) Rethinking plastic packaging—towards a circular economy
      (iii) Going beyond zero waste to landfill
      (iv) Reducing food waste
   (d) Sustainable sourcing.
      (i) Our approach to sustainable sourcing
      (ii) Transforming the palm oil industry
      (iii) Sustainable tea—leading the industry
      (iv) Transforming global food systems

3 Enhancing livelihoods. Goal: By 2020 we will enhance the livelihoods of millions of people as we grow our business
   (a) Fairness in the workplace.
      (i) Advancing human rights in our own operations
      (ii) Advancing human rights with suppliers and business partners
      (iii) Understanding our human rights impacts
      (iv) Fair compensation
      (v) Improving employee health, nutrition, and well-being
      (vi) Building a safer business
   (b) Opportunities for women.
      (i) Challenging harmful gender norms
      (ii) Advancing diversity and inclusion
      (iii) Promoting safety for women
      (iv) Enhancing women's access to training and skills
      (v) Enhancing entrepreneurial and life skills through our brands
      (vi) Expanding opportunities in our retail value chain
   (c) Inclusive business.
      (i) Connecting with smallholder farmers to enhance livelihoods
      (ii) Empowering small-scale retailers for growth
      (iii) Creating and sharing wealth

Clearly, striving to reach these goals involves all aspects of Unilever's business operations, supply chain, and global operations. Very few companies have such a clear, far-reaching initiative that bridges the desire to promote products that they produce while advancing a sustainability agenda.

Lessons from Walmart and Unilever

The two case studies clearly demonstrate approaches to corporate sustainability. In the case of Walmart, they are setting internal goals for their operations in a number of areas. In the case study, we focused largely on energy and the efforts the company is making to strive to get 100% of its energy from renewable sources. There is no doubt that the company is making serious efforts in this area and that they are changing policies and developing initiatives to reach their goals.

At the same time, Walmart, through their sustainability index, is changing the production of consumer goods by setting standards by which they expect their vendors to operate. Since it is the largest public company in the world, the index is a transformative development in seeking to make the business world more sustainable.

Unilever has also set sustainability goals within the company that are far reaching. What is interesting about Unilever's efforts is that they are examining the impact of their products on broader society. They are trying to make edible products healthier and personal care products better for the environment. At the same time, they are striving to promote public health education to try to improve the lives of others around the world.

Many have criticized Walmart and other companies that have taken on the sustainability initiatives. They suggest that the initiatives only address the things that the companies can manage and do not tackle difficult issues associated with their business model. Walmart, in particular, has been criticized for its impact on communities and its labor practices.

Sometimes, companies report on issues in their sustainability plans that may not put the organization in the best light. Unilever, for example in discussing water and sustainability notes that their water use has increased 15% since 2010. While this public reporting of negative progress on sustainability may seem like a poor idea, the reality is that many find this honesty refreshing. Indeed, it provides a basis by which improvements can be made and by which goals can be set.

In thinking about the initiatives of Walmart, Unilever, and other companies, it is worth asking if they are addressing the overall sustainability of their business operations. Can more be done in certain areas? What else could be done? If it appears that companies are not fully honest in sustainability reporting, they fall into a situation by which they could be accused of using sustainability to greenwash their unsustainable activities.

Can businesses with unsustainable products be sustainable?

One of the challenges in our modern era is how we move our highly energy intensive technological society into a more sustainable condition. We are not going to stop the use of coal, oil, nuclear energy, or other products that many consider unsustainable any time soon. Is it better to consistently critique the organizations involved with the production of products that we believe to be unsustainable, or work with the organizations to help them to become more sustainable? This is a question that each of us must answer individually.

Third party verification

If we produce a product and put it on the market, it is up to the consumer to trust that what we say about the item. We might use terms like "natural" or "green" or "organic" to describe it. Each of these terms has a vague or distinct meaning. We are asking the consumer to trust what we say about the product.

Because many consumers have seen products that have been subject to greenwashing, they are dubious about some of the claims that manufacturers might make about the "greenness" of the product. If they are distrustful about the product they are unlikely to buy it.

However, if we use a neutral organization, a third party, that is impartial about the merits of the product to evaluate it, consumers can trust that the product meets the standards of the verifying agency Third parties serve as a bridge between the consumer and the producer, the standard two parties involved with transactions.

Third parties are typically paid by manufacturers to evaluate their products. Labels like Energy Star, LEED Certified, and ISO Certified have distinct meanings with real verification behind the label.

Think about your own purchasing experiences. Would you feel better about buying something that is labeled All Natural or something that is certified USDA Organic?

Third party verification agencies also help organizations do the right thing in sustainability. Leaders of organizations may wish to move forward on sustainability initiatives, but they may not know the right way to go about starting an initiative. Third party verification organizations provide distinct guidelines that can help businesses meet sustainability objectives.

Moving to Portland versus the hard work of saving the planet

Portland, Oregon, is often considered the greenest city in the United States. Over the years, I have had a number of my sustainability and environmental science students tell me that after graduation they were going to move to Portland to be with like-minded people. I've always tried to convince them to change their mind.

Why?

If you want to make a difference on the planet, would you move to Portland? While Portland is a wonderful city and anyone would be lucky to live there, I am not sure I would move there if I were trying to change the world. Portland already gets it. It has bike lanes, green energy, an abundance of organic food, and a community where environmental equity infuses local decision-making. The important work on sustainability is not in Portland, but in other areas of the world where there are serious issues as to the long-term viability of the region.

I think that there is a similar analogy with corporations and other businesses. As sustainability or environmental experts we may not necessarily consider jobs in the corporate or business world. However, if we don't find ways to "green" our corporate world, we will maintain the status quo and not much will change.

So, move to beautiful and green Portland if you wish to join other like-minded individuals interested in sustainability. However, if you want to help your community or an organization make big strides toward sustainability, you don't go where others have gone before.

The Dow Jones Sustainability Index

The Dow Jones Sustainability Index has been criticized by some in the sustainability community for including companies that are not seen as particularly green. For example, British Petroleum was listed in the index up until the Deep Water Horizon Oil Spill in the Gulf of Mexico, when it was removed for "extraordinary events"—a clause in the rules of the Index that allow the owners to remove a company.

The Index also has invited controversial companies like Walmart and Chevron to be listed. The criticism of adding companies like this arises out of the issues associated with their products and services-products and services that are used by the many people on the planet in one way or another. Those who promote sustainability ideals suggest that listing companies like this within a sustainability index sends the wrong message and rewards overall bad behavior that leads to over-consumption and pollution.

(Continued)

However, others argue that industry group comparisons help to transform global corporate culture and improve sustainability within all industries. They argue that if corporations are going to exploit energy resources, they should be compared with each other to ensure that their practices are as green as possible. While their products may not be green, their business practices can be examined to ensure that they are doing as much as they can for global sustainability.

The cognitive dissonance of having major companies like Toyota listed in a sustainability index is uncomfortable for many in the sustainability world. They find the listing distasteful and inappropriate. However, pragmatists argue that the listing encourages good behavior by corporate leaders and improves practices. They feel that the corporation would have been active no matter what. Why not make their practices as green as possible?

How do you feel about this issue? Do you find it appropriate to reward companies for their improvements in sustainability or do you think that some companies should never be included within a sustainability index due to the products that they produce? Would you as a sustainability expert be willing to work with companies within some of the less green industry groups in order to help them improve their sustainability practices?

Halekulani Hotel: A new green standard in 5-star dining

Halekulani Hotel, on Waikiki Beach in Honolulu, Hawaii, is a high-end luxury property with discriminating clients that expect quality experiences. Along with well-appointed ocean-view rooms, the hotel has a luxurious spa and a Forbes Travel Guide five-star and AAA 5 Diamond restaurant (Figure 14.5).

"We listen to our customers and try to react to their wishes," said Peter Shaindlin, Chief Operating Officer, Hotels & Resorts of Halekulani. This approach inadvertently led Halekulani to be a leader in the green food world.

Several years ago, customers started to inquire whether or not the food in their five-star restaurant was free of genetically modified organisms (GMO). The issue of GMO food is big in Hawaii, where most of the papaya—one of Hawaii's iconic tropic export fruits—is produced using GMO plants that are resistant to the damaging ringspot virus that decimated the papaya industry in the 1980s.

"We find that luxury clients are interested in fresh and local food," said Shaindlin. "When we started to get requests for non-GMO food, it was an easy decision for us to try to find options for our menu."

Halekulani Executive Chef, Vikram Garg found several sources of non-GMO food and placed several dishes on the menu. They labeled which options were GMO-free. "We didn't realize it at the time, but we were the first hotel restaurant in the world to label GMO-free menu items in order to offer guests various dining options," said Shaindlin.

Figure 14.5 Peter Shaindlin (right) and Vikram Garg of the Halekulani Hotel on Waikiki Beach in Honolulu, Hawaii. (Photo courtesy of the Halekulani Hotel.)

Shaindlin was contacted by some in the pro-GMO community who complained about setting apart GMO food from other food. "I explained to them that we were reacting to the wishes of our customers," said Shaindlin. The green consumer helped Halekulani hotel set a new green standard for luxury dining.

15

Sustainability at Universities, Colleges, and Schools

Examining sustainability at universities provides an opportunity to holistically examine how institutions can become drivers of long-term sustainability within their communities (Figure 15.1). These institutions are transformative in our culture. They capture the imagination and innovation of our youth and help to guide them into ways that help to create a better future for all of us. Schools are the places where new ideas can be tried and evaluated. They also influence new generations of citizens.

Universities are often the driving force behind social change in the world. Just take a look at some of the large social movements in the last 50 years and you will find that most of them have their roots, at least in part, in universities. It is not a surprise that universities have been the center of a considerable amount of attention in the advancement of the sustainability movement.

Plus universities are like small cities. They have residents, a governing system, and they have expansive infrastructure systems that must be managed. They have power plants, roads, buildings, and landscapes. Some have farms and produce food. They have complex societies with differing access points to power and decision-making. Because of these issues, they are excellent places to test sustainability programs and conduct experiments on improvements in technology and infrastructure.

There are several issues to discuss within the context of universities and schools: curriculum, external benchmarking, internal initiatives, and student and faculty activism.

Curriculum at colleges and universities

Many of you are reading this book because you are enrolled in a course that has a focus on sustainability. As you have found out from the previous chapter on the history of sustainability, the field is relatively new. The word didn't really come into wide usage until the late 1990s. However, concepts of sustainability within the realms of environment, environmental economics, and social justice, were taught for decades on campuses all over the world. Yet, it wasn't until the late 1990s and early 2000s that degree programs in sustainability emerged, along with specific courses on sustainability.

Given the diversity of the field, sustainability curricula vary considerably from university to university (Figure 15.2). Often, the programs are interdisciplinary in that they have a

Introduction to Sustainability, Second Edition. Robert Brinkmann.

Companion website: www.wiley.com/go/Brinkmann/IntroductiontoSustainability

Figure 15.1 This is a picture I took of one of my sustainability classes at Hofstra University. Hofstra has many sustainability initiatives throughout campus and offer undergraduate and graduate degrees in sustainability.

Figure 15.2 The way sustainability is taught varies considerably around the world.

core set of required courses along with a range of courses that students take from other departments. Programs often have thematic tracking.

Some programs focus heavily on scientific, social science, humanities, or business themes of sustainability. The science themes may build expertise on things like energy, ecological systems, or water technology, with strong support from existing science courses. These tracks will prepare students to work in technical areas of sustainability. Social science tracks may build expertise on issues like environmental policy or environmental justice. Students who take this type of curriculum will be prepared to work in government, planning, or policy. Students who earn degrees with specialization in the humanities will take courses in writing, literature, film, or art. This prepares students to work in the creative fields. Finally, students who take sustainability business or economics tracks often take courses in finance, economics, business administration, and entrepreneurialism. These students often work in a variety of business operations related to sustainability and many end up in entrepreneurial careers. There are also many sustainability degree programs that offer more general degrees, where students can pick and choose their courses and design a curriculum that makes sense for their interests.

While there are a growing number of colleges and universities that offer degree programs, there are also many more that offer minors. Universities that offer the sustainability minor give students the opportunity to gain a degree in a major of their choice, while also getting some level of expertise and experience in sustainability topics. Many students majoring in the sciences, social sciences, and business choose to minor in sustainability because it is such a helpful minor in these times of rapid environmental and social change.

Some universities also provide the opportunity to take sustainability courses as part of their overall university general requirements for graduation. Because sustainability is such an interdisciplinary field, it is often used as an elective that students can take to cover requirements in the sciences, social sciences, humanities, or interdisciplinary studies.

Sustainability curriculum at K-12 schools

There are also a number of public and private primary, secondary, and high schools that have gotten involved with sustainability in some way. In most instances the involvement includes a range of initiatives that focus on improving the facilities and management of the schools via energy improvements, construction of green buildings, and better purchasing and grounds management. However, many are also involved in improving curricular and extracurricular activities.

Some schools have adopted curriculum around particular themes of sustainability such as local food, transportation, or energy production. Some schools have built sustainability laboratories that allow students to explore green energy production, plant gardens to produce their own food, or that allow students to get involved in their community on important local sustainability projects.

Take for example Learning Gate Community School near Tampa, Florida (Figure 15.3). This elementary school seeks to be the premier school for teaching students the importance of sustainability in their everyday lives. The school is in suburban Florida and has a garden, promotes zero waste, composts, and provides a number of life skills that will encourage students to be sustainable throughout their lives. Parents are very involved with

Figure 15.3 Many K-12 schools put a strong emphasis on the environment, including Learning Gate Community School in Florida. (Photo courtesy of Learning Gate.)

the school and help to transform their homes into more sustainable places. In other words, the school is not just educating students about the importance of sustainability, they are also transforming the lives of students and their families, and thus their communities. According to their mission, they are seeking "To promote academic excellence, community service and environmental responsibility through family and community partnerships." (http://www.learninggate.org).

One of the challenges with sustainability education at the non-university level is that many schools are tied to distinct outcomes set by national, state, and local educational organizations. This means that curricular flexibility is almost non-existent in many school districts and teachers have to have a regimented approach to teaching content in order to ensure that students have a mastery of materials on which they will be tested. This "teaching to the test" approach to education makes it difficult to add sustainability or other interdisciplinary topics into the curriculum without changing the nature of the outcomes assessments.

Plus, many have advocated greater focus on science, technology, engineering, and math (STEM) education for children. This is a very good thing for those of us interested in the science of sustainability. However, sometimes this is interpreted as focusing on the compartmentalized versions of science: chemistry, physics, math, etc. without looking at the interdisciplinary issues or broader societal impacts. Sustainability, by its very definition, is interdisciplinary and thus can be lost when schools focus too much attention on STEM education without application.

As a result of this "teaching to the test" approach to education, some are working to infuse sustainability within the state or national educational standards to ensure that themes of sustainability make their way into the classroom. In addition, many parents are choosing to send their students to alternative schools or are homeschooling their children to avoid a lock-step approach to curriculum. Many schools also offer clubs and other informal learning opportunities for students to get involved in activities that relate in some way to sustainability.

There are also external organizations and clubs, such as the Girl Scouts and Boy Scouts, that have embraced sustainability. The Girl Scouts, for example, have been educating their members about issues of sustainable agriculture through their advocacy of sustainable palm oil within the cookies they sell in their annual fundraising campaigns. The Boy Scouts offer a Merit Badge in sustainability that is based on themes of water, food, energy, consumerism, and community. So, while many schools may not have a strong focus on sustainability, there are ways to get young people engaged with sustainability outside of the school environment.

External benchmarking

A number of organizations provide some degree of external benchmarking for universities and schools. These organizations range from broad-based operations that focus on all aspects of sustainability, to more thematic organizations that seek to make a difference on one particular issue.

American Association for Sustainability in Higher Education

Perhaps the most well-known organization that promotes university sustainability is the American Association for Sustainability in Higher Education (AASHE). This group serves all stakeholders of universities, including facilities operations, students, faculty, administrators, and staff. They provide a great deal of information on best practices that can be undertaken by any of these stakeholders and they also provide some benchmarking tools for assessing sustainability initiatives. AASHE holds a conference each year that brings together students, faculty, staff, and administrators to discuss current issues. Most universities in the United States are members of this organization. If a university is a member, all faculty, students, and staff have access to the resources available on the AASHE Website. Check if your university is a member, and login and check out the materials.

The resources available on their website offer opportunities to learn about all aspects of university sustainability. They have information on curriculum and course content, information for those involved with facilities and all operations associated with the day-to-day life of a university, and information for administrators. The site is a good place to start if one is contemplating starting a campus sustainability initiative or working on a research paper or project on campus sustainability.

AASHE also has a university benchmarking system called STARS, which stands for Sustainability Tracking, Assessment, and Rating System. Like many rating systems, STARS gives credits within several categories to earn points. Based on the points earned, schools are granted Bronze, Silver, Gold, or Platinum ratings. Points are earned in four categories: academic, engagement, operations, and planning and administration. Points can be earned in the academic category within the context of sustainability curriculum and research. Points in the engagement category are earned by providing opportunities for campus and public engagement. Some ways that these points can be earned is by having sustainability information available during student orientation or by having university community partnerships or established community engagement operations. Points can be earned in the university operations category in several classifications, including air and climate, buildings, dining services, energy, grounds, purchasing, transportation, waste, and water. Finally, points can be earned in planning and administration in the themes of: (1) coordination, planning and governance, (2) diversity and affordability, (3) health, well-being and work, and (4) investment. Universities can also earn points within an innovation category for initiatives that do not fit the other categories.

As of this writing, there were 986 organizations that use the STARS rating tool throughout the world. Most of these institutions are in the United States. Take a look at the list of schools that are using STARS here: https://stars.aashe.org/institutions/participants-and-reports/. Is your school on this list? If not, is your school using another benchmarking tool to evaluate campus sustainability? While AASHE is a distinctly American approach to assessing sustainability, what approaches are used in other countries or regions to assess university sustainability and compare with others?

Presidents' Climate Leadership Commitments

Another organization that is very well known is the Presidents' Climate Leadership Commitments which is part of the group, Second Nature. The Commitments evolved from the American College and University Presidents' Climate Commitment (ACUPCC) (Figure 15.4). This organization consisted of college and university presidents who committed their universities to significantly reduce carbon emissions. As signatories, the universities conducted regular greenhouse gas inventories and made those inventories available to the public. They also produced climate action plans that detailed how they were going to reduce greenhouse gases on their campus. Over 700 college and university presidents have committed their universities to the ACUPCC plan.

The ACUPCC started about 15 years ago. In 2015, the organization rebranded into the Presidents' Climate Leadership Commitments that focuses on three main areas: carbon reduction, resilience, and climate change. The organization has helped to encourage sustainability especially climate neutrality, on campuses around the United States. However,

Figure 15.4 This is your author at the celebration announcing that the University of South Florida was a signatory on the American College and University Presidents Climate Commitment.

some campuses have left the system or are delayed on their reporting for some reason or another. Based on the requirements as outlined in the text box, there needs to be a substantial commitment from the college or university in order to be successful at reaching targets. The university must have internal resources to complete a greenhouse gas inventory and a climate action plan or they will need to work with a consulting firm in order to complete these requirements. In addition, the university must provide a plan for becoming climate neutral. This is a very difficult thing for an organization as complex as a university or college to achieve. Considering that renewable energy is only a small proportion of the energy budget of most universities, these organizations must think creatively about trying to become neutral.

Plus, some of the other initiatives that an organization is encouraged to develop are very expensive or difficult. For example, offsetting the carbon emissions for all air travel might be a difficult proposition for a research institution with faculty members and graduate students that travel frequently to attend conferences. Nevertheless, many of the goals are achievable and hundreds of organizations across the United States are actively involved in moving their campuses to climate neutrality.

The American College and University Presidents' Climate Commitment

Signatories to the American College and University Presidents' Climate Commitment did the following:

1 Initiate the development of a comprehensive plan to achieve climate neutrality as soon as possible.
   (a) Within two months of signing this document, create institutional structures to guide the development and implementation of the plan.
   (b) Within one year of signing this document, complete a comprehensive inventory of all greenhouse gas emissions (including emissions from electricity, heating, commuting, and air travel) and update the inventory every other year thereafter.
   (c) Within two years of signing this document, develop an institutional action plan for becoming climate neutral, which will include:
      (i) A target date for achieving climate neutrality as soon as possible.
      (ii) Interim targets for goals and actions that will lead to climate neutrality.
      (iii) Actions to make climate neutrality and sustainability a part of the curriculum and other educational experience for all students.
      (iv) Actions to expand research or other efforts necessary to achieve climate neutrality
      (v) Mechanisms for tracking progress on goals and actions.
2 Initiate two or more of the following tangible actions to reduce greenhouse gases while the more comprehensive plan is being developed.
   (a) Establish a policy that all new campus construction will be built to at least the US Green Building Council's LEED Silver standard or equivalent.
   (b) Adopt an energy efficient appliance purchasing policy requiring purchase of ENERGY STAR certified products in all areas for which such ratings exist.

(Continued)

(c) Establish a policy of offsetting all greenhouse gas emissions generated by air travel paid for by our institution.
(d) Encourage use of and provide access to public transportation for all faculty, staff, students, and visitors at our institution.
(e) Within one year of singing this document, begin purchasing or producing at least 15% of our institution's electricity consumption from renewable sources.
(f) Establish a policy or a committee that supports climate and sustainability shareholder proposals at companies where our institution's endowment is invested.
(g) Participate in the Waste Minimization component of the national RecycleMania competition, and adopt 3 or more associated measures to reduce waste.

3 Make the action plan, inventory, and periodic progress reports publicly available by submitting them to the ACUPCC Reporting System for posting and dissemination.

Other external benchmarking organizations

Other external benchmarking programs encourage student involvement within thematic areas. One of the most active of these programs is RecycleMania, which is managed by a non-profit group that is funded by the United States EPA, the College and University Recycling Coalition, and a number of corporate sponsors like The Coca Cola Company, SCA Tissue, Alcoa Foundation, and the America Forest and Paper Association.

RecycleMania encourages students to focus on recycling and waste within an eight-week period in spring. Recycling rates at the per capita basis are compared. Universities are given awards for the best recycling rates and the least amount of trash created. The program allows universities to compare their progress on a weekly basis with other universities in order to try to build friendly competition. However, the program helps to educate students, faculty, staff, and administrators on campus waste and recycling issues. The program seeks to encourage the development of sound recycling and waste management policies. Over 300 colleges and universities participated in the 2019 competition. The winners that year were Loyola Marymount and Knox College. Is your university a participant in RecycleMania? If so, how did your school rank in the competition? If your school does not participate, what kinds of efforts are underway on your campus to measure and benchmark recycling rates with other schools?

Campus Conservation Nationals (CCN) is another competition-based benchmarking system like RecyleMania. However, CCN is focused on education on energy and water conservation. Specifically, the goals of the organization are:

- Engage, educate, motivate and empower students to conserve resources in residence halls and other campus buildings.
- Foster a culture of conservation within campus communities, and propel campus sustainability initiatives.
- Enable students to teach each other conservation behaviors that they can employ on campus and in their future homes and workplaces.
- Enable students to develop leadership, community organizing and career development skills.

- Achieve measurable reduction in electricity and water use, preventing thousands of pounds of carbon dioxide from being emitted.
- Highlight the ability of behavior change tools such as competitions, commitments, and social norms, to conserve energy and water.

The CCN was developed in partnership with several organizations: The US Green Building Council, Lucid (a company that develops software for measuring energy and water use in buildings in order to provide real-time feedback on use and conservation initiatives), the National Wildlife Federation, and the Alliance to Save Energy.

CCN works by measuring energy reductions over a three-week period. The competition is done on individual campuses between buildings and between individual campuses. What this means is that individual residence halls can compete against each other within a single campus in areas of water or energy reduction. At the same time, campuses within particular regions can compete against each other.

The challenge is recorded within a dashboard that shows how individual buildings are doing. Students can enter data and track progress. They can also share strategies as to how they are achieving their goals. Savings are shown graphically in ways that demonstrate the energy and water savings that are taking place as a result of the students' efforts.

One of the particular challenges for the CCN is that many buildings on campuses do not have individual monitoring of energy and water. On my own campus, most buildings are clustered within single meters that measure energy or water use. At the present time, only a few residence halls can participate in CCN because of this metering issue. However, there is a movement to add meters to buildings on campuses throughout the world in order to better assess energy and water conservation strategies.

Tree Campus USA is a program established by the Arbor Day Foundation to promote protection of trees on university campuses and to advance education about the importance of trees (Figure 15.5). This program is similar to others established by the Foundation, including the Tree City USA program. According to the Arbor Day Foundation's website, trees on campuses reduce the amount of energy a campus needs to generate by providing shade, they reduce carbon dioxide in the atmosphere, and they provide a green space that promotes relaxation.

In order to be an official Tree Campus USA campus, it must have a campus tree advisory committee, develop a campus tree care plan, have a campus tree program with a budget, celebrate Arbor Day, and include trees within a service learning project. Over 250 campuses have earned Tree Campus USA designation. Is your campus a Tree Campus USA campus? What would it take for you to earn this designation? What do you do to celebrate Arbor Day on Campus? Does your campus provide any education about trees or the environment using the campus grounds?

Internal initiatives

In the last decade or two, many universities have developed a number of key internal initiatives that focus university resources and attention on sustainability. Campuses, due to their unique geographies and histories, vary significantly. A one size fits all approach toward campus benchmarking does not work for some of these institutions. An urban campus,

Figure 15.5 How does your campus grounds reflect sustainability?

like New York University, will have very different issues from a more rural campus like Michigan State University. Likewise, a small private university like Hofstra University, will have different sustainability issues than a large public university like the University of South Florida. Similarly, universities vary greatly from country to country and region to region. That is why it is important for each university to develop sustainability initiatives that work for their own unique situation.

Sustainability officers

Many universities have a sustainability officer. This person may be a faculty or staff member. If they are a staff member, they are likely housed within a facilities office or have strong interactions with university facilities. A university facilities office is where the action is on technical aspects of sustainability within the university infrastructure. If you want to make a difference on energy consumption, lighting, transit, landscaping, water, or other infrastructure issues, you need to get to know this office. If the sustainability officer for the university is a faculty member, they are probably more focused on curricular and educational issues. Obviously, sustainability crosses both facilities and education at universities and thus, regardless of where a sustainability officer might be housed, they have to be able to work with both facilities and educational stakeholders in advancing the sustainability goals of the institution.

Sustainability committees

Many universities also have sustainability committees that consist of faculty, staff, students, and administrators. These committees often drive the campus sustainability agenda and take ideas from campus stakeholders and try to bring them to fruition. They encourage the development of key programs such as a student green fees, new educational programs, and purchasing guidelines.

It must be noted that many university facilities departments are rather sophisticated on sustainability, particularly in the area of energy, water, landscaping, and food. For example, Ball State University built the largest geothermal power plant in the United States in recent years. They were able to take a coal power plant offline thereby significantly reducing their greenhouse gas emissions. Other universities have significantly reduced their unnatural landscaping and have planted native vegetation to replace the water- and fertilizer-dependent landscaping. Others have focused on developing green building standards. Regardless of approach, most universities are doing something to try to improve sustainability and make a difference to the lives of students, faculty, and staff on campus. They are serving as models for how to advance sustainability within their communities. When facilities departments work with sustainability committees and sustainability officers, they have the potential to make tremendous change within their communities.

As noted earlier, universities are also advancing a number of sustainability educational initiatives. A large number of them have developed sustainability degree programs at the undergraduate and graduate levels. They have also developed required courses or courses suitable for general education credit. Sustainability degree programs are advancing quite rapidly around the world in reaction to the need to better understand how we can maintain reasonable standards of living during a time of decreasing resources and changing technologies, while also advancing fairness for all.

Food service

One of the more personal issues we confront as individuals is our individual food choices (Figure 15.6). In our own homes, we can prepare whatever food we like. However, when we live and work on campus, we are at the mercy of the food service vendors to provide us with our dietary options. In recent years, there has been a growing interest in green and healthy food options on campus. Many students are choosing vegetarian or vegan diets and many also have special dietary needs such as gluten- or lactose-free diets. Still others have particular dietary needs based on their religion.

Figure 15.6 What we offer to students to eat on campus matters. How does your campus dining services reflect your university's commitment to sustainability?

Long gone are the days of bland campus food options. Campus dining offices are providing a variety of food options for students. Most strive to meet the dietary needs of students. They are usually happy to meet with students to discuss particular diets or to hear suggestions of how they could improve their options. Many dining services now have sustainability initiatives that include buying from local farmers, using sustainably certified seafood, and providing vegan and organic options. They are also striving to reduce packaging and waste. Some campuses have moved heavily into composting food waste (see text box on composting at St. Johns College).

Think about the food service on your own campus. Do you know of any sustainability initiatives underway in your dining halls? Are there vegetarian or vegan options? Does the office try to buy local products? Do they compost waste?

Campus composting at St. Johns University

St. Johns University in Queens, New York moved into composting in a big way (Figure 15.7). They now compost almost all of the kitchen waste produced on the campus.

Food waste from the kitchen is collected daily during the regular academic year by student employees who work under the direction of the university's sustainability officer. Tons of kitchen scraps and coffee grounds are gathered in large plastic barrels

Figure 15.7 A waste container at Hofstra University.

for transport to the composting bins. They mainly collect from the kitchen and campus coffee services since these are places where they can easily separate the compostable materials.

Once collected, they are transported to a mixing station near composting bins. The waste is mixed with wood chips from downed trees on campus and already created compost. The addition of processed compost helps to drive the decomposition process. Mixing is accomplished by using a mechanical loader. Once mixed, the compost is added to large aerated bins. It takes about a week to gather enough compostable material to fill a bin.

The bins are designed with a floor venting system. Fans on timers vent the bins for a few minutes every hour. The venting helps to create pore space within the compost that leads to faster decomposition. Moisture is added as needed and the temperature is checked regularly to ensure that appropriate decomposition is taking place.

Although St. Johns' academic year coincides with the coldest time of the year in New York City, compost managers are able to maintain the composting process throughout the year due to the great amounts of heat that are generated through the decomposition of the compost.

After a month of monitoring, the compost is finished with the decomposition process and can be added to the grounds. St. Johns maintains a vegetable garden where the compost is used as a fertilizer. Students manage the garden and the food is given to a local food kitchen for distribution to the poor. They also add the compost as mulch to landscaped areas of the campus grounds. They filter the compost to remove coarse

(Continued)

fragments for use as a lawn fertilizer. Some of the compost is used to make a nutrient "compost tea" that can be used as a liquid fertilizer for landscaping or potted plants.

This project has saved tons of waste from entering the landfill stream by keeping local waste on campus. It also reduces St. Johns' needs to purchase fertilizers and mulch for gardens and grounds. It employs several students and gives them opportunities to learn more about sustainability through their job. They also try to educate others on composting by giving tours and demonstrations on campus composting. How does St. Johns' composting initiative address issues of environment, economy, and equity?

Student and faculty activism

Students and faculty are helping to drive the sustainability agenda at universities through their activism. There are a several environmental or sustainability clubs at most universities that provide a platform for student activism and involvement. In some cases the activism is on campus, focused on improving practices that occur on campus. In other cases, students are working on local, state, or national issues such as global climate change or social justice.

One type of student activism that has gotten considerable attention in recent years is organized in part by 350.org, which was founded by writer and activist, Bill McKibben who has spent most of his life trying to educate people about the dangers of climate change and the likely impacts of our heavy use of carbon-based energy and other greenhouse sources.

350.org trains students on how to encourage carbon and greenhouse gas reductions on their campus. Currently a big focus of this effort is on divestment of university endowments from fossil fuel companies. The historical record demonstrates that some energy companies sought to deceive the public on the impacts of their products on the environment. As a result, 350.org is trying to get universities to divest their endowment portfolios of fossil fuel companies.

To date, this effort has met with limited success. Most university endowments are managed by investment firms that mix investments among a range of companies and it is difficult to pull funds from one particular business. Nevertheless, some universities have moved toward divestment and there is considerable pressure from student groups across the world to divest of fossil fuel firms. It is likely that more universities will move to strip their funds of these companies in the future.

While 350.org is a great example of national-level activism on climate change, there are many other organizations that engage with campuses on a regular basis. Sierra Club, the Nature Conservancy, People for the Ethical Treatment of Animals (PETA), and others regularly recruit student members and offer campus programs and talks. Plus, there are new emerging groups like Extinction Rebellion and Fridays for Future who have taken on new urgency as climate change is starting to change our environment. Many are excited by the individual actions of activists like Greta Thunberg who have made personal commitments to advance a sustainability agenda.

What organizations interact with your campus? Is there a student chapter of any of the major international or national organizations on your campus? If not, is there a local chapter of these organizations in your community?

Universities are also home to many homegrown student organizations that focus in some way on sustainability. Student clubs are managed by an office of students that ensures that the clubs are run according to university rules. Most campuses have several clubs that relate to sustainability. They may include those that are strictly focused on the natural world, such as a tree club or a wildlife club. Others may be more focused on environmental problems. Still others may focus on social justice or equity issues. The clubs sometimes come together to focus on big issues, such as climate change divestment or a student green fee.

The clubs provide an opportunity for students to share ideas and learn about specific things outside of the classroom. Because most student clubs have a service component, they give students a chance to interact with other students, faculty, and staff in meaningful ways, often off campus. Getting involved with student organizations gives students the opportunity to develop leadership skills, while also making a meaningful difference in their communities.

One of the most important innovations that emerged from student clubs in recent years is a student green fee. This is a fee that is tacked on to other student fees, such as an athletic fee, that students pay as part of their overall tuition. Green fees are used to fund sustainability initiatives on campuses all over the world. The initiatives may include the development of campus renewable energy, water conservation projects, green building initiatives, or educational programs.

Take a look at your own campus. What student clubs related to sustainability are present on campus? Take a look at their website. What is their purpose or mission? What kinds of activities are they involved with on a regular basis? What special events have they scheduled during the semester? How often and where do they meet? Which of the clubs interests you the most? Would you consider being an officer in a club? Why or why not? Does your campus have a green fee? Have students ever tried to get a green fee built into the costs of attending your school?

University faculty and staff are also very involved with sustainability issues. Many universities have sustainability committees that are populated by faculty staff, and students. In many cases, the committees have been very effective at implementing change on their campuses. They do this by building consensus and having collegial relationships among various campus stakeholders in order to change culture and policy. They often work in partnership with facilities and other administrative offices that are responsible for implementing change. Does your university have a faculty committee working on sustainability issues? What kinds of topics have they worked on over the last year or two?

Universities are the kinds of places where one person can make a difference. Individuals are behind some of the most important changes that take place on campus. It takes leaders to make positive change. Student, faculty, staff, and administrative leaders interested in sustainability issues are present on all campuses. Think about the leaders on your campus? Who is leading efforts on sustainability on your campus? What kinds of traits do they have in common? What are some things on your campus that you think that could be improved from a sustainability perspective? How can you get involved on your campus to become a leader to try to solve the problem?

Sustainability at my university: Hofstra University in Hempstead, New York

Several years ago, a student group called Students for a Greener Hofstra lobbied administrators at Hofstra University to do two main things: (1) start a degree program in sustainability and (2) create an office of sustainability on campus to oversee broad sustainability initiatives on campus. Since then, great things have happened and we are considered one of the greenest universities in the country.

Degree programs. Hofstra is now home to a degree program that allows students to earn a BA, BS, minor, and masters degree in sustainability. Given its location on Long Island a few miles from New York City, it tends to focus on urban and suburban sustainability issues like energy, climate change, food, urban and regional planning, equity, and economic development. However, the degree program is very flexible and allows students to take electives in a number of departments across campus including geology, biology, chemistry, sociology, anthropology, business, art, and engineering. The university tries to give students hands-on experiences by building community-based education within our curriculum as much as possible. Community engagement is supported by the Center for Civic Engagement which has developed formal partnerships with a number of community groups near campus. The university also brings major speakers like noted agriculturalist Will Allen or sustainability policy expert Van Jones on campus to interact with students. Each year it celebrates Earth Day, has sustainability programs around a campus Day of Dialogue, sponsors public debates on important sustainability topics, and hosts community groups, like the Long Island Food Coalition, on the campus.

Student groups. There are several main student groups working on sustainability issues on our campus. Each works on distinct issues, while also providing opportunities for students to interact and learn from each other.

Faculty and staff activism. A committee called the Environmental Priorities Committee is a regular committee of the Hofstra Faculty Senate. They work on a number of issues related to sustainability. In recent years they have worked on divestment of fossil fuels, procurement policies, and the promotion of electric cars. They work closely with staff and administrators to effect policy changes on campus.

Campus sustainability officer and sustainability initiatives. For many years, the campus has had a campus sustainability officer. She worked closely with faculty, staff, and students on a number of important projects. For example, the campus is reducing its energy use via transforming the campus light bulbs to high-efficiency LED lighting. She has also expanded the number of e-car charging stations, provided water bottle filling stations, and added opportunities for recycling. She also organizes participation in events such as Campus Conservation Nationals. Since the campus is an arboretum full of beautiful plants and trees, she works closely with our campus arboretum director on issues that pertain to the campus landscaping. The campus does use integrated pest management protocols to limit applications of pesticides and herbicides.

Like all universities, Hofstra University has a way to go to become a fully sustainable campus. However, important strides have been made in many areas. There is strong support from the university leadership to work on sustainability issues. Students and faculty are actively engaged with the issue and are working on sustainability issues on campus and in the community. Some key successes:

- Recycling of paper, plastic, and glass is available throughout campus
- The campus purchases green energy credits to account for 30% of our energy
- We produce a significant amount of energy via cogeneration
- Our campus food service has a policy to buy local first and they also only buy seafood approved by the Marine Stewardship Council. They also provide vegan and vegetarian options in the dining hall
- We are members of AASHE
- The library has books and journals dedicated to sustainability
- The sustainability officer is working on a greenhouse gas inventory
- We have a research center, The National Center for Suburban Studies that focuses its mission, in part, on sustainability
- While the university's endowment has not divested from fossil fuels, Hofstra is one of the few campuses in the United States where students met with the Board of Trustees to discuss the importance of divestment
- There are dozens of water bottle refill stations on campus where students can refill water bottles
- There are many faculty engaged with research on sustainability issues on campus

Students, faculty, and staff at Hofstra University work collaboratively to try to achieve great things on our campus. While there is always more to do, the university community is proud of the accomplishments so far. Each campus is unique and has its own sustainability path that it is following. Each place has its own distinct history, geography, student body, faculty, and administrative structure that influences how sustainability gets done. The examples of some of the initiatives at other universities around the country provide an example of interesting initiatives that are transforming campus communities. What story can you tell about your campus?

Building your own case study

In this book, I have used a number of case studies to demonstrate the effectiveness of a number of sustainability efforts. However, when looking at universities, it is appropriate for you to build your own case study. Answer the following questions:

1 Is your university a member of American Association for Sustainability in Higher Education (AASHE)? You can find this out by searching the organization's website at www.aashe.org. If so, create an account on their website. Any student who is at a university that belongs to AASHE can create a free account that gives you access to tons of information. What things can you find on the AASHE website that is of interest to you?

2 Does your university have a commitment to sustainability in its mission statement?
3 Does your library have books and research journals in sustainability?
4 Has your university's endowment divested of fossil fuels? Or have they been asked to divest?
5 Has your university completed a greenhouse gas inventory? If so, what are your annual carbon emissions?
6 Does your university have a sustainability officer? In what university department are they found? What types of things is s/he working on at the present time?
7 What types of sustainability initiatives are taking place on your campus? See if you can find something out about sustainable approaches to energy, food, water, landscaping, building, or social equity. Do you have a student garden? Does your food service offer vegan or vegetarian options? Do they purchase organic food or strive to purchase from local vendors? Does your campus have a landscaping policy that promotes sustainability? Do you have a green building policy? Does your campus produce renewable energy or purchase green energy credits?
8 Does your university offer a sustainability degree program? What are the requirements for the degree and what types of things can one do with the degree upon graduation? What sustainability-related classes are available for you to take on your campus?
9 Does your university offer opportunities for students to get involved with community-based learning via formal courses? Is there an office of community engagement on your campus? If there is a program on community engagement, what kinds of sustainability issues are highlighted?
10 What student clubs related to sustainability can you join? What do the clubs do to focus on sustainability?
11 What external organizations, like the Sierra Club or People for the Ethical Treatment of Animals (PETA), interact with your campus?
12 What faculty organizations are involved with sustainability on your campus? Are they part of the formal governance of your university?
13 Who are the student, faculty, staff, and administrative leaders working on sustainability?
14 What special events are taking place this year on your campus related to sustainability? Has your school hosted any major sustainability speakers recently? Do you celebrate Earth Day? Arbor Day? Any other special days related to sustainability?
15 Does your campus participate in RecyleMania or the Campus Conservation Nationals?
16 Does your campus have a green fee? If so, what kinds of initiatives does it fund and how is the fund managed?
17 How can you make a positive difference on your campus or in your community?

When you answer these questions, you'll have some idea as to what your university is doing on sustainability related topics. How do you think your university is doing overall? How do you think you compare with other universities in your region?

Sustainability at Oxford: a campus commitment

Oxford University in the United Kingdom has integrated sustainability into many aspects of the university (Figure 15.8). Besides having outstanding academic programs, the university infuses sustainability within their overall facilities management system (called estate

Figure 15.8 Venerable Oxford University has a myriad of sustainability initiatives.

management at Oxford). On many campuses around the world, small offices of sustainability are present.

The university has done extensive work on sustainability in recent years within the following areas:

1 Carbon and energy
2 Sustainable food
3 Reduce, reuse, recycle
4 Sustainable purchasing
5 Sustainable building
6 Environmental management
7 Water
8 Biodiversity

Each of these areas has distinct planning associated with it. For example, under the water category, the university is undergoing major efforts to conserve water. Plus, individual departments can be granted a Green Impact award for making major contributions to sustainability at Oxford in a variety of areas, including water conservation.

What type of strategic planning is taking place at your university related to sustainability? What types of goals have they implemented? Who is responsible for making sure that the goals are achieved?

Making school lunches healthier in the United States

While it may not seem like it, healthy school lunch programs definitely are part of the sustainability movement. What we eat impacts the environment. Since school lunch programs are heavily subsidized by state and federal governments, public policy drives the kinds of lunches that students eat.

In the US, school lunch programs have been around since the 1940s. They provide a source of nutrition for many young people around the country who go to school without lunches made at home. While the intent of the program is altruistic, many have criticized it for relying too heavily on lunch products that are not particularly healthy. Schools often get excess food that is purchased by the government in some way. Some have criticized the menus as too reliant on meats and fats.

Unfortunately, approximately 17% of young people in the United States are obese. School lunch programs are certainly responsible, in part, for the high rates of obesity. In addition, diabetes rates are increasing quickly among young people.

In the last several years, schools have been working hard to improve the nutrition of their lunch programs to try to address the obesity and diabetes problems. They are getting rid of vending machines, cutting sugars and fats, and adding more fresh fruits and vegetables. Unfortunately, unhealthy fatty foods are inexpensive and these changes can impact the budget of school lunch programs.

The former First Lady of the United States, Michelle Obama, made it one of her missions during her time in the White House to educate young people and families about the importance of nutrition. She installed a vegetable garden on the White House grounds and she speaks regularly in schools about nutrition and health. She has worked hard to improve the nutritional standards of lunches.

As a result of the work of Mrs. Obama and other food and garden advocates, schoolyard gardens are sprouting up all over the world. They provide an opportunity for students to:

- Connect with nature
- Learn about food and nutrition
- Learn about the science of gardening and agriculture
- Exercise out of doors.

The gardens also bring students and their parents together in the schoolyard settings. Some of the gardens allow students to take produce home to their families and many schools encourage family members to volunteer at the gardens. In some cases, the gardens become a new way for the school to build community on campus and with families of their students.

In Great Britain, the Growing Schools Garden organization, for example, focuses on encouraging outdoor learning for students around gardens. Their vision statement notes: "Every young person (0–19) should experience the world beyond the classroom as an essential part of learning and personal development, whatever their age, ability or circumstances." (http://www.thegrowingschoolsgarden.org.uk/). These types of programs get students thinking beyond the traditional classroom in ways that connect them with the broader issues of sustainability.

Does your school have a garden? Can you eat the food from it? Does your campus food service get any of its produce from campus gardens or farms? How healthy is your school lunch? How healthy is your diet?

The cow powered carbon neutral campus

Green Mountain College, a small liberal arts college in Vermont in the United States with about 750 students, purchases the majority of their energy from a plant that converts cow dung to methane. They became the first cow methane powered college campus in the world when they implemented this program in 2006. Since then, they have become the second climate neutral campus in the United States.

The development of this energy source is an important way that Vermont is dealing with excess cow manure from its extensive dairy herds. Vermont is famous for its high-quality cheese, butter, and ice cream. The state is taking waste products and turning it into usable energy. The campus gets approximately 1.2 million kWh annually from methane power. The benefit of using methane is that it is a much more potent greenhouse gas than carbon dioxide (20 times more potent to be exact). Thus, using the methane helps to avoid burning greenhouse-gas-producing fuel, but also burns a potent greenhouse gas to limit its impact on the environment.

To achieve climate neutrality, Green Mountain College has also focused heavily on energy conservation. One of the best ways to cut greenhouse gases on any college campus is to find ways to reduce energy consumption. This can be done by changing lighting, improving building insulation and windows, and limiting energy use during down times of the year. At the same time, the school has also looked to local renewable energy resources and is now using efficient woodchip burners to take advantage of the extensive forest resources in the state. The woodchip furnaces provide the majority of the heat in the winter on campus and about 20% of the electricity.

The efforts of Green Mountain College demonstrate that schools can achieve high sustainability standards if they set goals.

What is your campus doing to become climate neutral? Does your school purchase green credits from renewable energy generating sources? Does your campus generate renewable energy?

Whitman College builds wind turbines on campus farm

Whitman College is a small liberal arts college in Walla Walla, Washington in the United States that is home to approximately 1600 students. It has a strong environmental focus. As a result, students, faculty, staff, and administrators are working to promote alternative energy on campus. In 2013, 70 wind turbines were built on land owned by the college. Together with other property owners, there are 450 wind turbines that are part of a greater regional wind farm called the Stateline Wind Project. Whitman College is certainly the largest campus producer of wind energy in the United States.

Whitman is also focusing heavily on purchasing credits for green energy. What this means is that they may use dirty energy, but spend more to buy credits in support of the development of green energy sources. Their purchase of green energy accounts for 50% of the energy consumed on their campus. Whitman also generates solar energy and is looking to expand their solar energy use.

Whitman's unique situation allowed them to develop wind energy on property owned by the university, but not on the main campus. Many campuses have properties such as farms, retreat houses, or research facilities that are not part of the main campus. Does your university have such properties? How are these properties used? Can you visit them? How are they integrated into the sustainability mission of the campus? Are there any sustainability initiatives unique to those properties?

Stanford University: dumping the car for bikes

One of the most vexing issues on college campuses is parking. Faculty, staff, and students are all competing for precious parking spaces, particularly at popular class times. Some campuses charge the equivalent of hundreds of dollars to park. Yet, one school, Stanford University, is turning the page on car culture and creating a new campus culture around the bike.

The campus has institutionalized the bike in a big way. Here are some initiatives:

- The campus has a bike shop that has bike repair clinics and classes on how to ride bikes and how to register them on campus.
- There are several bike safety repair stands scattered across campus where bikers can make small repairs to their bikes.
- To promote safety, the campus provides helmets at a reduced cost to those who participate in a safety class. They also provide coupons for purchasing folding bikes and storage lockers to provide a secure environment for storage.
- They provide several locations where bicycle commuters can store clothes and take showers.

These initiatives, among others, have made Stanford University one of the most noted bicycle-friendly campuses in the United States. Right now, the campus is home to over 15 000 bicycles.

What is the bicycle culture like on your campus? Are there places where commuters can easily change clothes and shower near where they take classes or work? Are there abundant bicycle racks or storage areas? Are there safety courses or bike rental facilities on campus? Do you have a bike shop or bicycle club? How friendly are the surrounding streets off campus to bike users? Do you have a bike share program on campus or in your community?

Green fleets: The University of South Florida's biodiesel Bullrunner

Many campuses have purchasing policies for university owned vehicles that require them to purchase hybrids or electric vehicles. At the University of South Florida (USF), campus officials have gotten into biodiesel in a big way.

The USF campus is huge. It has dozens of building scattered across its approximately square mile sized campus. Parking is only allowed on the fringes of the campus and many students live in apartment buildings just off campus. As a result of its unique geography, the university instituted a campus shuttle service, called the Bullrunner, which travels around campus and the surrounding neighborhoods to bring students, faculty, and staff from nearby campus neighborhoods and edge parking lots to the heart of campus. The shuttle also provides opportunities for off-campus shopping and easily connects with the local regional transit options that allow riders to gain access to a number of areas within the Tampa Bay region. Because of the size of the campus, which is home to nearly 50 000 students, there are six distinct shuttle routes that run almost continuously.

Each year, tens of thousands of miles are covered by the shuttle buses. To limit the impact of USF's internal mass transit operations, the campus purchased 30 buses that burn biodiesel fuels. They have used a variety of biodiesel products, but mainly use soy-based renewable fuels. The shuttle service, while limiting the impact of cars on campus and providing mass transit options for employees, also promotes the development of renewable energy resources.

Does your campus have any fleet vehicle purchasing policy? Are campus owned vehicles green vehicles (hybrid or electric)? Of course the greenest vehicles are the ones you do not buy. Has your campus cut its campus fleet? What kinds of mass transit options are offered by your campus? Does your campus or local mass transit organization use renewable energy to power their vehicles? Does your campus mass transit connect with regional transit options? How often do you and your friends use campus or community mass transit? Do you have a car on campus? Why did you bring it? Do you really need to have a car on campus while you are going to school? How could you go shopping or run errands off campus without a car?

Community engagement at Portland State University

Portland State University (PSU) is big on community engagement. The university is one of the first to build the idea of engagement into the broader mission of the organization. Community engagement at universities is a mutually beneficial arrangement between a community and the university. Usually, the community of interest is the home campus community, as is the case with PSU. Often, community organizations work in partnership with faculty and students, sometimes in class settings, to address concerns of community groups, citizens, or governments.

PSU, in their university mission, states their vision of community engagement clearly (http://www.pdx.edu/portland-state-university-mission):

> PSU values its identity as an engaged university that promotes a reciprocal relationship between the community and the University in which knowledge serves the city and the city contributes to the knowledge of the University. We value our partnerships with other institutions, professional groups, the business community, and community organization, and the talents and expertise these partnerships bring to the University. We embrace our role as a responsible citizen of the city, the state, the region, and the global community and foster actions, programs, and scholarship that will lead to a sustainable future.

Clearly, within the main university mission statement, sustainability and community engagement are linked. Faculty at PSU are also evaluated, in part, by their community-based research and teaching. It doesn't matter what department you are in at PSU, community engagement is front and center in the mission of the university and thus the kinds of experiences one will have on campus.

Researchers engaged with sustainability at PSU have defined their main research objectives and they link closely to community engagement:

- Urban sustainability: building smart cities
- Ecosystem services: understanding nature's benefits
- Social determinants of health: connecting wellness, place, and equity.

The success of PSU's approach to community engagement and sustainability has taken root at universities across the world. Many universities now offer opportunities for students to get involved with community-based projects on campus and even earn credits for community initiatives. The examples of community engagement courses provided by PSU faculty inspired many university faculty members to create their own community-based courses. These classes provide learning opportunities for students while also helping community organizations achieve their goals.

Many universities now have offices of community engagement that link community needs with faculty research and teaching interests. Is community engagement part of your university's mission statement? Is sustainability in the mission statement like it is at Portland State? Does your university offer courses that require community engagement? Have you been in a course that requires community engagement? Are there opportunities to do class projects in your community at your school?

Green buildings on college campuses: University of Florida goes for gold

Many universities are embracing the green building movement. Indeed, many campuses today have a green building standard that requires new buildings to be rated to at least LEED Silver (see chapter on green building). However, few universities in the world have done more for the green building movement than the University of Florida (U of F). Based in Gainesville, Florida, the university requires all new buildings and major renovations to be built to LEED Gold standards. The Gold rating, as you will recall from the previous chapter on green building, is relatively difficult to achieve. Committing to going for the Gold rating requires a strong commitment from the university to create some of the greenest buildings ever constructed on college campuses. They are seeking to build all buildings to LEED Platinum standards in the future.

As a result of the efforts of university officials, U of F has the highest LEED building registrations than any other college campus. The campus is home to three LEED Platinum buildings, 26 Gold-rated buildings, 11 Silver-rated buildings, 14 certified buildings, and 15 registered buildings. This impressive array of buildings includes examples of some interesting technology, such as green roofs, reclaimed water, low-flow plumbing, waterless urinals, rain water harvesting, high efficiency lighting, energy efficient airflow and air conditioning systems, local and re-used building materials, and renewable energy use.

While green buildings have clear environmental benefits, they also provide an opportunity to educate the next generation of building consumers who are taking courses in the buildings. U of F has worked hard to use the buildings for educational purposes.

What is the green building policy on your campus? How many LEED certified buildings are on your campus? What kinds of green technology innovations are used in buildings on your campus? What about in your local community? Have you been in a LEED certified building? How was it different from a non-certified building?

Native and sustainable landscaping at one of the largest schools in the nation: Valencia College

Valencia College (VC) is home to approximately 100 000 students in the Orlando area. It is also home to one of the greenest campus landscaping policies in the US. They have been granted Tree Campus USA status for their efforts. Campus residents regularly celebrate Arbor Day. In addition, they have a campus landscape waste composting program and they utilize integrated pest management practices which significantly limits the use of harmful pesticides and herbicides.

What is interesting about VC's landscaping policies is that they have codified the protection of trees, the use of native plant species, and the protection of wildlife habitat. The policy is listed below and can be read here: http://valenciacollege.edu/sustainability/ campuses/:

> Landscape designs should be developed to provide landscaping that is relatively low cost and low maintenance and should emphasize simplicity, balance and ecological sensitivity. Designs should incorporate plant materials that are water wise, disease, pest and drought tolerant. The use of native plant material and natural plant arrangements is strongly encouraged, whenever possible. Natural landscaping is considered an important component of the design and the designer shall make every effort to incorporate existing natural landscapes into the design and to preserve any natural vegetation on the site. Landscape and site designs should preserve existing trees to the maximum extent possible. Proposed removal of existing trees shall be thoroughly evaluated before committing to a design strategy. The college encourages the preservation of wildlife habitat and the consideration of wildlife use in plant material selection. Landscape designs shall be based on the long term cost effectiveness and sustainability of the materials selected. The use of materials requiring excessive pruning, which drop noxious fruit or plant parts is discouraged. Landscape designs shall be cognizant of the need for a safe and secure environment for campus users. The use of inappropriate plant materials (e.g. poisonous, sharp needled, nuisance or invasive) is discouraged. Landscape and pedestrian area lighting shall be incorporated into the landscape design. In general, all designs shall promote the principles of low cost, safe, sustainable, cost efficient and low maintenance landscapes.

What is the landscaping policy at your school? Does your facilities department utilize integrated pest management? Does your university have a native plant policy? Does your university water the grounds? Is there an irrigation policy in place? How often does the university water the grounds? Do you have space for wildlife on campus? Is the habitat protected? When new buildings are constructed, are efforts made to protect trees?

Campus archaeology at Michigan State University

Universities are complex places that have multiple layers of history. Prior to them becoming campuses there were land uses that were in place that are recorded in the archaeological record. As we learned in the chapter on green buildings, the greenest buildings are the ones that you don't build and historic preservation is a key element of any organizational sustainability plan. Thus, on campuses, the preservation of historic buildings becomes a key component of campus sustainability. Universities often store tremendous amounts of information about buildings and data about university events and history within campus archives. However, there is also a great deal of information stored within the soil covering the campus.

At Michigan State University (MSU) in East Lansing, Michigan, archaeologists are working to preserve the past. Under the direction of Professor Lynne Goldstein, they have instituted an official campus archaeology program that infuses sustainability into the discussion of the history of the campus.

Goldstein and her team have found not only artifacts from the nearly 200-year history of the campus, but have also found Native American artifacts from those that lived in the area prior to the establishment of MSU. By finding, recording, and publishing this information in scholarly venues as well as social media outlets, Goldstein is able to educate the campus community about how different people lived during different eras of time.

Prior to contact with Europeans, the Native Americans that lived on the site were hunters and gatherers who lived in semi-permanent settlements. They hunted for game in the surrounding wooded landscape of the region and took advantage of nearby prairies and waterways for food sources. They also worked the ground to grow crops. They lived simply and certainly had an impact on the region. However, their overall footprint was very small.

Goldstein and her crew excavated the site of the earliest dormitory on campus, interestingly called Saints' Rest. This building was eventually abandoned and collapsed. However, Goldstein and her crew excavated the site and studied university archives to learn that some of the first students at MSU had a relatively large impact on the environment. They cut trees, hunted, and grew food. They heated the residence halls with wood cut from the surrounding forests. They had a much larger impact on the surrounding landscape than the Native Americans.

Of course, today's students have a much larger impact on the campuses. We now have extensively landscaped campuses that in many cases are fertilized and treated for pests. Our residence halls are often heated with fuels imported from many miles away, often across the world. We also import huge amounts of food from far distant lands. Students no longer hunt on campuses and rarely grow their food.

Each period of time provides a window into understanding the lifestyles of people who came before us. We can estimate their carbon footprint and calculate their impact on the environment. We can map the extent of their impacts and assess how they have changed the world. Think about your own campus. How has the campus sustainability changed through its history? What archaeological or archival information is available about people who lived on your campus grounds in the past? Campuses are always undergoing changes. What evidence are you leaving behind about your time on campus? What will this information tell future generations about the sustainability of your campus today?

Index

a

Abbey, Edward
- *Desert Solitaire* 10
- *The Monkey Wrench Gang* 10

acidification:
- ocean 39, 117, 121, 123, 304
- soil 40, 345

Adams, Abigail 183
Adams, Ansel 4
Adams, John 183
Agency for Toxic Substances and Disease Registry (ATSDR) 274
agriculture 12, 18, 26, 44, 55, 67, 70, 76, 84, 100, 149, 150, 173, 175, 176, 180, 181, 183–186, 191–193, 195, 199, 304, 311, 345, 348, 362, 382, 404, 420
air quality 4, 22, 52, 66, 81, 124, 179, 214, 218, 219, 241, 266, 268, 269, 290, 299, 311, 332, 369, 388, 421
Allen, Will 192, 416
American Association for Sustainability in Higher Education (AASHE) 405, 417
American College and University Presidents Climate Commitment (ACUPCC) 125–126, 406–407
Amtrak 231, 254
Anderson, Ray 85, 379–380
Anthropocene 25, 33, 53–55
Apalachicola River 1487
American Society of Heating, Refrigerating and Air Conditioning Engineers 208
Athabasca Tar Sands 271
atmospheric warming 96
aquifers 36, 146, 148–150, 159, 166, 167, 169

b

benchmarking 21–22, 57, 60, 67, 69, 75, 83, 202, 320–321, 401, 405, 408, 409
Bhopal 16, 303–304
biological oxygen demand (BOD) 275
biomass 95, 100–102, 365
Bloomberg, Michael 80, 132, 183
blue baby syndrome 45
Brandes, Oliver 158
Branson, Richard 243
Bronx Green Machine 183–184
Brooklyn Navy Yards 206
Brooks, David 158
Brown, Jerry 72
brownfields 81, 204–206, 223, 295–297, 300
Brundtland, Gro Harlem 16–18, 58, 332, 374
Brundtland Report 16–18, 58, 332, 374
Building Research Establishment Environment Assessment Method (BREEAM) 217–219, 228
Bullard, Robert 20, 293, 299
bus 76, 82, 205, 231, 236–237, 254–258, 423
bus rapid transit 254–255, 258
Bush, George W. 69
business sustainability 2, 6, 8, 15, 19, 26–27, 69, 74, 85–86, 201, 206, 219–220, 231, 283, 345, 358, 362, 366, 372–380, 397, 414

Introduction to Sustainability, Second Edition. Robert Brinkmann.

Companion website: www.wiley.com/go/Brinkmann/IntroductiontoSustainability

c

C.W. Bill Young Reservoir 147
Campus Conservation Nationals (CCN) 408, 416–417
Canada 4, 24, 36, 38, 64–65, 68–69, 89–91, 118, 119, 144, 148, 169–170, 177–178, 189, 238, 271, 284, 303, 319, 361
Canadian Index of Wellbeing 68
cap and trade 42
carbon cycle 29, 37–38, 46, 47, 99, 116–117
carbon efficiency 82–83
carbon footprint 21, 86–88, 231, 326, 426
Carson, Rachel 8–10, 291
  Silent Spring 9, 291
Carter, Majora 294, 299
Carter, Shawn Corey (Jay Z) 80
Cato, Molly Scott 345
  *Environment and Economy,*345
Chernobyl 16, 108–109, 131, 168
China 38–39, 60, 70–71, 90–97, 103, 106, 107, 118, 124, 157, 167, 170, 176–177, 181, 233, 238, 242, 252, 298, 303–304, 313, 317, 328, 335, 353–354, 351
Chinatown 24
classes of rocks:
  igneous 29–33, 40
  sedimentary 29–33, 36–37, 40, 91, 94, 97–98, 135
  metamorphic 29–33, 40
Clean Air Act 8, 13, 268–270, 290
Clean Water Act 13, 39, 268–271, 290
climate change 24, 26, 37–40, 42, 65, 69, 72, 80, 81, 91, 96, 99, 108, 112, 113, 116, 141, 148, 178, 242, 244, 256, 264, 272, 300, 304, 310, 312, 326, 333, 370, 379, 394, 414, 416
Clinton, Bill 185
Club of Rome 11
  The Limits to Growth 11
College and University Recycling Coalition 408
colony collapse disorder (CCD) 195
Commoner, Barry 11
  The Closing Circle 11
community development 84, 201, 204, 205, 218, 223, 226, 256, 295, 298, 321, 356, 358, 389
community engagement 73, 405, 416–417
community sponsored agriculture (CSA) 173, 188, 191
Compensation and Liability Act 272
Comprehensive Environmental Response, Compensation, and Liability Act of 1980 (aka Superfund, or CERCLA) 268
conservationist 6
Convention for the Protection of World Cultural and Natural Heritage 15
Convention on International Trade in Endangered Species of Wild Fauna and Flora (CITES) 15
Cordal, Isaac 140
  Waiting for Climate Change 140
Corporate Average Fuel Economy Standards 235
Cousteau, Jacques 12–13, 171, 334
  Silent World 12
Covanta 102
Cuomo, Andrew 96, 342

d

Dancing Rabbit Ecovillage 19
Darwin, Charles 46
dead zones 45, 249
Declaration of the United Nations Conference on the Human Environment 15, 332
Deep Water Horizon oil spill 93, 397
Deforestation 55, 100–101, 282, 300, 345, 346
Deming, W. Edwards 374–375
  Out of the Crisis 374–375
Denver, John 13
  Rocky Mountain High 13
desalination 147, 159
Diamond, Jared 345
  Collapse: How Societies Choose to Fail or Succeed 345
Dickens, Charles 287
Dow Jones Sustainability Index 385, 397

e

Earth Day 14, 416–417
Ehrlich, Paul 10
  The Population Bomb 10
electric cars 207, 231, 235–238, 242, 255, 416, 422, 423
Emerson, Ralph Waldo 4–5
Empire State Building 22, 224–25
Endangered Species Act 15, 18, 290, 351
energy 89–141
  nuclear 16, 90, 107–110, 131, 155–156, 243, 353, 365, 364, 396
  solar 23, 72–73, 79, 90, 105–106, 111–112, 131, 202, 209–210, 215, 220, 228, 243, 256, 365, 392, 421
  wind 72, 90, 103–108, 112, 131, 202, 209, 228–229, 237, 325, 365–366, 392, 421
energy efficiency 39, 66, 76, 89, 106, 110–112, 130, 208, 218, 222, 225, 235, 241, 255, 391–393, 424
Energy Star 208, 396
environmental activism 10, 13, 19, 24, 99, 136, 141, 293, 401, 414–416
Environmental Defense Fund 12
environmental impact statement 273, 351–352
environmental justice 19–20, 136, 138, 283–311, 353, 231, 403, 414–415
environmental policy 10, 13, 15, 27, 38, 64, 69, 76, 83, 121, 132, 138, 202, 244, 268, 271–272, 303, 333, 335–336, 349, 358, 384, 391, 403, 415–418, 425
Environmental Priorities Committee 311, 416
Environmental Protection Agency (EPA) 13, 39, 51, 69, 204, 273, 297
environmentalism 3, 12, 291, 379, 380
ephemeral streams 36, 246
erosion 32–33, 51–52, 100–101, 151, 170, 212, 247, 304, 355
ethanol 103, 235
eutrophication 44–45, 151, 153, 249, 251, 267, 274–275
Evangelical Environmental Network 136
Exxon Valdez Oil Spill 93, 240

f

fertilizer 8–9 18, 43–44, 150–151, 153–154, 161, 173, 178, 186, 251, 261–262, 267, 364, 413
Florida Everglades 45
Florida Green Building Coalition (FGBC) 75, 79–80, 320
Food 9, 10, 12, 17, 18, 26, 52, 57, 58, 61, 66, 67, 78, 79, 86, 99, 103, 112, 144, 151, 155, 173–177, 179–191, 193–199, 202, 224, 228, 232, 251, 259, 278, 279, 300, 301, 310, 317, 321–324, 326, 339, 340, 348, 354, 360, 363, 364, 366–368, 373, 387, 389, 393, 394, 397–399, 401, 403, 404, 411–413, 416–418, 420, 426
Food Not Bombs 198, 199
fossil fuels 37–40, 43, 90, 93, 95, 96, 102, 117, 121, 131, 157, 211, 235–237, 241, 255, 264, 285, 325, 345, 364, 416–418
freeganism 196–198
Friends of the Earth 12
Friends of the Everglades 12
fuel efficiency 235
Fukushima 108–109, 131, 155, 168, 262

g

Gaia Hypothesis 47–48
Garg, Vikram 398–399
Gates, Bill 391
General Motors streetcar conspiracy 253
genetically modified crops (GMOs): 176–177
Geologic Time Scale 53–54
global climate change 1, 2, 24–26, 37–40, 42, 91, 96, 99, 108, 112–141, 178, 242, 244, 246, 256, 272, 300, 304, 305, 326, 333, 377, 378, 414
globalization 16–18, 211, 287, 304, 311, 326, 327–332, 347, 348, 377, 389
Grand Canyon 6
Grant, Ulysses S. 5
Great Smog of 1952 8–10
green businesses 2, 206
green economics 352–353
green economies 26

green entrepreneurs 2
green fuels 242
green revolution 43, 176, 304, 329
green walls 184
greenhouse gas emissions 38, 39, 65, 66, 72, 81, 82, 122, 124–129, 131, 178, 235, 242, 244, 308, 340, 378, 379, 388, 393, 407, 408, 411
greenhouse gas inventories 124–127, 322, 406–407
greenhouse gas management 26, 127
Greenpeace 12
greenwashing 3, 57, 353, 380–382, 390, 396
gross domestic product (GDP) 22, 61, 68, 358
Gross National Happiness Index 67
Growing Schools Garden 420
guerilla gardening 196–197
Gulf Coast of the United States 4
Gulf of Mexico Dead Zone 45
Gurman, Stephen 158
Guyomard, Pierre 190

h

Halekulani Hotel 398–399
Hardin, Garret 10, 343
  The Tragedy of the Commons 10, 343
Hawking, Paul 379
  The Ecology of Commerce 379
  Confessions of a Radical Industrialist: Profits, People, Purpose: Doing Business by Respecting the Earth 379–380
Hazardous Materials Transportation Act 13
Highway Beautification Act 12
Hill, Julia Butterfly 19
Hofstra University 25, 137, 184, 187, 192, 220, 237, 288, 367, 402, 410, 413, 416, 417
Hofstra University Sustainability Club 416
Hudson River School of Art 4
hybrid cars 237–238
hydraulic fracturing (fracking) 96
hydrogen 34, 40–41, 257, 262, 363
hypoxia 32–33, 111

i

India 178, 181, 216, 233, 238, 303–304, 328, 348
indicators 21–22, 59, 60, 64, 66–68, 76, 81–84, 87, 134, 148, 242, 333, 358–361, 382, 385
industrial revolution 3–4, 33, 38, 40, 41, 117, 121, 165, 173, 175, 187, 328, 356
Intergovernmental Panel on Climate Change (IPCC) 121–122
International Council for Local Environmental Initiatives (ICLEI) 73, 75, 79, 319–320
International Federation of Organic Agricultural Movements 186

j

Jacobson, Joseph 374
Japan 9, 89, 90, 92, 93, 106, 108, 118, 178, 186, 191, 233, 235, 328, 330, 332, 361, 386
Jaroff, Bhavani 366
Johnson, Bea 280–281
Johnson, Hazel 19, 293, 299
Johnson, Ladybird 12
Jones, Van 416
Jurek, Scott 185

k

karst 36–37, 166, 167, 170, 221, 223
Katz's Delicatessen 251
Keystone XL Pipeline 271, 272
Kivalina 39, 40
Kyoto Protocol 38, 39, 333, 335

l

La Maison de Leontine 190
La Ruta del Cacao 340
Lambi Fund 101
Land Institute 12
landfills 20, 81, 95, 101, 102, 119, 128, 153, 154, 198, 211, 228, 250, 275, 279, 280, 285, 306, 315, 394
Lead-based Poisoning Act 13

Leadership in Energy and Environmental Design (LEED) 22, 79, 382
Leopold, Aldo 6, 47, 352
A Sand County Almanac 7, 352
light pollution 22, 207, 264–265
Linnean system 46
Lithification 31, 32
locavore 188–189
London Sustainable Development Commission (LSDC) 81–83
Long Island Sound 114, 154, 268
Love Canal 273–274
Lovelock, James 11, 47
Gaia: A New Look at Life on Earth 11

m

Manhasset Bay Protection Committee 268
Manhattan 24, 80, 81, 206, 251, 254, 257
Mann, Michael 131–132
Margulis, Lynn 48
Marine Protection, Research, and Sanctuaries Act of 1972 13
Markowitz, Gerald 293
McKibben, Bill 113, 414
The End of Nature 113–114
Climate Shift 135, 414
metropolitan statistical areas (MSA) 317
microcystin 284
midnight supper club 190
Millennium Development Goals 22, 57–60, 332
Millennium Ecosystem Assessment 354
Milorganite 275–276
Mississippi River 24, 45, 51, 145, 159, 168, 169
Montreal Protocol on Substances that deplete the Ozone Layer 16
Moses, Robert 256, 257
Muir, John 5–6, 287
Muller, Richard 133

n

Nader, Ralph 10
Unsafe at Any Speed (1965) 10
Naess, Arne 18
National Cave and Karst Research Institute (NCKRI) 221, 223
National Emissions Standards Act 13
National Environmental Policy Act (NEPA) 13, 268, 271–272
national forest 5–6
National Priorities List (Superfund) 273
natural gas 40, 89–92, 95–97, 108, 112, 117–119, 131, 235, 236, 254
natural systems 26, 29–55, 99, 176, 274, 345
neonicotinoid pesticides 196
Neshek, Path 185
Nevin, Rick 292
New York City 24, 34, 58, 69, 79–81, 182, 183, 206, 224, 225, 231, 251, 254, 256–258, 279, 281, 294, 295, 332, 334, 335, 367, 368, 413, 416
NIMBY syndrome (not in my backyard) 324–325
nitrate 43, 45–46, 162
nitrate poisoning 45
North America 4, 13, 24, 32, 34, 51, 95, 148, 167, 170, 191, 195, 196, 254, 382, 385
nutrients 41–45, 103, 150, 151, 153, 154, 241, 249–251, 259–262, 266–268, 271, 274–276, 284, 355, 414

o

Obama, Barack 12
Obama, Michelle 420
ocean acidification 39, 121–123, 171, 304
Ogallala Aquifer 148, 149
oil shale 93–94
organic matter 37, 92, 97, 116, 119, 135, 147, 250, 251
Organization for Economic Co-operation and Development 358, 361–362
Organization for European Economic Cooperation (OEEC) 361
Osuna, Raúl Cárdenas 139
Toro Labs: *One Degree Celsius* 139
overland flow (sheet flow) 36
overpopulation 10, 11

p

PassivHaus 291–220
perennial streams 36, 151
People for Community Recovery 294
People for the Ethical Treatment of Animals (PETA) 180, 185, 414
pH scale 39, 41
phosphorus 33, 42–45, 151, 251, 259–262, 284
Pinchot, Gifford 5–6
Pittaway, Simon 190
PlaNYC 79–81, 206
pollution:
  air 8, 39–40, 65, 70, 93, 98
  chemical 241, 259
  environmental 10, 120, 165, 305, 377
  heat 263–264
  noise 242, 265
  pharmaceutical 155, 168, 263
  visual 265–266
  water 51, 150–151, 154–155, 161, 179, 207, 246–250, 267, 270–271, 275, 321
Pope John Paul II 136
Preservationist 5–6
Prince Charles House 228
Prince William Sound 993
Protection of the Ozone Layer 16, 270

r

rail 109, 212, 223, 231, 232, 238–240, 245, 252–256
Ramsar Convention on Wetlands 14
recycling 23, 33, 36, 66, 82, 99, 102, 212, 250, 266, 277, 279–285, 306, 358, 362, 364, 367, 390, 400, 408, 416, 417
Recylemania 408, 418
Redford, Robert 13
Report of the World Commission on Environment and Development: Our Common Future 17
Resource Conservation and Recovery Act (RCRA) 13
Rio Declaration on Environment and Development 333, 335
Ritz, Stephen 183–184
rock cycle 29–33, 36, 40, 45, 55
Roosevelt, Theodore “Teddy” 6–7
Rosner, David 293
runoff 35, 40, 44, 51, 150, 151, 153, 154, 207, 219, 242, 248, 266, 267, 284, 341
Russia 38, 70, 89, 90, 92, 94–97, 118, 167, 178, 181, 238, 239, 253

s

Sacket, Jake 367
Safe Drinking Water Act 13
Salatin, Joel 186, 187
Schrader, John 334
Schumacher, E.F. 11
  Small is Beautiful: Economics as if People Mattered (1973) 11
sedimentation 32, 135, 151
Seeger, Pete 12–13
Senator Gaylord Nelson 14
Serifos 63, 197–198
Shaindlin, Peter 398–399
ship transport 239–241
Shiva, Vandana 18
Sierra Club 5, 414, 418
Sierra Nevada 4, 52, 53, 159
Sinatra, Frank 80
small farms 186
small house movement 226
smart grid 110–112, 364
Snow, John 165
Social Justice in an Open World 288
Solid Waste Disposal Act 13
Spurlock, Morgan 182
stakeholder 73, 202, 296, 315–316, 385
street sweeping 25, 246, 250–251
Students for a Greener Hofstra 416–417
suburbanization 49, 52
sulfur 33, 40–42, 93, 99, 127, 261, 269
sulfur cycle 40
sulfuric acid 40, 99
Superstorm Sandy 111, 132–133, 139
Surface Mining Control and Reclamation Act 13
Susanka Sarah 227

The Not So Big House: A Blueprint for the Way We Really Live 227
Sustainability Consortium 392
Sustainability Tracking, Assessment, and Rating System (STARS) 405, 406
sustainable building 202, 207, 217, 419
sustainable development 2, 17, 18, 21, 22, 59–60, 73, 81–84, 216, 319, 326–327, 330, 332, 333, 335–338
Sustainable Development Goals 59–60
sustainable landscaping 425
Sustainable South Bronx 293–294
Swoon: *Submerged Motherlands* 139

t

Tampa Bay Water 146, 147
tar sands 91–94, 271
Target Field 22, 79
Tennessee Valley Authority (TVA) 156
The Nature Conservancy 414
Thoreau, Henry David: *Walden* 4
Three Gorges Dam 70, 157, 302–303
Three Mile Island 108, 131
Toledo 284
total maximum daily loads (TMDL's) 267
total quality management (TQM) 373–374
Toxic Substances Control Act 13
transit oriented development (TOD) 254–255
transportation 13, 16, 23, 26, 31, 33, 55, 65, 76, 78, 79, 81, 95, 112, 128, 136, 179, 195, 201, 205, 212, 216, 218, 219, 231–258, 270, 279, 299, 316, 317, 329–331, 340, 350, 364, 369, 387, 391, 403, 405, 408
triple bottom line 2, 377

u

U.S. Green Building Council 22, 407, 409
U.S. National Forest Service 6
United Kingdom 8, 66, 103, 106, 217, 220, 228, 242, 309, 361, 386, 387, 418
United Nations 14, 15, 17, 22, 58–61, 73, 121, 145, 151, 288, 320, 328, 332, 335, 338, 362, 374, 382
United Nations Conference on Environment and Development (UNCED) 332
United States 3, 4, 6, 23, 24, 32, 36, 38–40, 42, 51, 52, 65, 67–69, 71, 78–80, 89, 92, 93, 95–97, 102, 103, 107–109, 117, 118, 124, 132, 136, 139, 144, 145, 148, 149, 156, 173, 176, 178, 183, 201, 204, 229, 232, 234, 235, 238, 242, 253, 257, 271, 272, 274, 282–284, 289, 293–295, 299–301, 304, 308, 317, 320, 323, 351, 366, 368, 385, 386, 391, 397, 407, 408, 419–421, 422
urban food movement 183
urbanization 4, 17, 18, 49, 55, 207, 247

v

vadose zones 36
Vienna Convention 16
viewsheds 266
Virgin Galactic 243
volatile organic compounds (VOCs) 214, 244

w

Wangchuck, King Jigme Singye 67
waste 9, 12, 13, 22, 26, 66, 72, 76, 78, 79, 81, 82, 94, 99, 101–103, 109, 126, 131, 143, 152, 155, 159, 165, 170, 178, 179, 196, 198, 199, 211, 212, 217, 218, 228, 236, 240, 249–251, 259–285, 293, 301, 305–307, 340, 364, 365, 370, 381, 391, 403, 408, 412414
water 4, 7, 9, 12, 22, 24–26, 31–36, 39–41, 43–45, 48, 49, 52, 53, 55, 61, 66, 72, 76, 78, 79, 81, 82, 86, 92–94, 96, 97, 99, 101, 103, 104, 106, 108, 115–117, 134, 140, 143–161, 165, 168–171, 198, 202–204, 207, 208, 218, 219, 223, 237, 240, 241, 246–251, 254, 262–268, 270, 271, 275–277, 283, 290, 294, 307, 320, 333, 340–344, 349, 350, 354, 355, 358, 360, 361, 364, 370, 393, 394, 396, 409, 416–419, 424–426
water cycle 34–36, 55, 143
water quality 4, 66, 145, 150, 151, 161, 212, 241, 268, 270, 271, 284, 341

Water Quality Standards Program 270
Williams-Steiger Occupational Safety and Health Act 13
World Business Council for Sustainable Development 148
World Commission on Environment and Development 17
World Wildlife Fund 12
Wright, Frank Lloyd 227

y

Yellowstone 5, 16
Yosemite 4, 5
Young Evangelicals for Climate Action 136–137
Yucca Mountain 109

z

Zigomanis, Michael 185